Urban Water Conflicts

Urban Water Series – UNESCO-IHP

ISSN 1749-0790

Series Editors:

Čedo Maksimović
Department of Civil and Environmental Engineering
Imperial College
London, United Kingdom

J. Alberto Tejada-Guibert
UNESCO International Hydrological Programme
Paris, France

Sarantuyaa Zandaryaa
UNESCO International Hydrological Programme
Paris, France

Urban Water Conflicts

Edited by

Bernard Barraqué

Centre International de Recherche sur l'Environnement et le Développement CNRS-CIRED, and AgroParisTech, Paris, France

UNESCO Publishing

United Nations
Educational, Scientific and
Cultural Organization

CRC Press
Taylor & Francis Group
Boca Raton London New York Leiden

CRC Press is an imprint of the
Taylor & Francis Group, an **informa** business

A BALKEMA BOOK

Cover illusration: Girls carrying water container in Africa, Boa Vista.
Courtesy of Shutterstock. Copyright: Sabino Parente.

Published jointly by

The United Nations Educational, Scientific and Cultural Organization (UNESCO)
7, Place de Fontenoy
75007 Paris, France
www.unesco.org/publishing

and

Taylor & Francis, The Netherlands
P.O. Box 447
2300 AK Leiden, The Netherlands
www.taylorandfrancis.com – www.balkema.nl – www.crcpress.com
Taylor & Francis is an imprint of the Taylor & Francis Group, an informa business, London, United Kingdom.

© UNESCO, 2012

Typeset by MPS Limited, a Macmillan Company, Chennai, India
Printed and bound in PrintSupport4U, Meppel, The Netherlands

ISBN UNESCO, paperback:	978-92-3-104121-1
ISBN Taylor & Francis, hardback:	978-0-415-49862-3
ISBN Taylor & Francis, paperback:	978-0-415-49863-0
ISBN Taylor & Francis e-book:	978-0-203-87702-9

Urban Water Series: ISSN 1749-0790

Volume 8

The designations employed and the presentation of material throughout this publication do not imply the expression of any opinion whatsoever on the part of UNESCO or Taylor & Francis concerning the legal status of any country, territory, city or area or of its authorities, or the delimitation of its frontiers or boundaries. The authors are responsible for the choice and the presentation of the facts contained in this book and for the opinions expressed therein, which are not necessarily those of UNESCO nor those of Taylor & Francis and do not commit the Organization.

British Library Cataloguing in Publication Data
A catalogue record for this book is available from the British Library

Library of Congress Cataloging-in-Publication Data

Urban water conflicts / edited by Bernard Barraqué. — 1st ed.
    p. cm. — (Urban water series—UNESCO IHP, ISSN 1749-0790; v. 7)
    "A Balkema book."
    Includes bibliographical references and index.
    ISBN 978-0-415-49862-3 (hardback : alk. paper) 1. Municipal water supply—Management—Case studies.
2. Urban sanitation—Management—Case studies. 3. Water resources development—Case studies.
I. Barraqué, Bernard. II. Series: Urban water series ; v. 7.

    HD4456.U73 2011
    363.6'1—dc23

                                                                        2011031359

Foreword

The current rate of population growth and urbanization is likely to exacerbate the difficulties that cities face when attempting to provide water and sanitation services to all their inhabitants. This situation is particularly acute in megacities of developing countries. In addition to decreasing availability of water resources and increasing demand, complex socioeconomic issues such as poverty, inequality, land use and tenure changes, inefficient legal and institutional frameworks, poor water governance, and lack of infrastructure and financial resources have serious implications for urban water management. Reforms in management and ownership of water and sanitation facilities through privatization and different forms of private-public partnership also have an impact on access to water and sanitation and, if not undertaken with consideration to local socioeconomic conditions, may lead to increasing conflicts over access to, and use of, water and sanitation services.

The collection of essays presented in this publication focus on urban water conflicts in an effort to examine controversial aspects of the management and, in many cases, mis-management of water resources in an urban setting. The essays present different cases of urban water conflicts that had arisen in large urban areas around the world over socioeconomic and institutional issues linked to access to water and sanitation services.

This book presents the results of the UNESCO project on 'Socioeconomic and Institutional Aspects in Urban Water Management', implemented during the Sixth Phase of UNESCO's International Hydrological Programme (2002–2007). Its production was the result of the dedicated work and efforts of the UNESCO-IHP Taskforce on Urban Water Conflicts with contributions from experts from various continents of the world. The contribution of Evan Vlachos (Colorado State University, USA) and Bernard Barraqué (Centre International de Recherche sur l'Environnement et le Développement – CNRS-CIRED, France) in coordinating the Taskforce was essential and is gratefully acknowledged. This publication, which is a volume in the *UNESCO-IHP Urban Water Series*, was prepared under the responsibility and coordination of J. Alberto Tejada-Guibert, Deputy-Secretary of IHP and Responsible Officer for the Urban Water Management Programme of IHP, and Sarantuyaa Zandaryaa, Programme Specialist in urban water management and water quality at UNESCO-IHP.

UNESCO extends its gratitude to all the contributors for their outstanding efforts, and is confident that the conclusions and recommendations presented in this volume will prove to be of value to urban water management practitioners, policy and decision-makers and educators alike throughout the world.

International Hydrological Programme (IHP)
United Nations Educational, Scientific and Cultural Organization (UNESCO)

Contents

List of Figures

List of Tables

List of Boxes

List of Acronyms

AASA	Aguas Argentinas Sociedad Anónima
AdP	Águas de Portugal (Portuguese water company, Portugal)
AEP	Adduction d'Eau Potable (Water Supply, France)
AESBE	Associação das Empresas de Saneamento Básico Estaduais (Association of State Water and Wastewater Companies, Brazil)
AMDP	Asset management and development plan
ANC	African National Congress (South Africa)
APF	Anti-privatization forum (South Africa)
ASSEMAE	Associação Nacional dos Serviços Municipais de Saneamento (National Association of Municipal Water and Wastewater Services, Brazil)
ATO	Ambito Territoriale Ottimale (Optimal Territorial Units, Italy)
ATTAC	Association pour la Taxation des Transactions pour l'Aide aux Citoyens (association for transactions' taxation and Citizen support)
AySA	Agua y Saneamiento Argentinos (National Water and Sanitation Utility of Argentina)
BAMR	Buenos Aires Metropolitan Region (Argentina)
BGW	Bundesverband der Deutschen Gas und Wasserwirtschaft e.V (German utilities association)
BMA	Barcelona Metropolitan Area (Spain)
BOT	Build-Operate-Transfer
BOTT	Build-Operate-Train and Transfer
BNH	Banco Nacional de Habitação (Brazil's National Housing Bank)
CAL	Companhia das Águas de Lisboa (Lisbon Water Company, Portugal)
CALS	Centre for Applied Legal Studies (South Africa)
CBOs	Community-based organizations
CDU	Community development unit
CE	Conseil d'Etat (Supreme Administrative Court, France)
CEDAE	Companhia Estadual Da Água e Estgotos (State water company of Rio de Janeiro, Brazil)
CEF	Caixa Econômica Federal (Brazil's National Savings Bank)
CESB	Companhia Estadual de Saneamento Básico (State Water and Wastewater Company, Brazil)

CFAL	Commission for the Inspection of Lisbon Waters
CCAEP	Cellule de Conseil aux AEP (Water Supplies' Support Unit, french-speaking Africa)
CIC	Carga Inicial de Conexión (Initial Connection Fee, Argentina)
CIS	Carga de Incorporación al Servicio (Yearly Connection Fee, Argentina)
CLO	Community liaison officers (South Africa)
CNC	Critical natural capital
COMPESA	Companhia Estadual de Saneamento (State water company of Pernambouco, Brazil)
COSATU	Coalition of South African Trade Unions
CSF	(European) Community Support Frameworks
CSO	Civil society organization (South Africa)
CSTM	Centrum voor Schone Technologie en Milieubeleid (Twente University, The Netherlands)
CVWA	Coachella Valley Water Authority (USA)
DA	Democratic Alliance (South Africa)
DJB	Delhi Jal Board (Water Utility of Delhi, India)
DVGW	Deutscher Verein des Gas- und Wasserfaches German Association of Gas and Water Engineers
DWAF	Department of Water Affairs Forestry (South Africa)
DWR	Department of Water Resources (South Africa)
EAL	Empresa de Águas de Lisboa (the first Lisbon Water Company, Portugal)
EFs	Environmental functions
EMSA	Empresa Sul Americana de Montagens (Brazil)
EPAL	Empresa Pública das Águas Livres (Lisbon Public Water Company, Portugal)
ETOSS	Ente Tripartito de Obras y Servicios Sanitarios (Institution in Charge of Water Services Regulation, Argentina)
EU	European Union
FCFA	Franc de la Communauté Financière Africaine (the currency of former French African Community)
FAE	Fundo de Água e Esgotos (Water and Sanitation Fund, Brazil)
FGTS	Fundo de Garantia do Tempo de Serviço (National fund for the compensation of labour severance, Brazil)
FWA	Frankfurter (Oder) Wasser- und Abwassergesellschaft mbH (Frankfurt/Oder Water and Wastewater Company, Germany)
GAC	Granulated activated carbon
GDP	Gross domestic product
GDR	German Democratic Republic (former)
GEAR	Growth Economic and Redistribution Programme
GES	Good ecological status
GNP-GDP	Gross national product – gross domestic product
GPD	Gallons per day (1 US Gallon = 3.785 litres)
GCS AEP	Groupement de Conseil et de Suivi des AEP (Water Supply Council and Support Board, French Africa)

GTZ	Gesellschaft für Technische Zusammenarbeit (German Agency for Technical Cooperation)
IBGE	Instituto Brasileiro de Geografia e Estatísticas (Brazilian Census and Territorial Statistics)
IFIs	International financing institutions
IHE	Institute for Higher Education
IHP	International Hydrological Programme of UNESCO
IID	Imperial irrigation district (USA)
IMF	International Monetary Fund
INA	Instituto Nacional del Agua (National Water Institute, Argentina)
INAG	Instituto da Água (national water institute)
IPCC	Intergovernmental Panel for Climate Change
ISD	Institutional and social development
JOWAM	Johannesburg Water Management (company, South Africa)
JW	Johannesburg Water (South Africa)
IRAR	Instituto Regulador de Águas e Resíduos (Water and Waste Regulatory Institute, Portugal)
LCD	Litres per capita per day
LOS	Levels of service
LWSC	Lusaka Water and Sewerage Company (Zambia)
MPG	Modelos Participativos de Gestión (Participatory Management Models, Argentina)
MAOT	Ministério do Ambiente e do Ordenamento do Território (Ministry of the Environment and Land-use Planning, Portugal)
MAF	Million acre feet (1 MAF = 1.234 km^3)
MCMA	Mexico City Metropolitan Area (Mexico)
MWB	Metro Water Board (Water Utility of Chennai, India)
MWD	Metropolitan Water District (Southern California, USA)
NBI	Necesidades Básicas Insatisfechas (Unfulfilled Basic Needs, Argentina)
NC	Natural capital
NGO	Non-governmental organization
NWASCO	National Water Supply and Sanitation Council (Zambia)
OGA	Operation Gcin'Amanzi (South Africa)
OPCT	Obras Públicas Contractadas a Terceros (Public Works Contracted Out, Argentina)
OSN	Obras Sanitarias de la Nación (National Water and Sanitation Company, Argentina)
PALYJA	Subsidiary of Lyonnaise des Eaux in Jakarta (Indonesia)
PAM Jaya	Perusahaan Air Minum Jaya (Jakarta's water utility, Indonesia)
PDM	Partenariat pour le Développement des Municipalités (Municipal Development Partnership, Cameroon)
PLANASA	Plano Nacional de Saneamento (National Plan for Water Supply and Sewerage, Brazil)
PMSS	Programa de modernização do setor saneamento (Water Sector Modernization Programme, Brazil)

PNAD	Pesquisa Nacional por Amostra de Domicílios (National Research based on a Housing Sample, Brazil)
PPP	Public-private partnership
PSCT	Public sewage collection and treatment
PSP	Private sector participation
PS-Eau	Programme Solidarité Eau (a French NGO on water and sanitation programs in developing countries)
PWS	Public water supply
RWE	Rheinische-Westfäliche Elektrizitätswerke (Electricity Company of North Rhine Westphalia, Germany)
SABESP	State water company of São Paulo (Brazil)
SANEANTINS	State water company of Tocantins (Brazil)
SANEPAR	State water company of Parana (Brazil)
SANEGRAN	Saneamento do Grande São Paulo (Greater Sao Paulo Sanitation System, Brazil)
SANEST	Saneamento da Costa do Estoril (Portugal)
SARS	South African Revenue Service
SDCWA	San Diego County Water Authority (USA)
SFS	Sanitation funding system
SHG	Self-help groups
SIMTEJO	Saneamento Integrado dos Municípios do Tejo e do Trancão (Wastewater Company of Greater Lisbon, Portugal)
SNIS	Sistema Nacional de Informações sobre Saneamento (National Information System on Sanitation, Brazil)
STEFI	Suivi Technique et Financier (Technical and Financial Follow Up, French Africa)
STP	Social tariff programme
SU	Servicio Universal (Universal service, Latin America)
SUMA	Servicio Universal y Medio Ambiente (Universal service and environment tax, Argentina)
TINA	There Is No Alternative (to privatization, Germany)
TPJ	Thames Pal Jaya (subsidiary of Thames Water in Jakarta, Indonesia)
TUWM	Total Urban Water Management
UAW	Unaccounted for water
USAID	United States Agency for International Development
USD	United States Dollar
USEPA	United States Environmental Protection Agency
UW	United Water (private company, USA)
UWC	Urban water conflicts
UWS	Urban water supply
VPL	Ventilated pit latrine
WB	World Bank
WDM	Water demand management
WFD	Water Framework Directive (European Union)
WRCB	Water Resources Control Board (USA)
WSS	Water and sanitation services
WWAP	World Water Assessment Programme

Glossary

Alluviums *Syn.* Alluvial deposit; clay, silt, sand, gravel, pebbles or other detrital material deposited by water.

Aquifer Geological formation capable of storing, transmitting and yielding exploitable quantities of water.

Best management practices (BMPs) Methods that have been determined to be the most effective, practical means of preventing or reducing pollution from non-point sources, such as pollutants carried by urban runoff.

Cesspits A pit for the reception of night-soil and refuse.

Cholera An acute infection of the small intestine caused by the bacterium *Vibrio cholerae* and characterized by extreme diarrhea with rapid and severe depletion of body fluids and salts.

Combined Sewer A sewer receiving both waste- and rainwater.

Combined Sewer Overflow (CSO) Discharge of a mixture of storm water and domestic waste when the flow capacity of a sewer system is exceeded during rainstorms.

Commodification Consideration of a good or a service which was not in the market area, as if it were a tradeable commodity, e.g. considering water as a market good (in particular potable water).

Consumerization The transformation of water from a common good of communities, or a public good paid by citizens (taxes), to a commercial public service paid by customers through bills, the service being provided by a local authority, public in-house company or even a private company, including the possibility of social tariffs.

Conventional/non-conventional water When water services are provided mostly collectively through a system of pipes, they are considered conventional. Non-conventional water includes grey- or waste water reuse, rainwater harvesting (at neighbourhood level) and eventually brackish- or seawater desalination.

Credible threat A threat that one's opponent has good grounds for believing will be carried out. The shift from a tension to a conflict occurs when one party implements a credible threat. There are several indicators. For instance, the use of the media, one party bringing the other before the courts, or the production of signs (such as a notice); finally, both parties may enter into a direct confrontation (verbal or physical).

Externality Outside effect, such as social and environmental benefits and costs, not included in the market price of goods and services being produced, or a consequence of an action that affects someone other than the agent undertaking that action, and for which the agent is neither compensated nor penalized. Externalities can be positive or negative.

Full cost pricing   A price covering operation and maintenance costs, plus a fair share of investments depreciation, plus a charge to cover environmental costs, resource costs and opportunity costs.

Greywater   Domestic wastewater generated from dish washing, laundry and bathing, excluding blackwater (i.e. wastewater from toilets).

Haussmann   French administrator responsible for the transformation of Paris from its ancient character to that which it still largely preserves. Although the aesthetic merits of his creations are open to dispute, there is no doubt that as a town planner he exerted great influence on cities all over the world.

Home rule *Syn.*   Local rule; the ability of a jurisdiction to adopt and enforce its own rules, policies and procedures related to carrying out its functions.

Hydrological cycle *Syn.*   Water cycle; succession of stages through which water passes from the atmosphere to the earth and returns to the atmosphere: evaporation from the land or sea or inland water, condensation to form clouds, precipitation, interception, infiltration, percolation, runoff and accumulation in the soil or in bodies of water, and re-evaporation.

Natural capital   Natural resources which economists can assess as an input in an economic activity. Once calculated, they can be paid for as 'ecosystem services'.

Not in my back yard (NIMBY)   Usually used to describe members of the public who are not opposed to an activity in principle (e.g. nuclear power or waste disposal), but do not want it located near their community.

Peri-urban   An area immediately adjoining a city or conurbation.

Public-private partnership   All the arrangements between the authorities responsible for public services provision and private companies, where management and technical tasks, and financial risks, are shared between authorities and private companies. It is not clear whether operators with private or commercial status fully owned by the public authorities are included in these partnerships. PPPs usually include: the concession (where the private company does the investment, takes the financial risk and manages the infrastructure); the management contract (company only does the management, no investment, no risk); the lease contract (company takes a limited risk in assuming the renewal of ageing assets); and the various types of mixed economy companies under commercial status.

Reverse osmosis   A treatment process used in water systems by adding pressure to force water through a semi-permeable membrane. Reverse osmosis removes most drinking water contaminants. Also used in wastewater treatment.

Runoff   That part of precipitation that appears as streamflow.

Separate sewer system   Sewer system receiving only wastewater and no rain.

Sewage *Syn.*   Wastewater; the waste and wastewater produced by residential and commercial sources and discharged into sewers.

Sewerage   The entire system of sewage collection, treatment and disposal.

Stakeholder   Any organization, governmental entity, or individual that has a stake in or may be impacted by a given approach to environmental regulation, pollution prevention, energy conservation, etc.

Stormwater   Runoff from buildings and land surfaces resulting from storm precipitation.

Sustainable development   A development able to sustain or self-support, by meeting criteria in three dimensions: environmental, economic and social. Political scientists increasingly tend to add a fourth dimension: multilevel institutional governability.

Urban drainage   A system of conveyance and storage elements serving to drain urban areas.

Urban Water   In this book, defined as the water carried by public water supply and sanitation services, plus stormwater on the city, plus water resources in case of an eventual interaction of the city with its hinterland or river basin.

Urban water cycle   A water cycle including all the components of the natural water cycle with the addition of urban flows from water services, such as the provision of potable water and collection and treatment of wastewater and stormwater.

Water Demand Management   An approach to meeting water demands by users not only through the expansion of supply, but by the control of demand. The notion does not necessarily refer to a classical economic approach (supply vs demand in a market), but to bring the dynamic interactive loop between supply and demand to a more sustainable equilibrium.

Water Framework Directive (EU)   Directive 2000/60/EC establishing a framework for the European Community action in the field of water policy. It aims to secure the ecological, quantitative and qualitative functions of water. It requires that all impacts on water be analysed and actions taken within river basin management plans.

List of Contributors

Karen Bakker
Professor, Department of Geography, University of British Columbia, Vancouver, Canada

Bernard Barraqué
Research Director, Centre International de Recherche sur l'Environnement et le Développement CNRS-CIRED, and AgroParisTechParis, France

Sarah Botton
Training and Research Project Manager, Agence Française de Développement (AFD), Marseille, France

Anne Bousquet
Capacity Building and Training Officer, Global Water Operators Partnerships Alliance, UN-HABITAT, Nairobi, Kenya

José Esteban Castro
Professor, School of Geography, Politics and Sociology, Newcastle University, Newcastle, UK

Elena Domene
Researcher, Barcelona Institute of Regional and Metropolitan Studies, Universitat Autònoma de Barcelona, Bellaterra, Spain

Sylvy Jaglin
Professor, University of Paris-Est Marne-la-Vallée, Latts, France

S. Janakarajan
Professor, Madras Institute of Development Studies, Chennai, India

Petri S. Juuti
Adjunct Professor (docent) and Senior Research Fellow, Department of History and Philosophy, University of Tampere, Finland

Tapio S. Katko
Adjunct Professor (docent) and Senior Research Fellow, Department of Chemistry and Bioengineering, Tampere University of Technology, Finland

Michelle Kooy
Urban Program Director, Mercy Corps, Jakarta, Indonesia

Marie Llorente
Researcher, Centre Scientifique et Technique du Bâtiment, Paris, France

Antonio Massarutto
Associate Professor, University of Udine, and Research Director, Centre for Research on Energy and Environmental Economics and Policy (IEFE), Bocconi University, Milano, Italy

Martin V. Melosi
Distinguished Professor, Department of History, University of Houston, Texas, USA

Gabriela Merlinsky,
Researcher, Research Institute 'Gino Germani', Faculty of Social Sciences, Universidad de Buenos Aires, Argentina

Matthias Naumann
Researcher, Leibniz Institute for Regional Development and Structural Planning (IRS), Erkner, Germany

Ana Lucia de Paiva Britto
Adjunct Professor, Post-Graduate Programme on Urban Studies (PROURB), Federal University of Rio de Janeiro, Brazil

João Pato
Postdoctoral Fellow, Institute of Social Sciences, University of Lisbon, Portugal

Christelle Pezon
Programme Officer, IRC International Water and Sanitation Centre, The Hague, The Netherlands

Stephanie Pincetl
Researcher, Institute of the Environment and Sustainability, University of California, Los Angeles, USA

Tiago Saraiva
Research Fellow, Institute of Social Sciences, University of Lisbon, Portugal

David Sauri
Professor, Institute of Environmental Science and Technology (ICTA) and Department of Geography, Universitat Autònoma de Barcelona, Bellaterra, Spain

Luísa Schmidt
Principal Researcher, Institute of Social Sciences, University of Lisbon, Portugal

Laïla Smith
Senior Programme Manager, Water and Sanitation (Africa) AusAID, Pretoria, South Africa

Ricardo Toledo Silva
Professor, School of Architecture and Urban Studies, University of São Paulo, Brazil

Markus Wissen
Professor, Institute of Political Science, University of Vienna, Austria

Sarantuyaa Zandaryaa
Programme Specialist, International Hydrological Programme, Division of Water Sciences, UNESCO, Paris, France

Marie-Hélène Zérah
Researcher and Head Urban Dynamics, Centre de Sciences Humaines, New Delhi, India, and Institute of Research for Development, Paris, France

Introduction

The book starts by providing the background and conceptual framework for understanding urban water conflicts, developed within the framework of the UNESCO-IHP Taskforce on Urban Water Conflicts. This introduction is followed by a chapter by Bernard Barraqué, Tapio Katko and Petri Juuti presenting a historical overview of urban water services in Europe and water-related conflicts. Based on a comparative and historical approach to water services provision, this contribution stresses the changing nature of water conflicts over the decades, through shifting technological, economic, political and social contexts. The European experience provides historical perspectives on private companies' initiatives in the initial development of piped system services and the role of local authorities, with the recent lessons of crises in the municipal model followed by commercial approaches to services, if not always private companies. A box by Kostas Chatzis summarizes conflicts raised by the initial tariff formulas in Paris in the nineteenth century. The European experience with water tariffs is further explored by Elena Domene and David Sauri in their chapter on the 1990s 'Barcelona water war', which illustrates a new social dimension to sustainability issues.

The contribution of Martin Melosi on the history of water supplies in the United States describes the same trends as those found in Western Europe: initial private initiatives are followed by municipal takeovers, with recent years showing a trend in return to private sector involvement. A box on water supply service in Montreal, Canada, by Dany Fougères provides a case study of successful public management and financing of water services in the 19th century.

The contribution by Christelle Pezon examines the historical failure of the concession contract model, and the subsequent development of the French water services delegation model, which was framed by a succession of Conseil d'Etat (supreme administrative court) cases. Among the various types of conflicts, emphasis is placed on those pitting local authorities against private companies. Resolution of these conflicts by the courts was the preferred solution in this context, and the court rulings thereafter dictated the form that governance of services would take. It would be relevant to extend this analysis of the court cases to other legal contexts, since the French model of *Public Service Delegation* is being experimented with and debated in various other countries. The contribution by Luísa Schmidt, Tiago Saraiva and João Pato covers the history of water services in Lisbon, detailing a century of conflicts between the city and the Portuguese government.

The contributions that follow provide cases from Latin America, a continent well-known for unrest related to water issues. The chapter by Ana Britto and Ricardo Toledo Silva provides a synthesis of urban water conflicts in Brazil, stressing the inter-governmental nature of the major disputes and issues of sustainability. Sarah Botton and Gabriela Merlinsky examine in detail the complex case of Buenos Aires and some factors related to the Bolivian water conflicts, explaining why the French-based multi-national corporation Suez S. A. lost two contracts in different situations. The contribution by José Esteban Castro provides an analysis of water unrest in Mexican urban areas, suggesting that an 'untold epistemic conflict' exists between engineers and natural scientists on the one hand, and critical social scientists on the other, creating an intellectual barrier to the better understanding of water conflicts.

The complex African situation is represented by two chapters: the first contribution by Laïla Smith explores the contrasting situations of Johannesburg and Cape Town in South Africa; the second by Sylvy Jaglin and Anne Bousquet examines the paradoxical role of NGOs in water services provision in small towns in sub-Saharan Africa.

The book contains two cases on water conflicts in Asia, which are linked in part to the very high population density of the region and rural-urban migration. Karen Bakker presents the case of Jakarta, where privatization under a weak governance system failed and resulted in a volatile situation. S. Janakarajan, Marie Llorente and Marie-Hélène Zérah examine the cases of two large metropolises in India, Delhi and Chennai, where water scarcity and lack of access to water aggravate an already unsatisfactory situation.

The chapter by Stephanie Pincetl provides a case of water scarcity in the United States, totally different from the Indian situation. The transfer of water from the Imperial Irrigation District to the City of San Diego illustrates conflicts over water resources between various water supplies, and with environmental protection and irrigation; and between southern California and the Colorado Delta area in Mexico.

The last two chapters illustrate various situations in Europe: Antonio Massarutto presents the Italian attempt to rationalize water services as public-but-commercial; and Matthias Naumann and Markus Wissen explore the Eastern German experience of a water services crisis in 'shrinking cities' after the re-unification of Germany.

The book ends with a more theoretical analysis of urban water conflicts from an ecological-economics point of view by Antonio Massarutto.

The case studies collected in this book cover all four types of cities described in the first chapter and illustrate a wide range of urban water conflicts. The aim of these case studies was not to provide solutions to urban water conflicts, but to describe the conflicts from the socioeconomic and institutional perspective of urban water management, while retaining diverging opinions on certain issues (for example, the role of the private sector). The book also aims to encourage consideration of the social sciences, such as history, geography, political science, sociology, anthropology, law and economics as a means to understanding urban water conflicts.

Bernard Barraqué

Chapter 1

Urban water conflicts: Background and conceptual framework

Bernard Barraqué[1] and Sarantuyaa Zandaryaa[2]

[1]Centre International de Recherche sur l'Environnement et le Développement CNRS-CIRED, Paris, France
[2]International Hydrological Programme (IHP), Division of Water Sciences, UNESCO, Paris, France

1.1 UNESCO-IHP TASKFORCE ON URBAN WATER CONFLICTS

Within the framework of the Sixth Phase of the International Hydrological Programme of UNESCO (IHP-VI, 2002–2007), one of the urban water projects under IHP-VI Focal Area 3.5 'Urban Areas and Rural Settlements' was dedicated to *socioeconomic and institutional aspects in urban water management*. The first meeting of the project implementation was held at UNESCO in Paris in March 2002, which defined the thematic framework of the project and topics to be addressed, and established a Taskforce on *Urban Water Conflicts* (also called the urban water conflicts group) to facilitate project implementation. The Taskforce held several meetings over the period 2002 to 2006 to develop key issues linked with water-related conflicts in urban settings.

This interdisciplinary project aimed to examine various aspects of water crises related to population growth, urbanization, patterns of development, increasing degradation in quantities and quality of water resources, and social impacts of costly solutions in urban contexts. In particular, the project aimed to develop typologies of urban water conflicts and methods to analyse them. A range of socioeconomic and institutional issues linked with urban water management, and those which may cause water-related conflicts in urban areas, were addressed, including:

- socioeconomic issues linked with urban water services, including issues related to access to safe drinking water as a human right
- socioeconomic issues (and water conflicts) related to urban water development and use
- integrated water and land-use policies to alleviate human-induced deterioration of the environment
- institutional development and participatory process for decision-making in urban water systems management
- issues related to legal aspects in urban water management and the evolution of trends in water regulations

- community participation in urban water projects and management through better communications between water professionals, decision-makers and local communities, and
- water and ethics.

These political, social, economic, institutional and cultural issues may become root causes of water-related conflicts in urban areas as a result of ineffective water management. Understanding and analysing urban water conflicts, therefore, requires an integrated approach to socioeconomic and institutional aspects of urban water management.

1.2 SOCIOECONOMIC AND INSTITUTIONAL ASPECTS OF URBAN WATER MANAGEMENT

Failure to consider socioeconomic and institutional issues in decision-making related to urban water management may have serious implications. One of the well-known examples of such failure is the privatization of the public water utilities of Cochabamba, Bolivia, in the late 1990s, which triggered protests over the poor results of the privatization and inadequate service quality that persisted thereafter.

Key socioeconomic and institutional issues related to different aspects of urban water management are briefly described below.

Access to safe drinking water as a human right. The most critical socioeconomic issue related to urban water management is access to safe drinking water, with most urban water conflicts arising from lack of access to water and sanitation services due to the unaffordability of high connection fees and water tariffs for low-income inhabitants. This issue has fundamental links with the human right to water and the challenge of combating poverty. Those who suffer most from inadequate management of urban water services are the urban poor and excluded social groups in peri-urban, informal settlements. Lack of access to safe drinking water and sanitation is a major barrier to improving the health and well-being of the urban poor and to reducing poverty in urban and peri-urban areas. As water is fundamental for life and health, the human right to safe drinking water is considered a pre-requisite for the realization of other human rights. The issue of access to water and sanitation as a human right has been extensively discussed and advocated at various international fora and levels, and the United Nations General Assembly and the Human Rights Council declared that access to water and sanitation is a human right in 2010 (UN, 2010). However, implementation of the human right to water and sanitation remains a major challenge, especially in urban areas.

Socioeconomic issues linked with urban water development. With rapid population growth and urbanization occurring at an unprecedented rate, urban water development has become a crucial issue, which requires consideration within the framework of sustainable development, in particular, with reference to its three primary dimensions: environmental, economic and social. Urban water development is intrinsically linked with complex socioeconomic and institutional issues, such as:

- institutional, legal, regulatory and policy frameworks for urban water management, involving, *inter alia*, issues related to water as a public good *versus* water as an economic good, water rights, water allocation, and environmental regulations and standards

- investment and financing of urban water infrastructure, characterized by high capital cost and long pay-back periods for both urban water services expansion and infrastructure maintenance
- centralization and decentralization of urban water management, requiring consideration at the appropriate decision-making level, taking into account issues such as administrative boundaries, resource availability, economies of scale, etc.
- water pricing and tariffs, which are directly linked with the quality of urban water services, maintenance of urban water infrastructure and operations, and affordability and cross-subsidizing of poor users, as well as public awareness of efficient water use and water conservation
- stakeholder and private sector participation, involving both public participation in decision-making related to urban water and public-private partnerships.

Institutional development for urban water management. Weak institutional framework of urban water systems is often linked with fragmentation, with different institutions being responsible for different aspects of water management and unclear division of responsibilities between the different levels of governments. Most urban water systems are excessively centralized and depend heavily on public funding. Outdated management practices and lack of adequate human capacities are additional causes of deficient urban water management.

Public-private partnership. Over the past two decades, different types of public-private partnerships have been considered in order to meet investment and financing requirements for urban water systems expansion. Operation and maintenance would also be improved by drawing upon private sector skills. However, experience has demonstrated the necessity for good regulatory and institutional frameworks. If these are lacking, public-private partnerships may have implications on service quality and accessibility, price affordability for certain groups of the urban population, the safety of drinking water, and even on access to water as a human right. Consequently, the advantages and disadvantages of private sector participation in water management and the potential conflicts associated with it need careful consideration, as shown in the review by the World Bank (Marin, 2009).

Participatory process for decision-making in urban water systems management. Individuals and communities should be involved in decisions related to urban water management that affect their lives. A participatory process for decision-making in urban water management is key to providing high quality, efficient and effective water services, and to ensuring principles of equity, accountability and transparency in urban water systems management. The participation process requires stakeholder participation and better communication between water professionals, decision-makers and local communities. Stakeholder participation in turn requires the empowerment of local people and communities.

Understanding urban water conflicts and their root causes remains a topic of much research interest, diverging opinions and debate. In particular, the establishment of an adequate institutional framework for urban water management and the choice of the appropriate level of decentralization of public services remain major issues, which depend on a number of factors. Similarly, the role of the private sector in the provision of public services, such as water and sanitation, is viewed differently depending on the socioeconomic context with varying results. Public participation in urban water management is necessary for formulating effective policies, minimizing or avoiding public confrontations and controversies, and involving stakeholders in the implementation and management of urban water projects. But it remains a complex process by which public needs, concerns and values are incorporated into decisions related to water services. Drawing on these issues, the UNESCO-IHP Taskforce examined various cases of urban water conflicts of different types and levels in various regions and countries so as to provide a scientifically sound basis for discussion and understanding of urban water conflicts and socioeconomic and institutional aspects of urban water management.

1.3 DEFINITION OF URBAN WATER: AN IMPURE PUBLIC GOOD

It is necessary to distinguish the notion of 'urban water' and its management from water resources management in general, and to study related conflicts separately from water resources allocation issues.

Urban water management has developed into a separate policy sector over the years, at least in developed countries, because of the growing importance of water services, namely drinking water supply, sewage collection and treatment, and urban drainage. Indeed, water services are at the core of our investigations. As described in the next contribution in this book, urban water is an artefact produced by the skills of engineers, flowing in pipes and hidden underground for the most part, in other words, separated from nature. Civil engineers first controlled quantities of urban water through hydraulic projects: elevation of water with pumps, long-distance water supply, and urban drainage with combined or separate sewers. Then came the time of sanitary engineers, applying technologies developed from the emergence of bacteriology at the end of the nineteenth century. Water quality then became the major issue, driving the development of urban water services. Ultimately, the separation of water services and water resources occurred through the creation of two types of water plants: water works at the incoming point, sewage treatment plant at the discharge point. For most urban residents, the relationship to water took place in limited areas between the tap and the sink.

However, if water resources and water services are different objects not only in terms of technology, but in the views of economists and lawyers, then the separation is not complete. On the one hand, there are still billions of people in the world who are not yet connected to piped water services and rely directly on the natural resource; their right to water is their water right (almost all legal systems give priority to domestic uses of the resource, even though this priority is poorly implemented). In addition, the global urbanization process results in very large metropolises with a dramatic impact on water resources, so that both issues of resources and services have to be re-integrated. We have reached the time of environmental engineering, where the goal of technological advances should be not only to protect human health and well-being from natural hazards and water-related risks, but also to ensure the protection of nature and the environment from the impacts of human activities. The uncertainties brought about by climate change increase the need for such re-integration.

The growing complexity of interactions between water services and water resources may lead some observers to oversimplify and amalgamate issues into the notion of 'water wars', particularly in the context of globalization. The typical reasoning is: *global water companies and international financial institutions support water privatization; if they prevail, ultimately the development of water markets will deprive the poorest of an essential good. Water then should be a public good in general.* However, very few countries have experienced water resources privatization (for example, as Chile did), and nowhere is there a population without access to any water. The problem is one of quality, but not of quantity for essential needs. In cities, the impacts of economic, industrial and human activities cause pollution of water resources, making water unsuitable for drinking. In any case, supporters of the privatization of urban water systems usually focus on public services and piped water systems, but are usually not interested in purchasing water resources or marketing them. Conflicts over resources in general are still essentially solved by the decisions of sovereign governments or international commissions, with little recourse to economic incentives. Conversely, conflicts over services are often linked to debates over tariff structures. These conflicts appear all the more important when a significant proportion of the urban population is not connected to water services. In many large cities in developing countries, initial water networks were developed during the colonial period, but these were limited to areas occupied by the colonists. Local populations relied upon traditional solutions, many of which were not adapted to high population densities, but which were usually free or cheap.

When independence took place, many new governments were inclined to develop large hydraulic projects to generate economic development. These projects developed water resources with public funds and made water a *de facto* public good. Access to water was largely subsidized, for example, for irrigation, whereas urban water services remained in the early stages of development and were inadequately maintained. When liberal policies replaced welfare approaches, international institutions exerted pressure on governments to introduce charging mechanisms and promoted privatization of the infrastructure. This policy met with increasing resistance from the public. Resistance to pay for water services is frequently highest in former colonial cities, as illustrated by the chapters on African cases in this book. The discrepancy between neighbourhoods in terms of service technology and quality is greatest where 'consumerization' trends are coupled with privatization of infrastructure. Therefore, it is no surprise if poor neighbourhoods refuse to pay for a service they consider as inappropriate.

Despite this resistance, water is an economic good. We are not free to decide if it is or not: when it is not scarce, water still mobilizes a lot of capital to be used for modern development, and calls to 'apply reason to choice', as economists define their discipline (Green, 2003). This does not equate with the notion that water is a market good, which it is clearly not. It is a mistake to confuse an economic good with a market good. Economic analysis does not ignore the collective dimension of some goods, and since the seminal work of Paul Samuelson (1954) has specified three types: *pure public goods* and two different sorts of impure public goods, *club or toll goods* and *common pool resources*. Thirty years ago, Ostrom and Ostrom set out a clear two-entry matrix with rivalry in consumption and excludability of benefits (see McGinnis, 1999). The matrix designates four types of goods. In the upper left-hand corner, one finds private or market goods. For example, when an orchard is fenced, people cannot

Rivalry in consumption

	Yes	No
Excludability of benefits — **Yes**	Private or market goods	Toll or club goods
Excludability of benefits — **No**	Common pool resources	Fully public goods

Figure 1.1 Market goods, pure and impure public goods

Source: After Ostrom and Ostrom (1977).

get in. Apples can be grown inside and sold on the market. If somebody buys them, nobody else can eat them unless they buy them from the owner first.

To drive on a restricted access highway, one must pay a toll. Those who cannot or will not pay the toll cannot have access. Similarly, if you want to play golf on certain courses, you must first join a club (upper-right corner of the table). Once you have paid the toll or membership dues, the fact that you use the highway or the golf course does not make it impossible for other members to do so. There is excludability, but no rivalry – at least, up to a certain point. Where too many people want to join at the same time a traffic jam will occur. In the club case, there is another solution: no more admittance. Meanwhile, on the highway, economists are tempted to introduce a corrective 'market' factor through a variation of the toll fee.

The bottom-right corner of the figure deals with the case of *fully public goods*. In some cases, it is impossible to exclude anybody from accessing a good, neither is this desirable. For example, education is deemed a public good in some countries; nobody is excluded from school, so it is therefore free and paid for by the citizens through taxes. Take the example of a lighthouse by the coast: one boat seeing the light does not diminish the light received by other boats, and when one boat sees it, all boats see it. There is no rivalry and no exclusion.

Lastly, let us consider a common pasture or forest which is over-exploited. There is rivalry between users, but it is difficult to exclude any user for both moral and practical reasons. In this case, Ostrom and Ostrom (1977) argue that there is a need for an institution to organize collective bargaining and a rational allocation of the resource, and that this solution is superior to splitting the commons into private plots of land.[1]

How can this framework be applied to water? Water resources are usually common properties. It is difficult, both for moral and practical reasons, to exclude users of a river or an aquifer that is over-drafted or polluted. Throughout history, the very fact that water flows freely led legal systems to give precedence to users' rights over

---

[1] Saying this, they follow the answer of Siegfried Ciriacy Wantrup to Garret Hardin and his *Tragedy of the Commons* (published in *Science magazine*, 1968). See Richard, Bishop and Andersen (eds.) *CIRIACY WANTRUP S.V.: Natural resources economics, selected papers*, Westview press, Boulder, 1985.

ownership rights, and to develop rules preventing water abuse. This is also why it is essentially wrong to argue that water is similar to oil or gold, opening the way to 'wars'. The latter two are ores, minerals, in other words, non-renewable resources, and their exploitation is usually subjected to mining codes, which differ from water laws.[2] Only deep aquifers of non-renewable water should be treated as minerals. Almost all water resources should be treated as common properties and managed by regulatory institutions where the community of users should reach agreements.

Conversely, water services, as a set of infrastructures, are club goods of a particular nature, since the club is ideally designed to supply all the demand and therefore be open to all water users in the full territory of a city and its suburbia (there are economies of scale and of scope, up to a certain point, and frequent club effects). Within this territory, a resident usually has to pay a basic connection fee, which corresponds to a small share of the renewal needs of the collective infrastructure. Once connected, a household will have to pay a price per unit of volume, which will be the same for all.

Because of technical and economic reasons, centralized public services do not extend to areas where connection to the public service systems might be too expensive or inefficient. For example, in some European countries a significant fraction of the population still relies on septic tanks. There is even a non-negligible fraction of population without connection to public water supply. People rely on private wells or cisterns and sometimes on 'group water schemes' (for example, in Ireland) or 'non-community water systems' (as in rural areas of the United States). In such cases, users form a common-pool resource institution with related constraints and no exclusion.

There is a large difference, however, between water resources with potential rivalry but little possibility to exclude (with small common-pool institutions to regulate access and allocation), and water services where there is no rivalry, but a possibility to exclude.

Yet, the situation is more complex in practice, and water has been eventually considered as a market good, or conversely as a fully public good. There have been very few cases of water market creations, except in Chile, where the neo-liberal economic model was adopted after the military coup and water resources were privatized. However, this model did not function very well[3] (Bauer, 2004). In California, US, the trading of Imperial Irrigation District's water to the City of San Diego proved a difficult and unique case: instead of an open market, the outcome is a complex agreement between a limited number of actors, including the farmers and the environmentalists around the Salton sea, and the institution in charge of 'wheeling' the water, the Metropolitan Water District of Southern California. It is probably in Australia that water markets or trading mechanisms are the most developed.

---

[2] Incidentally, it is because of the gold rush that in the American west, a unique legal system developed for water with the prior appropriation doctrine. It is however balanced by the notion of 'efficient and reasonable use': if one does not use his/her right to water, he/she may lose it.

[3] In a paper for the periodical RFF, Carl Bauer summarizes: 'The Chilean approach to "recognizing water as an economic good" has led to some important economic benefits, such as encouraging private investment and allowing more flexibility of water resources allocation. However, the legal and institutional consequences of this approach have hamstrung government efforts to respond to the growing social and environmental problems of water management, which the 1981 Water Code was not designed to address. This rigidity of the Chilean model, and hence its incompatibility with core aspects of IWRM, have been downplayed by the model's international proponents, who continue to argue that any flaws are secondary or can be readily corrected' (Bauer, 2003).

Conversely, in order to reduce existing scarcities, stimulate rural development and increase food self-sufficiency, governments have become increasingly involved in the water allocation process. Governments in Mediterranean countries such as Spain and in many socialist and developing countries declared some or all of water resources to be in the public domain, and this allowed them to build unprecedented storage capacities (to regulate water flows) and water transfers (to better use surpluses). In some cases, local communities had to be displaced and their traditional rights impaired. Ideally, they were compensated through relocation and the provision of cheap water surpluses generated by the dams, but were henceforth dependent on hydraulic institutions.

The same complexity occurs with water services: in areas with centralized systems, excluding those who cannot pay connection fees from public water services is morally debatable for public health reasons. One can then understand that governments in developing countries decide to make water a pure public good and to fund the expansion of services through taxes. This was also the case in nineteenth-century Europe, when outbreaks of waterborne diseases and their mortality extended beyond the poor working class. If the rich could also die from cholera, then excluding the poor from water services became a *de facto* impossibility, even though they could not pay. This was the fundamental reason why most water services were developed, during a long initial phase, under what could be termed *municipalism* – a particular kind of solidarity through public services. Moreover, municipalities were more creditworthy than private companies and could mobilize local taxes and national subsidies from state taxes, in addition to cheaper loans, bonds and other types of subscriptions, to extend the services to poorer neighbourhoods.

This is the situation today in developing countries, where a significant fraction of the population is poor, yet needs access to water. In this respect, mega-cities represent the biggest challenge since the end of the twentieth century. It is almost impossible to enforce the idea of club goods, or even to rationalize the use of potable water, as long as it is still rationed for a significant proportion of the population. Yet, there is a difference between developing world cities now and European cities then: the development of contemporary medicine reduces the risks of waterborne diseases for the rich, which may reduce their motivation to pay more taxes to support the extension of public water services, in particular, to areas where the landless and illegal immigrants live. This issue is developed by the Brazilian case in the book, and remains to be verified for other regions.

In the case of mega-cities an additional conceptual complexity arises. Even though it can be argued that water services and water resources are not the same thing in law and in economics, for the above-mentioned reasons, there are cases where the two overlap. When poor people are not connected to water services, they have to take water directly from the resources. In such cases, these water resources should ideally be maintained at a minimum in the state to which they were prior to the urbanization-industrialization process. In this case, the right to water equates with the fact that domestic use usually comes first in priority among water rights, together with the water needs of ecosystems. However, this is difficult to enforce. To provide a real example, in an Indian watershed where the tanning and textile industries had completely polluted the river and no drinking water network existed, people used water from either the river or the alluvial aquifer for drinking water and became sick as a result of pollution by heavy metals. The Supreme Court of India issued an injunction, but this was not really enforced. The solution, of course, is to provide treated water to the population. But the question then arises as to who will pay the associated costs, as the polluting companies may declare bankruptcy once subjected to the polluter-pays principle.

A second conceptual difficulty can be found in rich cities in developed countries discovering the limits of water treatment. There have been a growing number of cases where the separation of resources and services, through water and wastewater treatment plants (potabilization of water on one end and sewage treatment on the other), is challenged by the very progress of knowledge. More and more dangerous substances are discovered in water and it is becoming increasingly difficult to remove these in conventional water treatment facilities and still meet water quality standards and guidelines. In Germany, for example, it seems frequently preferable to compensate farmers for not polluting the resources than to add new treatment processes. In France, however, such projects of 'payments for environmental services' generate conflicts with farmers who fear being progressively driven out of business. In other places, large cities create quantity problems: peak water demand is so high that public water utilities need to buy water from other users. This is how water becomes a regional issue, and how potable water users become actors in the water resources allocation process. This also further complicates issues regarding water pricing.

Despite such complexities, our conceptualization leads to the need to distinguish between water rights (the allocation of water resources) and the right to water (the right to access to safe drinking water). Urban water conflicts primarily take place over the provision and cost of public water services, and include conflicts between these services and the regional environment.

1.4 A FEW METHODOLOGIES FOR ANALYSING URBAN WATER CONFLICTS

One way to analyse urban water conflicts is to use the above-mentioned conceptual framework as a basis; in other words, to develop a historical and comparative analysis over the contemporary period, starting from early periods of urban water systems development. The underlying hypothesis is that urban water conflicts that metropolises in developing countries are experiencing today are the same as those in the early stages of urban water systems development in developed countries, when only a fraction of the population could have connections to water services. Several contributions in the book have used this approach to analyse various water-related conflicts in urban settings.

In countries where the legal framework of water services has developed separately from water resources law in general, a systematic analysis of conflicts arbitrated by the courts over this contemporary period might be of particular interest. Although court cases may be numerous and their screening can be time-consuming, the same jurisprudence is repeated across many cases. Yet, over a long period of time, one can eventually discover an evolution in court rulings, reflecting progress in the definition and regulation of water services. Pezon's contribution to this book describes this legislative evolution in more detail.

Urban water is the product of a specific combination of geographical settings (resources availability), technologies, economics and finance, institutions and governance, and categories of water users. Conflicts may take place between all actors and non-actors involved in some of the dimensions listed above. For example: conflicts between the city and its hinterland, between engineers supporting alternative technical solutions, between the public and private sectors, between institutions at various levels of government, between the water policy network and water users, and so on. In fact, various conflicts could occur in different settings and be caused by different combinations of factors.

One possible approach to a typology of urban water conflicts could be the categorization of three types of cities (in developed, post-socialist and developing countries) based on the level of socio-economic development, and on the level of service and associated technologies. In developed countries, consumerization, that is, the commercial character of the service with cost recovery through metering and water bills, is widely accepted, and the rising issue is how to help the poor pay their bills, but not to abolish water metering. Conversely, water metering and billing is hotly debated in cities where a significant proportion of the population cannot afford connection to public water services. In developing countries' cities, it is often the lack of solidarity between elites, the middle class and poor rural migrants that precludes establishment of the basic trust needed to provide access to water services for all. In such situations, different neighbourhoods of the same urban area will depend upon differing water supply organizations and combinations of technologies (from centralized continuous supply systems to decentralized technologies, including vendors). Several authors[4] argue that instead of dreaming about a future expansion of networks, neighbourhood communities should improve their actual combination of decentralized solutions in the manner they find appropriate. This will obviously make inequalities in urban areas more evident, but there will be no alternative until *municipalism* prevails.

In post-socialist countries' cities, the situation is different again. Despite progress in public water supply and equal access, poor management, underpricing and decay of infrastructure have led to public distrust, which makes it very difficult to engage in global improvement of the service. For this reason, some authors (e.g. Davis and Whittington, 2004) have recommended improving water services only in certain rich neighbourhoods that would accept full-cost pricing against a high standard service (uninterrupted supply of potable quality water). The profit made in these areas would then be used to progressively improve levels of service for the poorer population. There are, however, reasons to think that this type of arrangement, which is the *de facto* situation in developing countries, would not be accepted in Eastern European cities, where the primary task would rather be to rebuild trust at the local/municipal level for global improvement of the service, despite the present experience of water consumption decline with the 'shrinking cities' phenomenon (see chapter by Naumann and Wissen in this book).

Mediterranean European cities represent a fourth type of city, characterized by: incomplete wastewater collection and treatment; a strong tradition of state water resources mobilization with large hydraulic infrastructure projects; low water pricing linked to this hydraulic tradition; and the development of subsidized irrigation resulting in irrational water allocation favouring agriculture and not urban areas or aquatic ecosystems, which in turn helps explain why urban residents refuse to pay the full costs of this policy.

This brief description of four types of cities (developed, post-socialist, developing and Mediterranean) demonstrates the complexity of the analysis of urban water

---

[4] See the work of Prof. Bunker Roy at Barefoot College, India; and also Zèrah (1997) and Maria (2007). www.barefootcollege.org

conflicts, and at any rate does not support a global 'one-size-fits-all' solution to the public or private provision of urban water services. Moreover, the frontiers between the different types of cities are being blurred by incoming new thinking in terms of sustainable development, further adding to this complexity. For example, the phenomenon of water demand collapse in Eastern Europe's 'shrinking cities' creates dramatic financial difficulties. But this situation might also occur in developed countries' cities, since the rising issue there is the long-term maintenance of ageing infrastructure in a context of water demand stagnation (implying reduced revenues for utilities). In the medium term, the part of the population living in the periphery of these urban areas might be tempted to disconnect from the network and adopt decentralized technologies, which have developed and improved recently, particularly in terms of energy usage. This preference for 'decentralized solutions' might, on the one hand, add to public services crises, but on the other, a new form of governance, where the two types of solutions complement one other, might be more suitable to local situations.

In addition, the emerging priority of restoring aquatic environment quality, as expressed in the European Water Framework Directive, asks all water users, including water and sanitation services, to develop new and more resilient forms of territorial management. A better state of equilibrium between the urban environment and natural environment can be reached through a choice of technologies following and adapted to prior definitions of territorial management strategies (and not the reverse). Of course, such a shift is unlikely to be conflict-free.

Depending on these various situations and factors, urban water conflicts usually involve:

1. Coverage of drinking water services and their quality in terms of equal access and continuity of service
2. Quality and coverage of wastewater collection and treatment
3. Urban hydrology (urban drainage, stormwater quality control)
4. Impacts of cities on the environment, in particular, over-exploitation of water resources, and excessive discharge of pollution
5. Financing investments needed for urban water infrastructure
6. Water tariffs, cost recovery and distributive effects
7. Public participation in decision-making related to water services, and
8. The territorial scale offering optimal compromise between infrastructure mutualization (which allows to average out costs of investments) and the loss of trust entailed by territorially concentrated management units.

Urban water conflicts may also relate to the 'triple bottom-line' of sustainability, comprising economic, environmental and social conflicts.

- *Economic conflicts* often relate to the financing of infrastructure and tariffs.
- *Environmental conflicts* relate to a city's footprint on its natural environment, as well as public health hazards resulting from inadequate water management.
- *Social conflicts* often relate to the affordability of water services and incompatibilities between land-use planning and good water management.

This can be expanded upon by distinguishing a fourth dimension of conflicts, as suggested by Barraqué et al. (2008), dealing with governance issues. Urban water conflicts arising from the interplay between different levels of government may be categorized as *political conflicts*. In developing countries, government at different levels is often characterized by rivalry rather than cooperation. This four-dimensional analysis of urban water conflicts has been adopted by Botton and Merlinsky with regard to studies of Buenos Aires and La Paz.

According to our colleagues in this book, urban water conflicts in large cities in India tend to arise mostly from scarcity issues: lack of access to clean water sources for the poor; issues concerning the quantity of water available for public services (reallocation of water rights); and issues of quality due to either over-exploitation or industrial pollution (decreasing quality of raw water implies growing treatment costs). They also viewed conflicts as 'the activation of a credible threat' at different levels, such as the media, public opinion, demonstrations, riots etc. (Janakarajan et al, see Chapter 13 in this book). Based on this notion of credible threat, another analysis of urban water conflicts consists of mapping levels of unrest in space and time through a systematic review of national and local press, as well as other media. This involves the reporting of payment strikes, complaints to the authorities, demonstrations and eventual outbursts of violence. The case of Mexico City clearly illustrates this approach, with a quantitative analysis of various forms of 'water unrest'.

During one of the meetings of the Taskforce on Urban Water Conflicts, Bakker (2004) proposed another approach to analysing urban water conflicts, described in Figure 1.2, consisting of two axes: governance of urban water services and technologies employed. The first axis proposes a continuum from individual to community to collective solutions to water services, and then to corporate solutions involving public utilities and eventually private capital. The second axis opposes artisan (crafted) and low-cost/decentralized solutions to centralized/industrial and often more expensive solutions, taking into account economies of scale. Conflicts would then occur in situations of change.

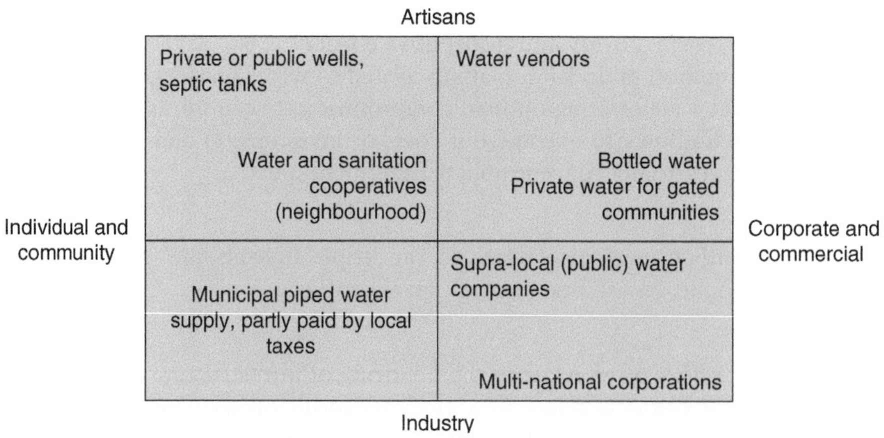

Figure 1.2 Governance of water services and technologies employed

Source: Courtesy of K. Bakker.

Urban water systems in cities in developed countries are usually placed in the bottom part of the table, while cities in the developing world mainly face issues of all four types.

This typology could be supplemented by a third dimension, representing the relationship between the city and its environment, with a double dimension: social and natural scarcities. The larger the city is in a given geographical context and the bigger the pressure on the environment, the more the city must rely either on more distant water resources and water transport, or on more treatment, with growing costs: desalination and wastewater reuse are at the end of the spectrum in terms of operation costs. Similarly, in large cities, wastewater systems often need costly technologies in order to reduce time and space requirements to treat large volumes of urban wastewater.

In other words, this third axis links the level of sanitary technology sophistication and the level of capital invested to the lack of territorial solutions. It allows analysis of an additional dimension of conflicts with the tools of environmental engineering: for example, instead of removing diffuse agricultural pollution from potable water, cities can opt to support the conversion of farmers to organic/extensive agriculture. They can also reduce the scale of water cycles, typically with decentralized solutions: replacing sophisticated technological solutions by land-use based ones, which may favour individual or neighborhood technologies and institutional downscaling.

1.5 CONCLUSIONS

Rapid urbanization, population growth and socioeconomic changes bring increasing challenges and complexities to urban water management. Increasing density in some cities, urban sprawl in others, and growing poverty and inequality in urban areas, especially in mega-cities in developing countries, not only imply increasing size and investment costs on the part of urban water systems, but also reinforce linkages between urban water management and broader socioeconomic and environmental issues, such as urbanization and its effects on the surrounding environment, comprehensive land-use planning and management, and socioeconomic changes. Inadequate management of these interactions leads to conflicts between different water users over competing and conflicting water demands, or conflicts between water users and other stakeholders.

As described in the above sections, urban water conflicts manifest themselves in different forms and dimensions. They involve fundamental water problems – those of quantity and quality – and a range of diverse stakeholders, including domestic water users, local governments, agriculture and farmers, and industry. One of the central problems associated with urban water unrest is the inability of the poor to pay for water services connections. Yet, when they are not connected, the poor frequently pay several times more for the same quantity of water, often from street vendors for questionable quality, compared to their richer neighbours with house connections to water services. This means that urban water conflicts are intrinsically linked with issues of poverty alleviation and equity, which make the problems more challenging and complex. The conflicts examined in the following parts of the book demonstrate the complexity of issues involved in urban water and the importance of the socioeconomic and institutional aspects of urban water management.

Systematic analyses of the interacting issues that contribute to urban water unrest are the first step towards adequately addressing the conflicts and their root causes. The main aim of the UNESCO-IHP Taskforce on Urban Water Conflicts has not been to provide ready-made solutions for local and national governments and international institutions, but to collect cases using various methods as a starting point for the development of future approaches to understanding and addressing urban water conflicts.

REFERENCES

Bakker, K. 2004. *An Uncooperative Commodity: Privatizing Water in England and Wales.* Oxford University Press.

Barraqué, B.R.M., Johnsson, F. and Nogueira de Paiva Britto, A.L. 2008. The development of water services and their interaction with water resources in European and Brazilian cities, *Hydrol. Earth Syst. Sci.,* 12: 1153–64. www.hydrol-earth-syst-sci.net/12/1153/2008/ (accessed on 9 May 2011).

Bauer C. 2003. *Marketing Water, Marketing Reform. Lessons from the Chilean experience.* Resources for the Future, 151, Summer 2003: 11–14.

Bauer, C. 2004. *Siren Song: Chilean Water Law As a Model for International Reform, Resources for the Future,* 179p.

Davis, J. and Whittington, D. 2004. Challenges for water sector reform in transition economies, *Water Policy,* 6: 381–95.

Green, C. 2003. *Handbook of Water Economics, Principles and Practice,* John Wiley, 424 p.

Maria, A. 2007. *Quels modèles techniques et institutionnels assureront l'accès du plus grand nombre aux services d'eau et d'assainissement dans les villes indiennes?,* Thesis, University Paris-Dauphine, 28 June.

Marin, P. 2009. *Public-Private Partnerships for Water Utilities. A Review of Experiences in Developing Countries.* World Bank – PPIAF, 208p.

Ostrom, V. and Ostrom, E. 1977. Public Goods and Public Choices, reproduced in M.D. McGinnis (ed.) 1999. *Polycentricity and Local Public Economies: Readings from the Workshop in Political Theory and Policy Analysis.* Ann Arbor, University of Michigan Press.

Samuelson, P. 1954. The Pure Theory of Public Expenditure, *Review of Economics and Statistics,* 36(4): 387–89.

United Nations (UN). 2010. United Nations General Assembly Resolution 64/292. The human right to water and sanitation. A/RES/64/292 (28 July 2010) http://www.un.org/depts/dhl/resguide/r64.shtml (accessed on 9 May 2011).

Zérah, M.-H. 1997. Inconstances de la distribution d'eau dans les villes du tiers-monde: le cas de Delhi. *Flux – Cahiers scientifiques internationaux Réseaux et Territoires,* 30 (October–December), 5–15.

Chapter 2

Urban water conflicts in recent European history: Changing interactions between technology, environment and society

Bernard Barraqué,[1] Petri S. Juuti[2] and Tapio S. Katko[3]

[1]Centre International de Recherche sur l'Environnement et le Développement CNRS-CIRED and AgroParisTech, Paris, France
[2]Department of History and Philosophy, University of Tampere, Finland
[3]Department of Chemistry and Bioengineering, Tampere University of Technology, Finland

2.1 INTRODUCTION

Two hundred years ago, European city-dwellers obtained their water from public fountains, public or private wells or water vendors. There were no public water supply systems. In Paris the first concession contract for a piped water system connecting private houses, granted in 1777 to the Perier brothers,[1] went bankrupt in 1788. In the nineteenth century, however, the skill of British 'mechanics' made it possible to offer in-house pressured water to urban areas, and later to rural areas. Other European countries soon followed the example, while engineers and hygienists also invented various sewer systems to handle both waste and rainwater away from the city. In the first half of the twentieth century, chemical engineering invented technologies for treating drinking water, as well as wastewater, which increased the autonomy of cities *vis-à-vis* their environment. Today, Europe is the only continent where the great majority of the population is connected to public water supply (PWS). Most of the population is also connected to centralized public sewage collection and treatment (PSCT), and those who are not live in low density areas, and can now rely on efficient decentralized on-site sanitation systems. In several countries, innovation in sewerage has enabled the development of environment-friendly stormwater control.

We argue in this chapter that it is important to keep in mind this two-centuries long experience, not only because of its global success, but also as a history of many different conflicts. Generally speaking, it is a story at odds with the ongoing globalization debate over urban water management: the open conflict between the World Bank, economists and water companies on the one hand, and the alter globalist movement on the other, about the respective merits of public and private sector intervention in water services. This debate is present in Europe, in the light of proposals made by the

---

[1] Their project was to pump water from the Seine river into a reservoir with a steam engine and to distribute it under pressure to private subscribers.

European Commission to open up the provision of 'services of general *economic* interest' to the private sector and to competition. Another issue is the potential impact on the budget of poorer households of 'full cost pricing' of water services, advocated by the European Union's (EU) Water Framework Directive (WFD).

However, further attention to the history of water services in Europe leads to the discovery that the public vs private debate is at least inseparable from, and probably not as important as another issue, pertaining to the relative centralization or decentralization of water services provision. After the initial development of water supply via concessions to private companies, most urban municipalities in Europe decided to take over the responsibility of extending infrastructure to the whole population, and to operate and maintain it over the long term. However, in many cases this was only possible as a result of financial help from the upper levels of government. Even today, as some European countries question this level of municipal involvement, and reorganize services at the supra-local level, it is still important to analyse and discuss the historical decentralization of water services in a non-European context. This in turn permits understanding of how environmental engineering has come to challenge the long autonomy of the water services sector.

This implies a link with the three dimensions of sustainability (economic, environmental and ethical), and the related conflicts that arose across the decades, to the conflicts of power between various levels of government. But it is also important to follow the historical development of water supply and wastewater control technologies, as well as the changes in technological paradigms and engineering disciplines, which were involved in many of the conflicts.

If we go back to the beginning of public water services, we can broadly sketch their development in three stages. In the nineteenth century, or rather until popularization of the discoveries of Sir Edwin Chadwick (1800–1890), Florence Nightingale (1820–1910), Louis Pasteur (1822–1895) and Robert Koch (1843–1910), some hygienists supported the idea that urban water should be drawn from natural environments at a distance from the cities, where it was contaminated. Yet, private water suppliers in those early days usually relied on nearby rivers, hoping that flowing water would reduce risks better than stagnant water. The most important issue was to 'wash the city clean'.

Large metropolises in particular wished to source potable water from increasingly further away, and would obtain support from governments to this end. At the end of the nineteenth century, however, this initial strategy began meeting with resistance in some places, because of competition with local uses or other cities. Eventually though, the related conflicts were resolved: the discovery of bacteriology resulted in the invention of water treatment, which then allowed local authorities to provide good sanitary water from much closer sources, including the rivers upon which the cities were located. Municipalities then took the lead, and in many cases terminated contracts with private companies which were often unable to connect the whole population.

Once water services became a mature industry, municipalities also had to face issues of long-term systems maintenance and capital reproduction. Some were then led to create or recreate utilities with an industrial and commercial status, or to delegate services to private companies. At the end of the twentieth century, the issue of cost recovery put in question some of the technological choices of the chemical engineering period, and new strategies inspired by environmental and ecological engineering were formulated to try to solve the sustainability issue.

2.2 GOVERNMENTS INTERVENE TO PROVIDE LONG DISTANCE SUPPLY OF CLEAN WATER

When industrialization started in Europe, prevailing legal systems were still largely inherited from the feudal period, wherein water rights usually differed between land-lords and peasants. Communities had the right to use water for domestic and husbandry purposes, and landlords had a right to use and abuse (alienate, export, destroy) water, provided they respected the inalienable rights of communities. In expanding cities, migrants from the countryside imported their customs, and requested free and good quality water from public fountains. Growing needs could eventually be met through some sort of water transfer from a distant and 'pure' natural environ-ment, but this implied the consent of landlords. Later, with the formation of nation-states, national government sovereignty over water permitted the settling of disputes between cities and the country.

Domestic water supply was initially provided by private companies, who tried to recover their initial investment from bills paid by customers. These companies were frequently industrial rather than financial in nature: their aim was to install infrastruc-ture and get repaid for it, not to operate services over the long term. Indeed, they were tempted to sell the initial network to local investors, to enable them to recommence the process in another city. However, few people were ready or able to pay water bills in the nineteenth century. This created growing conflicts between companies, users and municipalities: lack of proper payment for their services meant companies often postponed investment in extending or maintaining service, which in turn attracted critiques, and even legal prosecution, from municipalities. The latter were increasingly tempted to terminate concession contracts and recover the services, with the intention of running them under direct labour. In some cases, national governments sided with private companies to maintain their operations. This was the case in Lisbon: central government distrusted the municipality's capacity to operate the service. This was also the case in France (*see* Chapter 5 in this book). But in most cases, governments permit-ted municipalization, and even supported it financially. To understand this, one has to recall that most initial private ventures were undertaken by foreign English or French companies, a situation which attracted broad criticism. A clear example is the Italian law of 1903 which imposed municipal management for water, gas and public trans-portation services. A municipal takeover also took place in Berlin, while in Britain, a systematic and precocious takeover by local authorities occurred, eventually referred to as 'water and gas socialism' – an ironic expression indicating a certain level of con-flict (Saunders, 1983). Indeed, use of the term 'social municipalism' would be better adapted than 'municipal socialism'. Interestingly enough, such discussion went little-noticed in the Nordic countries. In any case, municipalities managed to develop serv-ices and make them at least temporarily trustworthy, expanding and self-financing.

One important factor was public financing: local governments obtained 'cheap money', particularly from the early popular savings banks which they controlled. Government bonds were also attractive to the public. In New England (Anderson, 1988) and Finland, fire insurance companies contributed significantly to the development of water services, as the widespread installation of standpipes proved effective in fighting fires.[2]

In Britain, consumption-based bills by private companies were partly or totally replaced by the payment of rates on property values or local taxes by all citizens, which provided revenues independent of actual water consumption. Today, the payment of water supply through rates is still largely dominant in Britain, and in many countries, sewerage has long been paid through local taxes, all the more so where the systems are combined.[3] In addition, governments would subsidize projects for national health policy reasons. Public procurement was then bound to experience greater success than private, as in the example of Glasgow:

> direct municipal provision seemed to offer several advantages to the city. The existing private company had [. . .] outdated infrastructure [and] consequently was unable to cope with the demands of the rapidly growing population [. . .] Moreover, the company was not in a position to raise the necessary capital for improvements, unlike the Town Council, whose extensive community assets made it eminently creditworthy. Public accountability meant that unpredictable market forces could be over-ridden, and a stable service provided [. . .] Loch Katrine was located in the Perthshire highlands, some 55 km from Glasgow, and thus well away from the polluted city [. . .] The official opening by Queen Victoria on an appropriately wet autumn day in 1859 was an event of enormous significance for Glasgow [. . .] Loch Katrine was unquestionably the prime municipal showpiece for the city, combining the wonders of Victorian technology with the nurturing quality of pure Highland water (Maver, 2000: 90–91).

To ensure the whole population was connected, many cities had to take over the utilities created by private companies, and force the population to connect to PWS (*see* Box 4.1, adapted from Fougères (2004) on the Montreal case in Chapter 4 in this book), and, even more, to PSCT. This is another reason for the choice of payment by taxes or rates. Indeed, creating a local tax or rates on property values provided more money and also cross-subsidies from the richer to the poorer.

In Paris, this municipal move was only partially successful: part of the population preferred the taste of river water sold by vendors, and other people trusted only their own wells, even though urban densification increased the risks of contamination of surface and groundwater. Private companies in the suburbs, and the city itself, also sold water partly pumped from the river. It was Baron Haussmann who made water

---

[2] This issue became increasingly topical immediately after the Great Fire of Turku in 1827. The General Fire Assistance Company of the Grand Duchy of Finland was established in 1832, and then funded cities to establish water works. The quite advantageous loans from this insurance company (averaging about 6% in the second half of the nineteenth century) played a large role in the expansion of city water works. But there were also other important forms of funding, in particular, taxes from spirit distilleries. In each locality a company was given the exclusive right to distill spirits against the payment of a liquor tax. From this tax, little by little, capital was raised over time for the establishment of a water works – about 10% of the total required – most came through taxes and quite substantial donations and willed sums. Loans were also taken from local banks where necessary. A loan from the fire insurance company was nevertheless generally the largest single source of funding, and the interest charged was clearly lower than that of other creditors (Juuti and Katko, 2005: 61–62).

[3] Today, a growing number of countries like Finland, France, Germany and Sweden have legalized the inclusion of wastewater charges in water bills, or created direct charges. This will be explained below.

supply a municipal monopoly, with water coming from distant sources (a project of the engineer Belgrand). But when Haussmann decided to enlarge the capital by annexing surrounding suburbs, he had to strike a deal with *Compagnie Générale des Eaux*, which had a pre-existing contract to bill private customers there. While the city would retain the management of all infrastructure and the provision of free water at public fountains, Générale des Eaux retained a management contract to connect and bill those people who had, or would subscribe to, an in-house connection. In most cities then, billing was rejected by the population, and conflicts developed when operators had to provide the same water for free as 'public' water, and at a cost as 'private' water: where should the frontier be? Metering was still in its infancy, and several systems were used (see Box 2.1).

Box 2.1 Water conflicts involving domestic users in nineteenth-century Paris

During the nineteenth century, three formulas were developed to supply water directly in homes. First was the 'free tap' subscription, or fixed fee, with no measurement of water consumed: the subscriber pays a fixed charge according to several parameters, such as number of persons and animals, number of steam engines, garden surface, and after 1880, number of taps in the housing unit. In the second system, water was delivered through a gauge: the subscriber received a fixed quantity of water each day, thanks to a limiting diaphragm placed on the connection. After 1876, a third system was developed: the meter, which allowed charging for water in direct proportion to the actual volume consumed.

Each subscription mode generated or induced specific conflicts between the various actors around the water supply. These included: the municipal water service in charge of producing water and extending the system; the Compagnie Générale des Eaux, who after 1860 were in charge of recovering bills or fees; the subscribers, who in Paris comprised land and building owners; and the tenants, who constituted the majority of consumers.

With the 'free tap', the chief source of conflict between the water suppliers and subscribers was the selling of water by subscribers to other users. Gauge subscriptions were associated mainly with frauds dictated by the size of the diaphragm. With the support, or at least the tolerance of the billing service, some customers modified the diaphragm's diameter to receive more water than they were entitled to. The development of metering put an end to such conflicts, but generated new ones. Initial frauds developed around the new device, taking advantage of the relative unreliability of the first models rolled out in 1876. These were unable to meter small flows, so when taps were barely opened, the meter was not put to work, thus putting the supplier at a disadvantage. This apparently generated opportunistic behaviour from some customers, who ran water undetected. Even after the invention of increasingly reliable meters, this type of fraud seems to have continued, combined with the tapping of water directly from the water system before it reached the connection. In a 1887 case, the court ruled that this kind of illegal connection constituted theft, rather than just a cheat on the quantity of the good provided. In another case, twenty years later, the appeal court reinforced this ruling, stating that a subscriber who manipulated his meter to avoid registration of part of the water supplied by the utility was guilty of theft.

Metering also impacted the relationship between tenants and landowners-subscribers and induced new conflicts. Prior to metering, a landowner was unconcerned about the amount of water used by tenants, since it did not affect his own bill. With the arrival of metering, landowners were tempted to control and limit the amount of water used by tenants. From then and up to the present day, metering has remained largely a collective matter at the building level.

Source: K. Chatzis (2006).

The technical solution of delivering untreated water from distant sources remained dominant in the New World, and was also extended to the rest of the world in the aftermath of the Second World War. This was due to the co-occurrence of international financing institutions offering cheap money to newly formed Nation-States, and various (Keynesian or socialist) forms of support to governments intervening in infrastructure provision. In the 1950s and 1960s large hydraulic projects were increasingly devoted not only to cities, but also to irrigated agriculture destined for the export market. Today, many states in developing countries still base their water policy on large water transfers, so as to indirectly subsidize the production of irrigated cash crops for the world market. In some cases, the maximum extractable water resources have been reached, while irrigation is often *de facto* privileged at the expense of public water supply.

In short, government intervention allowed the transfer of sufficient water from distant sources, through largely subsidized aqueducts. Local authorities began taking over infrastructure, initially created by private companies. The frequent financing of extensions by local taxes or rates reduced tensions between urban populations and water suppliers, making public services acceptable. But municipalities were now dependent upon higher levels of government, and distant water sources, which eventually would become unsustainable. The innovation of water treatment helped to alleviate this situation.

2.3 FROM QUANTITY AT A DISTANCE TO QUALITY CLOSE AT HAND

In the early twentieth century, in the heart of industrial Europe, growing population densities and reduced natural resources increased competition for pure water resources, while the development of bio-chemical analyses showed the growing extent of contamination. The issue was shifting from quantity to quality. In the end, it was decided that, whatever source it came from, surface water should be filtered (end of the nineteenth century), and later chlorinated, ozonised or disinfected through GAC (granulated activated carbon) beds (around the First World War). Groundwater (e.g. in Finland), was still not filtered, if treated at all.

But once water began to be treated, the question arose of why it should be transported away. Taking it from the river just upstream of the city would not make a difference in terms of public health and would save a lot of investment. Moreover, surface water intakes usually provided significantly more water than underground intakes, implying economies of scale. Consequently, large European cities changed strategy from investments aimed at increasing available quantities of water to those aimed at improving its quality. This of course, resulted in a significant rise in operation costs. However, the concurrent rise of the middle classes saw the status of domestic delivery of pressurized water change from a luxury good to a normal commodity. As a result, customers accepted to pay water bills, which covered the operation costs.

This is exactly what occurred in Paris a century ago. At the time, the idea that the city should get water from distant sources prevailed (having originated in Baron Haussmann's distrust of filtration, which conversely was already in use in Britain). The work of engineer Belgrand turned towards securing longer distance sources of water (around 100km away from Paris). It was even expected that the capital city would have to obtain water from the Loire, which would require central government licensing – a project that dates back to the seventeenth century, when water was needed for Versailles. However, the Loire is characterized by very low flows in summer, precisely when water demands would be highest. Moreover, the potential donor region around Orleans strongly opposed the transfer.

Then, in 1890, an engineer named Duvillard presented a project to draw water from Lake Geneva, in other words, 440km away from Paris. Despite its impressive scale, and the international status of the lake, the project was technically simple, even at that time, and there was plenty of water available in the Rhone. Proponents of the project soon produced all sorts of arguments to convince the Paris city council and the French government that a 'capital of the world' would need at least 1,000 litres per capita per day (lcd), that is, five times more than the highest standards of the time. Paris would have more luxurious fountains, improved street cleaning,[4] better domestic comfort and hygiene. Besides, such a quantity of water would extend navigation possibilities during drought periods, help flush wastewater away to the Seine and then to the sea, and make other water resources available for local economic development. In the end, they argued, a huge transfer would make Paris PWS reliable forever, and the bigger its size, the cheaper each cubic metre.[5]

But while Paris council debated, a disease epidemic broke out, and it was found that one of the distant natural intake points (the Loing springs) was to blame. This showed that even distant, 'pure' water could be contaminated and should be filtered and treated. In 1902, Paul Brousse, one of the founding fathers of so-called 'municipal socialism', inaugurated the new water filtration plant in Ivry, just upstream of Paris,[6] with a subsequent decision taken after the First World War to institute chlorination. A long-lasting choice was being made. Water demands were growing incrementally at the time and the big jump embodied by Duvillard's project appeared too risky. Finally, the Lake Geneva aqueduct was discarded by the Paris City Council for national defence reasons: what would happen in the future if the Germans conquered the aqueduct and cut off the supply? Beyond the geopolitical anecdote, investments focused on quality replaced those on quantity, just as chemical (sanitary) engineering supplemented civil engineering. Today, the city of Paris still gets 50% of its drinking water from distant springs, but the suburbs rely almost exclusively on surface water abstractions.

---

[4] Paris is one of the few cities in the world where streets are washed clean: for political and business reasons linked with Haussmann's decision to merge suburban communes and extend Paris from twelve to twenty *arrondissements* (boroughs), it was decided that water for public purposes would be produced by the city and delivered for free through a public network, while a second network would serve domestic and other private needs with the payment of water bills in return. Paris still has two PWS systems, one potable and the other non-potable by today's standards. The non-potable network produces hardly filtered Seine water to flush the sewers, to supply the lakes in the Boulogne and Vincennes parks, and to clean the streets. Other public uses such as fire hydrants and garden watering have ceased because of unreliability, lower pressure, sprinkler clogging and so on.

[5] It is interesting to note that the same type of arguments have recently been raised by the Franco-Spanish partnership to transfer water from the Rhone to Barcelona (350km). This example also demonstrates the indirect impact of irrigation: Mediterranean cities have to obtain water from ever further because all local surface water is provided quasi-free to farmers by central government projects, and groundwater is usually ignored.

[6] It was recently redesigned, and serves as a showcase for French know-how in 'slow filtration'. However, it will close down due to the fall in water demand.

After the Second World War, the growth of the capital and increased per capita consumption led to expectations of water scarcity. However, the proposed solution was not to build an aqueduct, but rather to enhance low flows. The prefect of the Seine *département* (county) took advantage of a severe flood to obtain agreement for the construction of three large upstream reservoirs on the Seine, the Marne and the Aube, which would increase summer flows and meet Paris water demands even during very serious droughts (as occurred in 1976). Interestingly, a fourth upstream reservoir and a new aqueduct to Paris were planned by Mayor Chirac's councillors in the early 1990s. However these were abandoned along with other planned aqueducts for the same reasons: water demand in Paris decreased by 13% between 1990 and 1996 (Cambon-Grau, in Barbier, 2000), and giant water supply companies stated that they must and can purify potable water anyway: why then abandon instream intake points? These latter developments form part of the third stage of the water industry with the birth of environmental engineering culture, comprising integrated river management and water conservation. Anticipating further developments, we believe that the Paris case illustrates how PWS used a combination of technologies issued from civil engineering, sanitary (chemical) engineering and, more recently, environmental engineering to reduce their risks.

As in Paris, the invention of water treatment allowed many cities to trust and turn to nearby surface water, and thus complete their networks and serve the population through a primarily local solution. This is one of the chief rationales for the generalization of municipal control of utilities, at least in temperate climates. Eventually, the best solution would be to establish inter-municipal joint boards in order to attain economies of scale. Treating the water also meant increased operational costs, supporting the notion of water as a universal service to be paid for by water bills, which would in turn enhance the financial autonomy of the utilities. Billing rather than taxation was better accepted later in the twentieth century as urbanites became accustomed to having tap water. After the Second World War, the rise in river pollution was deemed increasingly unacceptable, and in-stream users such as fishermen increasingly took city sewers and industry to court. Cities were constrained to build sewage treatment plants to reduce discharges, and the related services frequently became local utilities, once most urbanites would be connected.

In a way, the two types of plants (drinking water treatment and sewage works) materialized the frontiers of the 'networked city' (Tarr and Dupuy, 1988), enabling the development of an institutional, legal and economic system quite separate from the issue of water resources allocation, and one that would eventually be extended to rural areas. This was the domain of local public services. Conversely, in the Mediterranean part of Europe, central governments retained a strong involvement in water policy, sponsoring the development of regional bulk water transfer institutions. Local authorities were then provided with very cheap water to distribute to their population, which in turn helped maintain the tradition of inexpensive but relatively unreliable and incomplete services. Today, there is paradoxically less open criticism of service quality in these countries than elsewhere in Europe (e.g. in UK and France), but water users express a sort of 'silent distrust'. Raising water prices or rates is very difficult, because authorities first need to earn the confidence of their citizens by improving the service.

Despite these differences, water and wastewater services eventually developed into a mature industry in Europe, with a shared consequence: the initial infrastructure, which had frequently been subsidized, would eventually need replacement, while at the same time environmental performance had to improve. As such, the increased financial needs created a new set of conflicts.

2.4 THE CRISIS OF MUNICIPAL WATER SUPPLY SERVICES

France, Germany, the Netherlands and the Nordic countries all expended efforts on city sewage collection from the 1950s onwards, and on sewage treatment from the 1970s. To make investment funding easier, it was decided to change the status of PSCT from an imposed administrative service, for public health reasons, to a commercial service similar to PWS, and to include payments within water bills. This caused bills to jump or increase dramatically (depending on the degree of subsidy PSCT receives from governments). For example, in Finland the costs of wastewater treatment were covered mainly by sharply increased prices. In Germany, the billed sewage cost more than doubled water bills. In France, the development of sewage works was made through earmarked levies on drinking water, which then went into a kind of mutual banking system: the well-known *Agences de l'eau*.

But during the same period, PWS had to face the issue of renewing aging infrastructure without the support of government subsidies. This is the fundamental reason why municipalism had to evolve towards legal private status: under traditional public accounting it was neither practised in depreciating assets nor making renewal provisions, while private accounting could. PWS, and later PSCT, slowly shifted towards commercial status, adopting depreciation and provision practices, and this of course meant another rise in water bills. Moreover, governments today are influenced by economists who support full or at least fair cost-pricing,[7] and phase subsidies out. In turn, water bills rise dramatically: in France, bills doubled on average between 1990 and 2004. This has brought about unexpected consequences: an increasing number of large users, such as manufacturing industry and services, either changed their processes or invested in leakage control. This explains the recent stagnation of water volumes sold. In some countries, even domestic users have reduced their demand for PWS, through changes in fixtures and domestic equipment, different garden designs, and the use of rainfall storage or other alternative sources of water for non-drinking uses. Such cases have been identified in several places in France, Belgium and Germany (Montginoul, 2005; Cornut, 2000). Similar instances have also occurred more recently, in Central and Eastern Europe (Juuti and Katko, 2005: 230–31). Eventually, these new attitudes will threaten the financial balance of public services.

Simultaneously, water suppliers are discovering that it is going to be ever harder to permanently comply with drinking water standards at reasonable costs. The control of eco-toxicologists over standard production tends to privilege a traditional 'no-risk' strategy (Lave, 1981), without taking into account the implied costs. For example, in Europe, the lowering of the lead content from 50 to 10 µg/l implies removal of all lead pipes; this will cost up to US\$35 billion while there is no observable evidence for lead poisoning from water under the former standard.[8]

---

[7] Fair cost-pricing means that the principle of cost recovery is accepted but only partly implemented, in other words, bills include part of the capital depreciation cost.

[8] Only some European countries are concerned: France, Italy, Spain and the UK. Other countries, Nordic ones in particular, had few lead pipes and had already phased them out.

The multiplication of drinking water criteria[9] is making the situation increasingly complex. For example, chlorination byproducts induce a very small probability of cancer. There are many other such examples. Year after year, the media report a growing proportion of people receiving non-compliant water, even though treatment is improving over the long term (for a history of drinking water criteria and the present result, see e.g. Okun, 1996). This is creating growing distrust among water consumers, and in some cases resulting in open conflict.

To lower the risk of being unable to make acceptable renewal choices, local authorities in charge of water supplies turned to a new strategy: land-use control on areas or catchments where groundwater is abstracted by PWS. This often implies a change to organic, or at least no-nitrate, no-pesticide agriculture and compensation programmes for farmers (Brouwer, Heinz and Zabel, 2003). This policy has turned out to be cheaper than sophisticated water treatment and results in a positive-sum (win-win) game with farmers. Yet, a new and more sustainable equilibrium still seems a long way off. Worse, drinking water criteria are still being regularly reinforced as new risks are discovered, and spiralling treatment costs and their negative effects are ongoing.

In any case, contrary to what some 'hydro-schizoid' water engineers seem to think in Spain (Llamas, 2001), the present time is not appropriate for large-scale, long-distance water transfers (Barraqué, 2000). The events that took place in California starting in the 1970s are now occurring generally in Europe: it is getting ever harder to build dams, because environmental movements which oppose their ecological impacts have been joined by economists and liberals who advocate full-cost payment of water infrastructures by their beneficiaries. The new rationale seems to be: 'conserve first and manage the demand, there is no cheap money in sight for water transfers'. Copenhagen is not going to buy water from Sweden, Bari isn't going to receive Albanian water, and London will have to reduce leaks drastically before it can obtain water from Scotland. In 1980, a project to transfer water from Umeå, Sweden to Vaasa, on the western coast of Finland through a submerged pipeline on the sea bottom was explored, but rejected. Many other elaborate projects on other continents are also dying like 'hydro-dinosaurs'.

However, water transfers or treatment technologies are still needed. The deciding factor is economic analysis. For instance, long-distance rock tunnels for raw water transfer were completed in 1982 for the Helsinki metropolitan area in Finland (Juuti and Katko, 2004: 28), and the southern region of Sweden in 1985 (Isgård, 1998: 79). There is now a boom in desalination in Mediterranean countries to produce additional water for summer tourists.

New York, like many US and Canadian cities, could follow a different path from its European counterparts, largely because of the abundance of available clean water. The city uses a lot of water and over time has taken it from further and further away, while protecting water intake points through extensive land-use control. However, the US metropolis currently lags behind Europe in relation to filtration, as clean natural resources are not immune to cryptosporidium, a type of protozoa, and other new (lethal) substances. A US EPA panel of experts concluded that New York should not be given any further derogation with regard to the need to filter and treat water

---

[9] Bacteriological criteria now form just a small part of drinking water standards; heavy metals and micro-organic compounds bring the total to sixty-three criteria in Europe and eighty-four in the US.

extensively, while city engineers have argued that increased land-use control would suf-fice (Okun et al, 1997; Ashendorff et al, 1997). In the end, the city obtained a 'Filtration Avoidance Determination' (FAD), and the success of the contracts with farmers in the Catskills and upper Delaware allowed EPA to renew this FAD. But the controversy is serious, since the new treatments requested are likely to seriously increase water prices.[10] This might, in turn, cause serious social impacts, plus an eventual water demand collapse, leading to a further question: why not just pump the water in the Hudson River, and forget about water from Canada? It seems, however, that the issue may prove highly controversial, effectively suspending any such major change.

In summary, we could say that population density and environmental issues in Europe have led the water industry to progressively supplement the quantity approach (hydraulics and civil engineering), with a quality approach (water treatment and sanitary engineering), and more recently with resources protection and demand management (environmental engineering). But in turn, this questions the viability of pure municipal procurement, and calls for various types of averaging out.

2.5 EUROPEAN WATER SERVICES AND THE THREE ES OF SUSTAINABILITY

Can the Europeans afford their water policy? This was a growing issue throughout the 1990s with the first estimations of cumulated investments implied by anti-eutrophication directives adopted in 1991 (directives on urban waste water and on nitrates from agriculture), on top of older directives. Water policy has been increasingly criticized by the public, the press and politicians. Paradoxically, criticism has been much stronger in northern European member states where infrastructure is complete, than in south-ern ones, where equipment is still being installed and water prices are still far from matching costs (which in turn means government subsidies or infrastructure degradation). This shows that *rationalization* (both in terms of economics and environment) cannot really start until *rationing* is over.

The issue developed further in discussions between certain member states and the European Commission. It was felt that supplementing traditional standard-setting based on emissions control (pollution discharges, drinking water criteria), with another type of regulation based on the desired ecological quality of the aquatic environment, would be over-costly. Moreover, compliance with all these regulations at the same time would prove impossible, leaving governments open to being sued at any time by their own citizens for non-compliance to legislation. As a result, a projected directive on ecological river quality was halted, until a framework directive could be set up and discussed, so as to introduce coherence and cost-savings. The WFD (EEC 2000/60) was issued in October 2000, but caused some to wonder if the European Commission, Parliament and Council had not set an impossible target of 'cleaning all water in fifteen years', as the US Congress had in 1972. Furthermore, the European Commission and Council had initially given their official support for full-cost pricing, without having undertaken a feasibility study. Finally, the WFD only requires calculating and publicizing the gap between full-cost and existing pricing.

---

[10] In particular, if metering is introduced to replace the outdated frontage rates system, negative redistribu-tive effects could be serious, as pointed by Netzer et al (2001).

However, the issue of costs may have been biased. The costs of implementing regulations have generally been calculated on the basis of technologies that, although mastered, were developed by sanitary engineers before the rise of environmental issues and under a different rationale *vis-à-vis* risks and costs. The high costs incurred today might then be due to incapacity to develop environmentally innovative technologies and, above all, strategies other than 'supply-side' or 'end-of-pipe': demand-side management, land-use control and integrated planning (see Moss, 2000). Moreover, southern member states, with different climate and population patterns, may need to develop specific technologies rather than copying northern states, even though they receive support from them via European structural funds. However, it will be difficult to induce southern member states to adopt this new technological stance before a severe crisis demonstrates the clear need for it. These issues were addressed in comparative research on the costs forecast for compliance with then most costly directive on urban wastewater treatment, issued in 1991,[11] on urban wastewater collection and treatment (UWWD). The Eurowater partnership of European water policy analysts has addressed the issue of future water services (i.e. PWS and PSCT) in the light of the UN's definition of sustainability, asking: how do we reconcile economic, environmental and ethical/equity sustainability criteria? This is referred to as the 'three Es approach':

- economically: how is the enormous capital accumulated in water services technologies maintained and reproduced over the long term? If we do invest enough, what will be the impact on water bills?
- environmentally: which extra investments are needed for the sake of public health and the environment? When accumulated costs are too high, what are the alternatives to classical systems?
- ethics and equity: if all 'sustainability costs' (long-term reproduction of infrastructure, environmental protection and users costs) are passed on to consumers, can they afford it, and is this politically acceptable?

The two first issues are in fact interrelated, through the limited capacity of governments and cities to face unsubsidized investment. Investments to replace ageing infrastructure and new environmental investments do overlap, but both have to be made without subsidies. For example, if England and Wales were to rebuild their water services infrastructure completely, they would have to spend £189 billion! Some of these assets can depreciate over a hundred years, but others in thirty or even ten years. Under the present centralized regulation system, OFWAT, the UK water regulator, can calculate what investments should be made and when, and which price should

---

[11] See Chapter 8 in Correia (1998) vol. 2: Eurowater is the name of an original partnership funded by DG Environment, LAWA in Germany, the NRA in Britain, and the Gulbenkian Foundation in Lisbon. The partners were the WRc in England, River Basin Administration centre in Delft TU (Netherlands), Ecologic, an environmental policy consultancy in Berlin, LATTS, a social sciences laboratory in Ecole Nationale des Ponts et Chaussées, under the leadership of Francisco Nunes Correia, then hydrology and environmental policy professor in Lisbon's civil engineering faculty, later minister of the environment in his country.

result. Other countries are also trying to do this despite decentralized organization and the small size of undertakings. However, many experts believe that most European utilities are in the quiet process of 'eating' the initial capital, in particular for sewers. Furthermore, UK water charges are reasonable when compared to other countries.

To make privatization of English and Welsh utilities attractive (i.e. reduce the general public's strong opposition to liberalization), the UK government cancelled the debt of the previous regional water authorities, then offered a so-called 'green dowry' to help the new companies deal with the 'nasty European directives'. What anyone other than the Thatcherites would have called subsidies amounted to £6.4 billion, in other words, more than the French or German governments had given away to their water services over the last twenty years. In her thesis, Bakker (2003) shows that the British 1989 privatization put the water industry into a fundamentally unsustainable situation, the effects of which will be increasingly realized in the coming years, more than twenty years after it took place. This is why the issue of cost recovery raised by the new WFD is so unclear: if you want to seriously verify economic sustainability, you need to go back thirty or fifty years to see how subsidies during that period have influenced today's price. If subsidies are removed now, effects on water prices will not be felt for another twenty years at least.

We have also found that water supplies or authorities – under various modes according to the politico-historical culture – have organized averaging out or cross-subsidy mechanisms to limit the impact on water prices of heavy and lumpy but long-lasting investments. Some have done this through spatial integration.[12] Indeed, there is a concentration process in effect across Europe, which tends to dispossess local authorities of their former responsibilities regarding water services. In Italy the law now bans local direct labour procurement, and promotes the integration of *servizio idrico integrato* at the level of provinces, although it has met with resistance from local authorities. Conversely in the Netherlands, since the Second World War, a significant and voluntary concentration of water services has taken place between local authorities, while commercial forms of water services provision were adopted. Today, only ten PWS mixed-economy companies remain, with twenty-seven water boards in charge of PSCT.

Another strategy exploits temporal averaging out (earmarked funds, water banks such as the French *Agences de l'Eau*, modernization of public accounting to allow for depreciation, etc.). The German model makes use of an original institution, the *Stadtwerk* – a single, formally private but publicly owned company that runs several technical networks in a given city. This model was criticized by the World Bank, but on the basis of incorrect data (Barraqué, 1998). The Stadtwerk often supplies electricity, transportation, water and gas services, while sewerage is taken care of by another municipal organization (Juuti and Katko, 2005: 82).

Lastly, social forms of averaging are well known, even if not always presented as such. For example, paying for wastewater via local taxes makes this service dearer for those who have a large house, and are usually richer. Under municipalism, water

---

[12] This is what took place in Britain: before centralization created the ten regional authorities in 1974, later privatized in 1989, a sustained concentration process had been ongoing since the Second World War.

supplies did not pay much attention to the detailed breakdown of potable water uses and distributive effects of tariffs, since they wanted the best quality in unlimited quantities to serve all purposes (in a *commonwealth* vision). Modern water supplies are still reluctant to really study distributive effects: they want water services to be customer-based (through billing instead of taxing) for financial reasons, and so assert that this way is more equitable. However, the English and French examples show that commodification is dangerous, even for water companies themselves, since consumers will not readily accept that a good service is costly, and are even less likely to accept that water conservation implies in the short term an increase in the unit price (per cubic metre), if not in the yearly bill. Hence, the opening of the traditionally closed PWS policy community to a range of newcomers, in particular, the public in general, makes water engineers feel awkward and insecure. But is there any alternative if the stake is the general confidence placed by the public in the service and those who provide it?

2.6 THE NEW SOCIAL ISSUE OF SUSTAINABILITY

It is therefore the third issue, the third E of Eurowater's sustainability – the ethics/ equity dimension – which is the most crucial today: is social and political acceptance still possible in the long run, or are we heading towards a collapse of the networked model for water? Until water supplies were forced to adopt demand management and pricing closer to real costs, water services were broadly accepted. However, this model rested on an 'out of sight, out of mind' basis, as Martin Melosi terms it. Now the limitations of the supply-sided model force managers to study demand like any other business, in other words, through the use of marketing. The breakdown of drinking water demand and the evolution of sector-differentiated uses have to be studied. But studies alone may not be sufficient, and utilities may well have to involve the public in their research. For example, claims that metering is moreequitable and efficient (because water conservation is rewarded) may seem problematic to residents of central cities in condominiums, since water use is largely determined by appliances, and there is little elasticity to (existing) pricing (Barraqué, 2011). Then, the additional cost of separate metering and billing often exceeds the benefits of water conservation, while it can have detrimental social effects (Netzer et al, 2001).

Even though local econometric studies show little elasticity of domestic demand to price, there is a slow reduction in use per capita taking place in the developed world. In the US, this may be obtained through information policies and subsidies to individual conservation measures (Dickinson, 2000). The point is for changes to take place slowly – to let people adapt to higher water prices over sufficient time. In northern Europe, smaller consumption and higher prices reduce the scope of change compared to North America. Yet, there has been such a reduction in water usage (mostly from large users and outdoor uses) that the only way to maintain the cost recovery principle is to raise unit prices. One way to ease this problem is to introduce a non-volumetric basic charge. Due to the lack of available information for users, in particular concerning the difference between short-term and long-term sustainability, and the real causes of (formal) privatization, the situation has generated a growing distrust of infrastructure systems, which had reached their equilibrium under a 'municipalist' type of welfare. In some cities in France this distrust is growing to the point where citizens resort to legal processes. But in many places, they look for alternative sources of supply such as private wells, water harvesting, etc.

Southern Europe illustrates this point differently, since the non-completion of infrastructure prevents prices rising to cost recovery levels, and this generates irrational allocation. Almeria was thus supposed to get water from the Pyrenees or the Rhone, at a cost of ten times what the city would pay to buy this water from farmers. Through a model initially developed under the dictatorship, but now involving European subsidies, Spanish farmers overexploit aquifers and request increasing volumes of surface water transfers to grow more crops, part of which will be bought as surplus by the EU. The farms themselves are largely owned by rich northern agribusiness. If not coerced by this uneconomical set up, farmers would make more money without working, just reselling 'their' water to cities, as occurred recently in California. This is related elsewhere in this book, and shows the adverse consequences of potential 'water markets'.

The growing orientation of water services towards demand-side management is very dangerous if not done within a collective re-learning process, far from the 'out of sight, out of mind' tradition. For example, there is a trend in Europe to encourage the generalization of individual water metering and billing, even in flats in small buildings, for the sake of better equity, enabling people to trade-off between saving and paying, and so on. Yet in most experiments, individual metering does not induce a significant change in indoor consumption patterns, in particular, after a few months. Moreover, in many cases the savings of the most thrifty are offset by the yearly cost of the meter itself (depreciation, reading and billing separately). Consumers are consequently frustrated and angry on seeing water bills rise after being told they would make savings. Distrust is growing again. According to Rajala and Katko (2004) installing meters in individual apartments might be feasible in new houses, but is too expensive in old dwellings.

One of the most interesting recent cases is the Government of Flanders's decision to strictly implement the Rio Agenda 21, and give an initial water volume away for free. For practical reasons, the free volume was set at 15 m^3 per capita per year, while extra volumes would be charged such that water supplies would make the same income. The consequences have been studied by the social and economic council of the Flanders region (Van Humbeeck, 1998). Firstly, water supplies have seen their total volumes shrink, because increasing block tariffs have made the extra volumes very expensive. They suspect that people reinvest in cisterns and private wells for watering gardens and so on, but in the meantime, they are obliged to raise unit prices. Secondly, it turns out that richer families pay a little less than before, while poorer ones pay a little more. This is due to specific socio-demographic conditions: richer families are larger than poorer ones, and extra children do not take enough water to imply a bigger bill (remember that the free initial volume is per person). This quite complex case illustrates how little we know about domestic water uses. Another illustration can be found with the Barcelona water bills strike (*see* Chapter 3 in this book). If we are going to tackle the present crisis of water services, it can't be just with an economist's toolbox and a sense of moral justice. We need anthropologists, sociologists, historians and geographers, as well as interdisciplinary approaches. Attitudes are so variable and culture-specific that it seems dangerous to relate eventual evolutions in water demand solely to price elasticity.

2.7 CONCLUSION

There are increasing discussions on how to reach a better compromise between the economic, environmental and ethical dimensions of water services sustainability in the EU and the US, and the development of new tools and indicators can give us some confidence for the future. However, what is missing at present is an indicator that would translate the degree of overall confidence among the general public into the systems and water services policy community.

The real threat to water services is illustrated by the situation in Eastern Europe and large developing cities. When utilities are not fully reliable and operational in terms of quantity, continuity and quality (in particular sanitary quality), the compensation strategies adopted by various groups of users tend to *increase* the uncertainty and unreliability of services. In such cases, attempts to apply new economic approaches, suited to the mature systems of developed countries, may just be catastrophic. At present, privatization and even public-private partnerships at state or regional levels demonstrate their limitations with regard to the universalization of water services. Ultimately, it may prove impossible to improve deficient utilities in developing countries, and their unsatisfactory condition will remain, along with bottled mineral water. Several chapters in this book raise this issue. Zérah (1997) showed the aggravation of social fragmentation by deficient water services in the case of New Delhi – the wealthier develop systems to provide permanent, albeit expensive water; the poorer the families, the more dependent they are upon vendors and unsafe raw water. But we could then argue: is there any significant difference between this example and the kind of poor services and private alternatives found on Greek islands?[13] If confidence in PWS in developed countries goes on shrinking, and there is a significant decrease in water volumes sold, we may end up with growing irregularities in service, such that the fragmentation and conflicts that characterize the systems of large cities in the developing world, would prevail globally in the long run (Barraqué, 2001).

In Durban (RSA), the special chip cards produced by *Lyonnaise des Eaux* to buy good potable water from fountains have met with some success: the company makes money, users are freed from gastroenteritis, and are also able to escape traditional tribal control over wells. In Buenos Aires, the same company proposed the delivery of cheap bulk water to the entry of *barrios*, instead of desperately trying to meter each user and fight for bill recovery. However, both examples highlight a debatable point: are poor, developing country populations destined to remain tied these alternative partial services? Europeans could argue that the fundamental reason for these alternative types of PWS is the lack of a municipality with sufficient legitimacy and capacity to build reciprocal confidence between the utility and water users.

So why don't we take a look back at the future of municipalism? In our view, the only way to avoid the above mentioned *Lyonnaise des Eaux* dilemma is municipal involvement, as in the good old days before local welfare and public economy were thrown into the dustbin of liberal economics. The achievement of municipalism was to channel the savings of the upper and middle classes into financing a long-term solidarity system for all, based on public control of urban services, sometimes with the

---

[13] In some Greek islands, people rely first on their wells and cisterns, and turn to PWS for additional uses. In these cases, demand for public water is highly variable, which is bad for reliability of the service.

participation of the private sector, but without privatization of infrastructure. Besides, municipality-owned water undertakings can adopt modern accounting practices with depreciations to cover their long-term infrastructure costs. In many countries, such undertakings buy goods, services and works from the private sector based on continuous competition, without tying their hands with long-term operational contracts (Hukka and Katko, 2004). In most countries, such arrangements have been in place since the beginning of the modern systems in the late 1800s (Juuti and Katko, 2005: 71). The newest development is concerned with the settlement of disputes between large cities and surrounding rural areas: it turns out to be cheaper for utilities to help farmers to reduce or prevent diffuse pollution than to remove nitrates and pesticides from the raw water (Brouwer et al, 2003).

Indeed, the territorial dimension of this issue is more important than the simple public vs. private debate. If we keep hope for the cities of developing countries, we will have to invent similar mechanisms at appropriate territorial levels, depending on the national/local citizenship traditions and community cultures. Such a 'subsidiary' system would certainly offer better guarantees for national and international public investors, which in turn would result in access to cheaper money for water systems. This seems to be the lesson taught by the historical developments of European water services.

REFERENCES

Anderson, Letty. 1988. "Fire and Disease: The Development of Water Supply Systems in New England, 1870–1900." In Joel Tarr and Gabriel Dupuy (eds.), *Technology and the Rise of the Networked City in Europe and America*. Philadelphia, PA: Temple University Press.

Ashendorff, A., Principe, M., Seeley, A. and Beckhardt, L. et al 1997. Watershed protection for New York City's supply. *Journal of the American Water Works Association*, 89(3) March.

Bakker, K. 2003. *An Uncooperative Commodity: privatising water in England and Wales*. Oxford: Oxford University Press.

Barbier, J.M. 2000. Evolution des consommations d'eau. *TSM – Génie Urbain – Génie Rural*, periodical of the AGHTM, 2.

Barraqué, B. 1998. Europäisches Antwort auf John Briscoes Bewertung der Deutschen Wasserwirtschaft. *GWF Wasser-Abwasser*, 139(6).

Barraqué, B. 2000. Are hydrodinosaurs sustainable? The case of the Rhone-to-Barcelona water transfer. E. Vlachos and F.N. Correia (eds) *Shared water systems and transboundary issues, with special emphasis on the iberian peninsula*. Lisbon: Luso-American Foundation and Colorado State University.

Barraqué, B. 2001. De l'eau dans le gaz à l'usine à gaz. *Hydrotop 2001, colloque scientifique et technique*, Marseille, 24–26 April, Recueil des Communications, C-068.

Barraqué B. 2011. Is Individual Metering Socially Sustainable? The Case of Multifamily Housing in France, in *Water Alternatives* 4(2), pp. 223–244. *www.water-alternatives.org*

Brouwer, F., Heinz, I. and Zabel, T. (eds) 2003. *Goernance of Water-related Conflicts in Agriculture, New Directions in Agri-environmental and Water Policies in the EU*. Dordrecht: Kluwer Academic Publishers.

Chatzis K. 2006. Brève histoire des compteurs d'eau à Paris, 1880–1930. (archives). *Terrains & travaux*. n° 11 pp. 159–178.

Cornut, P. 2000. *La circulation de l'eau potable en Belgique: enjeux sociaux de la gestion d'une ressource naturelle*, Thèse de Doctorat, Université Libre de Bruxelles.

Correia, F.N. (ed) 1998. *Eurowater, Institutional mechanisms and Selected Issues in Water Management in Europe*. 2 vols. Balkema, Rotterdam.

Dickinson, M.-A. 2000. Water conservation in the United States: a decade of progress. A. Estevan and V. Viñuales (eds) *La efficiencia del agua en las ciudades*. Bakeaz y Fundacion Ecologiay Desarrollo.

Fougères, D. 2004. *Approvisionnement en eau à Montreal: Du privé au public 1796–1865*. Sillery: Septentrion.

Hukka, J. and Katko, T. 2004. 'Liberalisation of the water sector' – a way to market economy or to monopoly market? *Water & Wastewater International*, 19(9): 23–25.

Isgård, E. 1998. I vattumannets tecken – Svensk VA_teknik från trärör till kväverening Ohlson & Winfors (in Swedish).

Juuti, P. and Katko, T. 2004. From a Few to All: Long-term Development of Water and Environmental Services in Finland. KehräMedia Inc.

Juuti, P. and Katko, T. (eds) 2005. Water, Time and European cities. History matters for the futures. Printed in EU. Available: htpp://www.WaterTime.net, also http://tampub.uta.fi/index.php?Aihealue_Id=20.

Lave, L. 1981. *The Strategy of Social Regulation*. Brookings Institution.

Llamas, R. 2001. *Las aguas subterraneas en España*. Fundación Marcelino Botín.

Maver, I. 2000. *Glasgow*, Town and city histories, Edinburgh University Press.

Montginoul, M., Rinaudo, J.-D., Lunet de Lajonquière, Y., Garin, P. and Marchal, J.-P. 2005. Simulating the impact of water pricing on households behaviour: the temptation of using untreated water. *Water Policy* (World Water Council), 7(5).

Moss, T. 2000. Unearthing water flows, uncovering social relations: Introducing new waste water technologies in Berlin. *Journal of Urban Technology*, 7(1).

Netzer, D., Schill, M. and Dunn, S. 2001. Changing Water and Sewer Finance, distributional impacts and effects on the Viability of Affordable Housing. *APA journal*, 67(4) Autumn.

Okun, D.A. 1996. From cholera to cancer to cryptosporidiosis. *Journal of Environmental engineering*, June.

Okun, D.A., Craun, G., Edzwald, J., Gilbert, J., Rose, J.B. 1997. New York City: to filter or not to filter? *Journal of the American Water Works Association*, 89(3) March.

Rajala, R.P., and Katko, T.S. 2004. Household water consumption and demand management in Finland. *Urban Water Journal*, 1(1): 17–26.

Saunders, P. 1983. *The Regional State*, a Review of Literature and Agenda for Research, Sussex University Papers.

Tarr, J., and Dupuy, G. (eds) 1988. *Technology and the Rise of the Networked City in Europe and in America*. Philadelphia PA: Temple University Press.

Van Humbeeck, P. 1998. *An assessment of the distributive effects of the wastewater charge and drinking-water tariffs reform on households in the Flanders Region in Belgium*, Report of the SERV (Sociaal-Economische Raad Van Vlandern), May.

Zérah, M.H. 1997. Inconstances de la distribution d'eau dans les villes du tiers-monde: le cas de Delhi. *Flux, cahiers scientifiques internationaux Réseaux et Territoires*, 30 (October–December): 5–15.

Chapter 3

Water, public responsibility and equity: The Barcelona 'water war' of the 1990s

Elena Domene[1] and David Sauri[2]

[1]*Barcelona Institute of Regional and Metropolitan Studies, Universitat Autònoma de Barcelona, Bellaterra, Spain*
[2]*Institute of Environmental Science and Technology (ICTA) and Department of Geography, Universitat Autònoma de Barcelona, Bellaterra, Spain*

The history of water supply in Barcelona and its metropolitan region has long been characterized by the constant search for resources to meet an ever-growing demand from industrial and especially residential uses in a context of recurring hydrological stress. Up to 1954, water resources consumed in the city of Barcelona came from groundwater captured in the nearby Besòs and Llobregat fluvial plains. In 1955, the first surface water intake from the Llobregat river was inaugurated, and in the mid-1960s the supply network grew to incorporate the first inter-basin transfer with water coming from the Ter river, some 100km north of Barcelona. Water supply in the city of Barcelona was privatized at the end of the nineteenth century, collectivized during part of the Spanish Civil War, and privatized again during the Franco regime. Agbar, the private water company still in charge of Barcelona's water supply, and purchased by French company 'Suez' in October 2009, has become one of the leading conglomerates in the utilities sector of Spain and (until recently) Latin America. Although the water supply is private, the legal nature of water is still public and it is perceived as a public good among the Catalan population. In this chapter, we focus on the dynamics of the conflict that occurred in the Barcelona Metropolitan Area (BMA) after the incorporation of several water taxes into the water bill in the early 1990s. The so-called 'Water War' of Barcelona exemplifies the popular view stating that water must be a universal basic service/good accessible to everybody. This case also highlights how water policies based on demand offer a number of economic and environmental gains, but must also face distributional impacts.

Between 1987 and 1993, average prices for domestic use in BMA rose from €0.59 to €1.23/m^3 (an increase of 108%). Upon approval of the Water Quality Bill by the National Parliament in 1981, sewage collection and treatment costs were transferred to users (industries and domestic) through the 'Water Treatment Tax'. In 1991 the Catalan government decided to substantially raise the water bill by adding a bundle of new taxes. Some of these were directly related to the European Wastewater Directive mandating all municipalities above 2,000 inhabitants to treat their wastewater by 2002. Another directive, the so called 'Hydraulic Tax', addressed the financing of supply and flood control works within the area. Both taxes were collected by the Catalan administration. A new municipal tax on garbage disposal, which had previously remained largely unpaid, was also incorporated into the water bill, following the correlation between water consumption and garbage production. Finally, other taxes were the result of increasing financial constraints. The part of the bill related to direct water consumption (supply fee) was organized along a fixed fee quota and an increasing block-rate structure. In total, the water bill in the city of Barcelona incorporated eight different items.

The introduction of these new taxes not only increased the price of the water bill, making Barcelona the third most expensive city in Spain in this regard (the Canary Islands also have very high prices), but also altered the balance between the supply fee (charged by the water company) and taxes (charged by local, Catalan and Spanish administrations). In 1987, 61% corresponded to a strict supply fee and 39% to taxes. In 1993 this proportion was reversed with 46% allocated to the water fee and the remainder going to taxes. Moreover, the new taxes were implemented without any prior dissemination of information to consumers about the fate of the money collected by the different administrations.

Discontent with this policy resulted in a tax revolt organized at the end of 1991 by neighbourhood community groups in Barcelona and its peripheral cities, which consisted of paying the water fees but not the taxes. The area potentially affected by the tax revolt covered a total of twenty-three municipalities and a total population of 2.6 million people. A citizenship platform against high water bills brought together trade unions, consumer groups and the Federation of Neighbourhood Associations of Catalonia. Much of the work was undertaken by the citizenship platform (through neighbourhood associations), but they also had the support of certain green NGOs and the advice of legal experts. At the end of the 1990s more than 80,000 families objected to the payment of taxes. To avoid disconnection from the service, each family deposited the fee for the supply service (plus the corresponding VAT) in a special bank account, but nothing more. A typical profile of a family joining the protest corresponded to four members, with a relatively low income, whose water bill reached around €84 every month. Retired people and people living alone also joined the revolt. Not surprisingly, the penetration of protests was greater in neighbourhoods and municipalities characterized by a strong tradition of urban struggles dating back to the Franco years.

The responsibility principle according to which every user must bear a proportion of the cost of improving the water cycle may violate some basic equity principles, such as the following (Sauri et al, 1998):

1.   different users are treated differently
2.   differential impact of pricing
3.   non-equitative use of taxes.

A citizen of Barcelona may have felt treated unfairly with respect to other users and other territories. In Barcelona, taxes for domestic water were higher than in other cities of Spain, and four times comparatively higher than those charged in the industrial sector. Moreover, during the period 1991–96 it was reported that golf courses remained exempt from the hydraulic tax with the argument that water for golf courses could legally be considered as water for agricultural irrigation, which was exempt from the tax. The disproportionate fraction of cost of environmental modernization paid by lower-income segments of the metropolitan population appears to have been the gist of the problem. For example, the bills of people in the lower-income neigh-bourhoods of the city may have been financing the construction of flood control infra-structures for residents with much higher incomes towards the coastal plains and hills north of Barcelona. Although the water bill in Barcelona barely represents 3% of the total expenditure of households, its cost may still be significant for low-income house-holds (Tello, 2000).

Another presumed violation of the equity principle is that large families with four or more members must pay more than small families, simply because the block-rate struc-ture does not take into account the number of members per household. For instance, in the Barcelona area (data for 1996), the first $18 \, m^3$ (for three months) were billed at €0.26/m^3; from $19 \, m$ to $42 \, m^3$, the price increased to €0.5/m^3; and for a consumption higher than $43 \, m^3$ per term, the price reached €0.66/m^3. In 1995, four families went to the Catalan Higher Court of Justice denouncing the unfair situation of large families. Two years later, the Court ruled that prices should be modified taking into account family size. At the other end of the spectrum, small families were also paying higher average prices per cubic metre than large ones, since the fixed service fee (quite high) was independent from consumption. In this way, the apparently progressive element of having three different consumption tariffs was erased, and greater consumption in fact implied a reduction of average prices.

Concerned with these inequities, the Federation of Neighbourhood Associations of Catalonia published a manifesto with different proposals. It demanded, for example, a simplification of Catalan legislation related to the water cycle, as well as political negotiation of water prices with popular representatives. As regards the composition of the water bill, the main claims related to the elimination of some of the new taxes such as the 'hydraulic tax'; keeping the 'Water Treatment Tax' price of 1991 down for the remainder of the political term (until 1996) for the domestic sector; the removal of all taxes not directly related to water consumption; the re-consideration of the applied block-rate structure; an increase in pricing for the industrial sector; the removal of the invoice minimum fixed at $18 \, m^3$; and the reunification of prices for the entire Catalan territory. Moreover, community groups asked for more transparency and information concerning the final use of the collected taxes.

Citizen activism took many different forms during the almost ten years of struggle. Community groups began collecting signatures opposing the rise in taxes. As this measure failed to fulfil their expectations, the subsequent strategy was the creation of a social platform, calling for public meetings, resulting in an attendance of around 10,000 in the most populous neighbourhoods (Morera and Perxacs, 2000). Other actions included blocking traffic and public demonstrations. The most notorious action was, as previously mentioned, the refusal to pay the taxes included in the bill. The evolution of what was dubbed in the media the 'Water War' was followed closely by the Catalan press, radio and television.

The Catalan Administration agreed to modulate the taxes and apply certain meas-ures to ease the economic impact on less favoured families. It offered a tentative response as soon as 1992 by agreeing to lower the proposed increase in the 'water treatment tax' from the expected 6.5% to 5%. In addition, the charge for hydraulic infrastructure would be frozen (but not reduced) from 1992 to 1995. In 1993, the Catalan Department of the Environment agreed not to raise the charge above inflation. In May 1994, the regional government agreed to remove the additional 20% in the water treatment tax paid by residents in the Barcelona agglomeration.

In September 1994, an agreement was signed between the public administration and the organizations that had initially joined the platform, with the exception of the Federation of Neighbourhood Associations of Catalonia. The latter continued opposing the rise, but this time alone. The agreement foresaw reductions in charges and taxes with the final price per cubic metre decreasing to around 4% or €1.17/m^3. Some tariff differentiation was also introduced according to household characteristics, such as location, apartment size, number of taps and so forth. In 1996, a 'social price' was introduced to take into consideration large and poor families by subsidizing part of the cost.

The ten-year-long popular tax revolt prompted the enactment of a new law on Water Taxation and the creation of the 'Catalan Water Agency'. This new regulatory institution would concentrate executive powers on the water cycle which were formerly shared between two different departments, the Department of Public Works and the Department of Environment. Beyond this, the most important feature of the new Law on Water Resources was the reorganization and simplification of all taxes belonging to the hydrological cycle. Thus, a unique tax entitled the 'Cánon de l'aigua' (Water Tax) was created, and extended to the entire Catalan territory. Since 1999, prices and taxes have followed a progressive tariff structure, with substantial increases for the largest consumption blocks. However, issues of efficiency and fairness still loom large, and are the object of many debates between community groups, water companies and public administrations.

Today, water tariffs are expected to increase in order to meet the requirements of the *full cost recovery* principle of the Water Framework Directive by 2010. There is institutional (and academic) consensus for internalizing the costs of environmental improvement. However, spreading these costs evenly across all users does not seem to meet similar agreement at the popular level. It is also interesting to note that if water supply alternatives, such as water transfers, have raised significant social opposition in Barcelona and other areas of Spain, the same has occured with demand management alternatives such as prices and taxes. This can be partly explained by the lack of knowledge among a broad section of the population concerning the urban water cycle and its associated costs. Moreover, Barcelona has never suffered restrictions and the perception of scarcity only appears in periods of water shortages (although these have been relatively frequent over the last ten years). Another reason for social opposition could be the general belief among Spanish citizens that water is a public good and should be subsidized. In fact, in some municipalities of Catalonia and Spain, water was not paid for until recently. Therefore, a tremendous political effort is required, taking account of information, participation and equity issues, in order to avoid future conflicts. As with the Flanders sewage tax reform, this case illustrates how difficult it is to abruptly increase water prices. The inclusion in water bills of elements not directly related to water consumption, such as a domestic solid waste tax, incited feelings of being robbed. Despite certain improvements, growing block tariffs and a high-level fixed fee still raise equity issues. Conservation and environmental protection do not necessarily equate with social justice, and more sociological analysis is needed to really improve water tariffs.

REFERENCES

Morerea, E. and Perxac, H. 2000. La guerra de l'aigua. Biblio 3W. *Revista Bibliográfica de Geografía y Ciencias Sociales.Universidad de Barcelona*, 253, 17 October 2000.

Tello, E. 2000. 'La guerra del agua' en Barcelona. Alternativas económico-ecológicas para un desafío socioambiental. A. Estevan and V. Viñuales (eds) *La eficiencia del agua en las ciudades*. Bilbao: Bakeaz/Fundación Ecología y Desarrollo, pp. 277–98.

Sauri, D., Dura-Guimera, A. and Muñoz, F. 1998. Sostenibilidad y conflictos distribucionales: el caso de las tasas sobre el agua en el area metropolitana de Barcelona. Communication presented at the Congreso Ibérico sobre Panificación y Gestion de Aguas, 14–19, September, Spain.

Chapter 4

Full circle? Public responsibility *versus* privatization of water supplies in the United States[1]

Martin V. Melosi

Department of History, University of Houston, Texas, USA

For much of the eighteenth and early nineteenth centuries, urban Americans acquired water through their own devices, from water merchants, or from public wells (some purchased by the local government). Beginning in the 1830s, many cities and towns developed centralized water systems managed and owned by the municipalities themselves. From that time until the late twentieth century, water was generally treated as a public good and providing it was regarded as a public responsibility, based on the assumption that market forces could not be depended upon to furnish services necessary to society (Jacobsen, 2000: 3, 13, 22). At the same time, freshwater was a commodity to be bought and sold, whether controlled by private or public entities. Gail Radford suggested the implications of developing public water systems:

> Mundane as it might seem, providing water represented a sharp break for cities, which had previously confined themselves to supplying relatively indivisible public goods, such as police and fire protection, that did not lend themselves to the commodity form – that is, to being socially defined as objects bought and sold in markets. Water, by contrast, was generally charged for according to use. In a sense then, water opened Pandora's box. The widespread reliance on municipal provision of this vital substance enhanced the plausibility of following the same course for other goods (Radford, 2003: 872).

In the 1990s, however, privatization of the delivery of water to American cities appeared to be a viable option for the first time in many decades. How could such a long-standing commitment to a pioneering municipal service be challenged? And, more generally, what accounted for the change from private to public service and possibly back to private service again?

It may appear simplistic to create a scenario based on changes from private to public and then public to private water supply systems. As Charles Jacobson and Joel Tarr

---

[1] An earlier version of this chapter was originally presented at 'Urban Infrastructure in Transition: What Can We Learn from History?', The 6th International Summer Academy on Technology Studies, Deutschlandsberg, Austria, 2004.

stated regarding infrastructure and city services, 'Although it is widely believed that today's movement toward privatization represents the first major shift from public to private supply of infrastructure, history provides examples of many shifts in both directions'. They added that 'a simple distinction' between what is 'public' and what is 'private' does not really 'encompass the range of arrangements that has existed with respect to the ownership, financing, and operation of facilities'. These might include plans where a government agency builds and operates a facility, contracts out the construction, or contracts out the operation. Funding provisions might rely on user fees, taxes and assessments to abutters, bonds, or a combination of some or all of these (Jacobson and Tarr, 1996: 2).

It should be noted that this chapter focuses on broad national trends in water supply only, not on all city services. A reasonable argument can be constructed using 'public' and 'private' construed quite generally to identify the biggest and most obvious changes over time. Furthermore, the privatization movement of the 1990s and beyond emerged out of circumstances very different from those in earlier times. Of particular significance was the globalization of the water industry, which changed the organizational structure of water services in many parts of the world, and presented a major challenge to local or regional approaches to water delivery in the United States.

4.1 URBAN WATER SUPPLY BEFORE 1830

Prior to the 1830s, many American cities faced the threat of fires, and suffered from poor sanitary conditions and the looming prospect of epidemic disease. A plentiful and pure water supply was a valued resource in such a setting. While some of the earliest city-wide water supply systems appeared at this time, few communities could boast of well-developed sanitation technologies of the order of those constructed several decades later. Much of the responsibility for obtaining water rested in the hands of the individual, who acquired it from wells or nearby watercourses. In low-density areas in particular, these methods proved to be adequate, even efficient, and thus resisted change, obsolescence or outright replacement. Such practices were often publicly regulated but rarely publicly managed or owned.

As the population grew larger, the number of structures increased, new technologies like the flush toilet (or water closet) came into use, water sources became polluted or infected, and/or local water supplies literally dried up. As a consequence, the traditional methods of acquiring water became less workable. The result – albeit somewhat slowly implemented – was the appearance of water-supply protosystems that placed emphasis on more sophisticated means of acquiring and delivering water than buckets and wells, were increasingly capital-intensive, and were publicly regulated and often publicly operated.[2]

European (especially English) experiences with water-supply systems influenced US cities. Philadelphia became the first to complete a sophisticated waterworks and municipal distribution system in 1801, but this constituted an anomaly, which did not

---

[2] 'Protosystem' connotes an original system or 'first in rank or time' as opposed to a primitive system. For a more thorough discussion of water supplies and waterworks in the United States, see Jacobson and Tarr (1996: 7) and Melosi (2000).

Table 4.1 American cities with waterworks, 1800–1830

Year	# Works	# Cities	Cities with Works	Public	Private
1800	17	33	51%	1	16
1810	27	46	59%	5	22
1820	31	61	51%	5	26
1830	45	90	50%	9	36

Cities = population of 2,500 or more.
Source: US Bureau of Census (1960: Part A, pp. 1–14, 1–15, Table 8) and Waterman (1934: 6).

spark an immediate nationwide trend.[3] Prior to the mid-nineteenth century, only about half of the major cities and towns had some type of waterworks and these were overwhelmingly private as Table 4.1 shows (Croes, 1885: 4–69; Eddy, 1932: 82).

4.2 THE RISE OF THE PUBLIC WATER UTILITY, 1830–1920

Beginning in the 1830s, the scale of urban growth in the United States, the persistent fear of fire, increasing demand for water, and vague notions connecting waste with sickness led to the construction of several city-wide water supply systems. As Charles Jacobson stated, '[W]aterworks represented a critical element in a distinctively growth-oriented American style of city-building, elements of which have survived to the present day' (Jacobsen, 2000: 3). Between 1830 and 1880, American cities underwent their first 'sanitary awakening', a time when prevailing public health ideas attributed disease transmission to 'miasmas', that is, decaying matter, foul smells and bad air. Sanitarians, engineers and city officials linked the new water systems to the goals of environmental sanitation, using sensory purity tests to seek out what they believed to be safe sources of water and to protect supplies from human and animal wastes. This strategy was crude and scientifically inaccurate – bacteria, not filth, transmitted disease – but waste removal and concern for water purity nonetheless had a salutary impact on making the delivery of water a high priority. It also placed the responsibility for public improvement in human hands. The number of waterworks multiplied at an accelerated rate from forty-five in 1830 to 9,850 in 1924 (Melosi, 2000: 73–89, 120).

Leaders, especially in large cities, concluded that control of the sanitary quality of water services would be difficult if the supply remained private. They also increasingly came to believe that a public water supply could be profitable for city government, and

---

[3] After examining various options, the city leaders chose the proposal of English-born engineer Benjamin Henry Latrobe, who recommended building a steam-powered pumping plant that would distribute water to the city from the protected Schuylkill River located more than 1 mile away. Latrobe began the task in 1799 and completed it in 1801. In 1811 the city's Watering Committee replaced the original plant, pumping water to a reservoir atop Fairmount Rise and then releasing the water by gravity to the city. The Fairmount Waterworks served Philadelphia until 1911. See Gibson (1988: 2–40), Jackson (1989: 635) and McMahon (1988: 25–26).

Table 4.2 Public v. private ownership of waterworks, 1830–1924

Year	# Works	Public	Private	% Public	% Private
1830	45	9	36	20	80
1840	65	23	42	35.4	64.6
1850	84	33	51	39.3	60.7
1860	137	57	80	41.7	58.3
1870	244	116	128	47.5	52.5
1880	599	293	306	48.9	51.1
1890	1879	806	1073	42.9	57.1
1896	3197*	1690	1490	52.9	46.6
1924	9850	6900	2950	70	30

*Includes seventeen undocumented systems.
Source: Waterman (1934: 6).

would keep a valuable resource from being controlled by businessmen. While many water companies had been profitable, capital investment in the modern systems escalated, and operating costs rose (Jacobson, Klepper and Tarr, 1985: 9). In addition, many private companies were accused of being inefficient or charging excessive rates. In essence, the push for municipal ownership had as much to do with the desire to influence the growth of cities as to settle disputes with private companies over specific deficiencies.[4] Thus, private owners came increasingly under pressure to sell their assets as several communities gradually phased out private service. Major cities tended to support public systems earlier and more uniformly than any other class of cities. They also tended to invest more heavily in water supply and distribution than had privately owned companies. Whereas in 1830 only 50% of cities had public systems, in 1897 forty-one of the fifty largest cities (or 82%) had public systems; in 1924 70% of all cities went public. As Table 4.2 shows, the most dramatic increase in public ownership occurred in the mid-1890s, during the Progressive Era, when promotion of government action in several spheres intensified (Committee on Municipal Administration, 1898: 726–27; Griffith, 1983: 180; Jacobson and Tarr, 1996: 8).

The desire of city leaders to convert private systems into public, or to build new public systems, rested on more than the will to do so. The central issue was the ability of cities to incur debt to fund major projects and to sustain the high costs of operation. As the nineteenth century unfolded, city finances underwent changes in scope and complexity that ultimately made the development of public water supply systems possible (Anderson, 1980: 106, 108, 112; Tarr, 1984: 26, 30). The urban bureaucracy itself experienced substantial change, making it more responsive to developing city-wide sanitary services. Professional bureaucrats became firmly entrenched in municipal government, and helped shift power away from state capitals to city halls. Beginning slowly in the 1870s, several cities made efforts to move away from state interference in their affairs by demanding more 'home rule'. The movement

---

[4] For a general discussion of municipal ownership as a public issue, see Griffith (1983: 86–87).

took many forms, including efforts to increase the appointive power of mayors and to gain control of various service departments (Finegold, 1995: 15; Teaford, 1984: 7).

'Home rule' – granted by legislatures or constitutions – proved viable in several states with large cities. In some cases, the cities demonstrated political clout, which they could wield at the state level. In Colorado, Denver was granted some home rule powers in 1889, but this action was essentially a rubber stamp for powers the city had already accrued. When a new political party entered office, Denver temporarily faced the institution of state boards, which cut into its local authority. In a quite different case, Louisiana granted statutory home rule to all cities in the state in 1896, except New Orleans. In some states with small cities, or where public service standards were high, legislatures often retained the right of special legislation, but wielded it carefully. By the end of the nineteenth century, the success of reform efforts in the cities and states made for a favourable political setting for greater home rule. More home rule did not insure political and financial stability for cities, but it did allow some latitude in setting local priorities, or at least in responding to perceived local needs (Glaab and Brown, 1976: 174–76; Griffith, 1983: 215, 128; Teaford, 1984: 105, 122).

By the late nineteenth century, faith in environmental sanitation as the primary weapon against disease lost followers as the 'germ theory' of disease replaced the 'filth theory'.[5] Bacteriology placed more emphasis on finding cures for disease as opposed to prevention, which had been the mainstay of sanitary reform since the 1840s. The commitment to develop elaborate urban infrastructure for water services was not deterred by changing notions of health and disease, since the need for pure and plentiful water was essential to city life. The technology to ensure the quality of water supply changed with more emphasis on chemical testing, treating and filtering water. By 1920, many American cities could boast plentiful sources of pure water, and water systems that took greater account of how to confront water-borne epidemics (Melosi, 2000: 117–48).

In the era of bacteriology, water supplies increasingly relied upon centralized organizational structures and capital-intensive technical innovations, which had been developing since the 1830s. The prevailing goal in the late nineteenth and early twentieth centuries was the transformation of evolving systems into more comprehensive public, city-wide systems that could provide permanent solutions to the delivery of water. The price of public-water infrastructure was high, but many officials and citizens came to believe it was worth the expense (Jacobson, 2000: 33–34, 61, 69).

Metering water usage became a powerful management tool for administering the water supply in public systems. Ostensibly employed as a way to set rates, the use of water meters was equally important as a means to check waste and anticipate future expansion of the system. By 1920, metering had made notable strides. While only about 30% of cities metered at the pump, more than 600 of 1,000 cities surveyed metered at the tap; 279 cities metered all taps (*American City*, 1920: 614–20; 1921: 42–49).

---

[5] The germ theory purported that microscopic organisms – or bacteria – were the cause of epidemic disease. A contagionist disease paradigm was replacing a noncontagionist paradigm.

Box 4.1 Same story in Quebec? Water supply in Montreal: 1801–63

In 1801, a group of private investors set up a drinking water service, under the name 'Company of the owners of Montreal waters'. In 1845, the company was in crisis and was municipalized, its capital being taken over by the Montreal City Corporation. But this passage under public procurement did not radically alter the situation of water supply. Another change was needed, this time in the financing mechanism. This took place in 1851, and by 1865 each housing unit was connected.

Indeed, after the municipal takeover, service interruptions went on repeatedly and for long durations, while the infrastructure system accumulated a modest 13 km or a little more. The penetration rate into homes remained quite low, and, as with the private company during the previous period, the city services delayed or even opposed new connection demands: financial profitability was not insured, and the stock of necessary material was not always sufficient. At this point, urban hydraulic elements were mostly imported from Britain.

Thus, when the system became a municipal service, the majority of citizens still went to public fountains or the banks of the Saint-Laurent river to fetch water. Yet, demographic growth, urban expansion and contamination of water abstraction locations, made these supply solutions dangerous and sometimes impossible.

The 1845 municipalization took place in the context of an emergency situation. Moreover, city managers were convinced that piped water supply was the future, and that public procurement could better extend and improve the service than private entrepreneurs. Yet, something else had to occur for significant results to be achieved. The new legal system not only allowed the city corporation to replace the company, but granted it sovereign powers over water supply: from 1851, it could compel citizens to connect to the service. In counterpart, it had to respect certain constraints: payment of water through rates proportional to the renting value of the housing unit and fixed by law (as in the UK), and the obligation to serve every house. The subscriber thus became the 'obliged' of public service through this rate system, but he/she could then take the water quantity desired for no extra charge, which favoured large and poor families.

This public service approach has been maintained up till the present day, with a project to involve the Suez company rejected.

Source: Fougères (2004).

4.3 EXPANSION OF WATER SUPPLY SYSTEMS, 1920–1945

From the end of the First World War to the end of the Second World War, neither the quality nor character of water-supply services underwent substantial change. The challenge for municipal officials, engineers, planners and sanitarians was to adapt those services to urban growth, increasingly characterized by metropolitization and suburbanization, on the one hand, and demand in numerous small towns and rural communities, on the other. Decision-making during this period was complicated by two major disruptions to American life: the Great Depression and the Second World War. Despite the fluctuations of the economy from the 1920s to the 1940s, national trends in the construction and expansion of waterworks continued to indicate steady growth. Many of the new systems were rudimentary ones in numerous small communities. In 1940, there were approximately 14,500 waterworks in the United States (Fuller, 1927: 1588; Turneaure and Russell, 1948: 9). Although the rate of growth was strongest from the 1890s through to the early 1920s, increases in the 1930s were significant due to the infusion of federal funds during the New Deal (*American City,* 1925, February: 185–91; March: 309–23; April: 435–45; May: 555–65; June: 665–77; July: 47–59; Davis, 1933: 92).

The relative stability of the waterworks business during the interwar years also saw some significant changes in the management of water supply systems. The need for greater cooperation between political entities in the acquisition and delivery of water was becoming obvious, especially in response to metropolitan and suburban growth patterns in major cities. In some parts of the country, special water districts sprouted up in the 1920s, especially for the development and delivery of water (Siems, 1925: 644–45).

Without question, the greatest change in the development, extension and financing of water supply systems in the interwar years came with the new role of the federal government. Management, however, most often remained in the hands of local – or regional – public authorities.[6] During the New Deal, the Public Works Administration financed between 2,400 and 2,600 water projects with a price tag of approximately US$312 million – half of the total expenditures for waterworks for all levels of government. The Federal Emergency Relief Administration, the Civil Works Administration and the Works Progress Administration spent another US$112 million for work relief on municipal water projects. Smaller communities realized the greatest impact of these funds; for the first time they were able to finance public systems, treatment facilities and distribution networks. In fact, almost three-quarters of the projects financed went to communities of less than 1,000 people. While federal support stimulated development of new waterworks and provided resources for improving others, wartime priorities ultimately shifted federal funds away from local sanitary services (Armstrong, Robinson and Hoy, 1976: 231–32; Daniels, 1975: 9; Public Works Administration, 1939: 170, 173–78; *Scientific American*, 1944: 18).

4.4 METROPOLITAN EXPANSION AND NEW DEMANDS ON WATER SUPPLIES, 1945–1970

Relentless growth on the periphery and deterioration of the central city characterized post-Second World War urban conditions, and placed increasingly stiff demands on the providers of water supply. Concern over decaying infrastructure, especially at the urban core, raised important questions about the permanence of the sanitary systems devised and implemented in the nineteenth and early-twentieth centuries. An array of mounting social ills – characterized as an 'urban crisis' – increasingly shifted attention away from physical problems. The last of a series of *Fortune* articles on infrastructure (December, 1958) stated flatly that water supply and sewerage 'remain a signal failure in public works' (Thompson, 1958: 102). This assessment was harsh, but many older water supply and sewerage systems were in decline by the mid-1940s. The Committee

---

[6] The states, more than the municipalities or the federal government, were the centres of action for new legislation to control stream pollution. See Besselievre (1931: 325–44), Hinderlider and Meeker (1926: 606–08), Micklin (1970: 131), Monger (1926: 790), Skinner, (1939: 1332), Tarr (1985: 1059, 1064), Vesilind (1981: 26) and Warrick (1933: 496). For regional systems, see Elkind (1998).

on Public Information for the American Water Works Association reported in 1960 that of the approximately 18,000 functioning water facilities in the US, one-in-five had a deficient supply, two-in-five had inadequate transmission capacity, one-in-three had defective pumping, and two-in-five had weaknesses in their treating capacity (Hanna, 1961: 22).

Decisions about improving water supply systems had to be made within a framework of rapid urban growth, increasing water usage and growing financial pressures on cities. Further concentration of industry in metropolitan and unincorporated areas also increased the need for more water, as did demands for service in unincorporated residential communities (Bollens and Schmandt, 1970: 176; Thompson, 1958: 102).

New waterworks continued to come on line, especially in the expanding metropolitan periphery and in smaller cities and towns no longer able to depend on private wells and rudimentary water systems. In 1945 there were approximately 15,400 waterworks in the United States supplying about 12 billion gallons per day to 94 million people. By 1965, there were more than 20,000 waterworks supplying 20 billion gallons per day to approximately 160 million people. By the mid-1960s, 83.4% of water-supply facilities (in cities with 25,000 or more population) were publicly owned. Between 1956 and 1965, US$10 billion was spent for new construction and additions in the United States. The annual value of the water placed waterworks within the nation's top ten largest industries (Babbitt and Doland, 1949: 40; Fair, Geyer and Okun, 1977: 14; US Department of Commerce, 1975: 619, 621).

Distribution problems resulted from the location of water facilities at the cities' cores, which often serviced the larger metropolitan areas and outlying suburban communities. It was frequently in the interest of the central city to extend water lines to the suburbs to maintain a healthy economic climate in the metropolitan area. For suburbs, growth was impossible without adequate services. In some cases there was reluctance on the part of central cities to extend distribution lines outward, if there was no guarantee of future annexation. Often, real estate developers or alternative public entities constructed pipelines beyond the existing city limits to make outlying suburbs attractive to future annexation. In the 1960s, the central plant in Chicago supplied water on a contract basis to approximately sixty suburban communities. The number of special districts and other administrative arrangements were increasing in number in response to the need for water. From the vantage point of the total water system, the cost of distribution represented as much as two-thirds of a utility's investment (Bolton, 1959: 67–68; Fleming, 1967: 94–95; Larson, 1968: 1316; Water Resources Council, 1968: 5-1-3).

For water-supply systems and other city services, the postwar economic boom and the dynamic expansion of metropolitan America obscured the chronic deterioration of infrastructure and the inability of cities to keep pace with sanitary needs. Water supply systems were failing to live up to expectations and foretold an unsettling fear of a new era of adversity.

4.5 FROM INFRASTRUCTURE CRISIS TO PRIVATIZATION, 1970–2004

In the wake of the so-called 'infrastructure crisis' of the late twentieth century, water supply systems avoided the direst predictions about decay and deterioration. A 1987 report stated that a national water supply 'infrastructure gap' of the magnitude that would require a substantial federal subsidy did not exist. Urban water supply systems as a whole, it concluded, 'do not constitute a national problem' (National Council on Public Works Improvement, 1987: 37–38). This assessment was based on comparisons with other components of the nation's infrastructure. Water needs appeared modest when compared with highway repair and replacement estimated in the mid-1980s to reach a twenty-year 'needs level' of approximately US$2 trillion. Studies set price tags of US$125 billion for water-supply repairs, expansions and improvements. The relatively small, but hardly insignificant, number masked problems that had been building for years. Some experts, looking beyond the statistics, charged that many drinking water systems were outdated, faced massive leaks, were poorly maintained and relied on pipes 100 or more years old (Ausubel and Herman, 1988: 265; Everett, 1996: 91; Grigg, 1986: 7–8).

Broadening federal regulatory authority over water pollution and the tightening of water-quality standards were the first steps in recognizing the severity and complex nature of water pollution in the 1970s, but added additional financial pressures to managing water systems at the local and regional level (Environment and Natural Resources Policy Division, 1980: 14; Luken and Pechan, 1977: 4; McClain, 1994: 2.1–2.2). The financing of water supply in the 1970s and 1980s remained largely at the local level. Statistics from the early 1980s indicate that state and local governments were primarily responsible for 83% of expenditure for municipal water supplies. Federal funds for water projects were on the decline in the 1970s, and capital spending by all governments for water resources had fallen by 60% from the late 1960s to the late 1980s (American Public Works Association, 1982: 11, 32; National Council on Public Works Improvement, 1987).

Regionalization of the water industry in the United States attracted considerable attention, especially the Metropolitan Water District in California and the Metropolitan Sanitary District of Greater Chicago (Grigg, 1986: 1, 85; Holtz and Sebastian, 1978: 71; National Council on Public Works Improvement, 1987). Moreover, efforts by several multinational companies to privatize water-supply delivery and treatment globally gained significant attention in the 1990s. American waterworks remained largely public ventures managed at the local level in the first decade of the twenty-first century, but privatization was a trend to reckon with (Ross and Levine, 1996: 261).

In recent years, various observers have come to believe that freshwater will be the most contested commodity of the twenty-first century, as oil has been in the twentieth (Gaura, 2002: A3; Gleick, 2002: E8; Swomley, 2000: 6). Deep concern about this turn of events on a global scale grew out of several converging issues.

First, some observers raised the spectre of a 'freshwater crisis', in much the same way as an 'energy crisis' was proclaimed in the 1970s. In an article published in 2000, social ethics professor John M. Swomley predicted that a water crisis 'looming on the horizon' could reach 'dire proportions within the next ten to thirty years'. It is unclear on what basis he made such a presumption (or if he has the expertise to do so) (Swomley, 2000: 8). An article in a 2002 issue of *Nation* also sounded an alarm:

> The world is running out of freshwater. Humanity is polluting, diverting and depleting the wellspring of life at a startling rate. With every passing day, our demand for freshwater outpaces its availability and thousands more people are

put at risk. Already, the social, political and economic impacts of water scarcity are rapidly becoming a destabilizing force, with water-related conflicts springing up around the globe. Quite simply, unless we dramatically change our ways, between one-half and two-thirds of humanity will be living with severe freshwater shortages within the next quarter-century (Barlow and Clarke, 2002: 11).

Second, beyond the issue of scarcity there was growing unease that freshwater was being commodified, that is, being treated more as an economic as opposed to a social and environmental good.[7] From this vantage point, water is not just another commodity or consumer product, but – as one writer noted – 'a shared resource and a public trust'. A United Nations committee asserted that access to safe and affordable water must be a human right (Krisberg, 2003: 15; Lenze, 2003: C-12). Others echoed the notion that commodifying fresh water was ethically wrong. On a practical level, treating water simply as a product leads to choosing the most profitable markets for providing water service, leaving some areas – especially poor communities and those located on the urban margins – without adequate service (Barlow and Clarke, 2002: 13–14; Gleick et al, 2004: 5; Knickerbocker, 2002: 1; Lee, 2001: A1). Such concerns, although raising legitimate questions about equity, failed to take into account the point that water historically has been treated as a product as well as a public good, and that the actions of multinational water companies did not initiate the commodification of water.

Third, people, especially in industrialized nations, have come to expect water delivered efficiently and at low cost. However, local governments and regional authorities often face budgetary hardships – including reduced federal funding – and the increased cost of compliance with environmental regulations to the extent that many historically profitable water supply systems are difficult to maintain in the pubic arena (Gleick, 2002: E8; Grunsky, 2001: 17–18; Hyman, 1992: 52; Lanza, 1992: 1; Runyan, 2003: 36–37). Local leaders must frequently choose between maintaining services 'in-house' that may also have political benefits *versus* substantial government spending that may have political costs. Given the predicament of many local authorities, private companies are increasingly pursuing opportunities to manage or to own local waterworks.[8]

In the United States, 'privatization' most often means governments contracting with private companies to provide specific public services. For example, a public-private partnership was established between Harrington Park, New Jersey and United Water Resources through which the city maintained ownership of the water utility, while the company managed the facilities. Selling off assets or complete liquidation of public holdings is also possible in some instances (Lopez-de-Silanes, Shleifer and Vishny, 1997: 447, 468; Nichols, 1996: 8A; Schundler, 1997: 45). From a business perspective, water supply systems often represent a 'hot investment'. Johan Bastin, with the European Bank for Reconstruction and Development in London, was reported as

---

[7] This, of course, is ahistorical since water was commodified long before the onset of the twenty-first century. On the recent claim of the commodification of water, see Peter Gleick et al (2004).

[8] In a few cases the reverse has occurred. For example, city officials in Marysville, Ohio were preparing to initiate eminent domain proceedings against Ohio Water Service Company in 1990 in an effort to purchase the private company. See Ball (1990: 10).

saying: 'Water is the last infrastructure frontier for private investors' (Fleming, 1998: A1). Thus, the most recent efforts of private water companies to penetrate the American market do not signal commodification of water *per se*, but rising expectations about new economic opportunities.

Fourth, critics are skeptical of claims that privatizing water supplies could revitalize systems, make them more efficient, and deliver the product at a reasonable cost once a city's rate-setting ability is shifted to a private company. They are also concerned as to whether the private market can deal with issues related to the public good in addition to focusing on profits – most likely to be taken out of the community. They are particularly weary of multinational companies with no local ties that most often constitute the driving force behind recent efforts at privatization of water-supply systems (Fleming, 1967: A1; Gleick, et al, 2004: 7; Schundler, 1997: 45).[9] As Maria Alicia Gaura stated in a 2002 edition of the *San Francisco Chronicle*, 'The transformation of water delivery from prosaic necessity to hot investment trend has startled many US ratepayers, who never dreamed that stockholders in Europe would be wringing profits from their water bills' (Gaura, 2002: A3).

Globalization of freshwater service adds a significant layer of apprehension to the privatization trend in government. The rising influence of international water companies and their pursuit of local opportunities around the world do not take us 'full circle' from individual and private water supplies before 1830, to public utilities established by the late nineteenth century, and back again to private providers. They take us into a new era entirely.

Where public-private competition over water supply, waterworks and treatment plants has been largely a local matter in the past, the potential impact of multinational – or transnational – water companies controlling vast numbers of systems represents a unique situation. Control over water supplies and water delivery is not a change from water as a public service to water as a commodity, but a fundamental erosion of local authority well beyond more traditional tensions between city and region, city and state, and the city and the federal government.

At the turn of the new century, privatization of water systems is much more widespread in Europe than in the United States. In 2003 only 5% of the water systems in the US were privately owned, and only about 15% of the population was served by corporate water. Of the 94% of water systems that are publicly controlled (about 5,000), most are municipal.[10] Between 1997 and 2003, however, the number of publicly owned systems operating under long-term contracts by private companies increased from 400 to 1,100.[11] The Center for Public Integrity – a non-profit advocacy group based in Washington DC – estimated that by 2020, 65% to 75% of public waterworks in Europe and North America would be controlled by private companies, with Africa and Asia not far behind (Cook, 2002: 19–20; Elie, 2003: 1; Haarmeyer, 1993: B5; Jehl, 2003: 14; Lenze, 2003: 1).

---

[9] See also Shleifer (1998: 14) and Runyan (2003: 38).

[10] Estimates do vary. Some observers argue that only about 85% of waterworks are publicly owned. See Gaura (2002: A3).

[11] In many cases private water companies agree upon long-term contracts (20–25 years) to manage and operate a particular city's waterworks. This has been more typical in recent years than outright purchases.

Leading the way to this potential sea change are ten major corporations,[12] several subsidiaries, and some smaller companies delivering water and wastewater services. The prospect of a long-term contract to monopolize a key resource has attracted substantial corporate attention. For their part, the World Bank and the International Monetary Fund provide backing to many of the larger ventures, especially in developing countries. Representatives of the World Bank have argued that governments in developing countries are too poor and too much in debt to subsidize water and sanitation services with public funds. International trade accords – such as the North American Free Trade Agreement – also incorporate provisions for governments to turn control of freshwater supplies over to global trade institutions, helping private companies gain access to those supplies. The bottled-water industry (Culligan, for example, is owned by Veolia) must also be included among water-for-profit enterprises, selling more than 90 billion litres of bottled water in 2002 alone. One report has noted that the annual profits of the water industry in recent years surpass those in the pharmaceutical sector and reaches about 40% of the oil sector, although only about 5% of the world's water is privately owned.

Two French companies dominate the international water industry: Veolia Environment and Suez.[13] Veolia, formerly Vivendi Environment, grew out of Generale des Eaux, which had been established by Napoleon III in 1852. Its first contract called for supplying water to the city of Lyon. Suez purchased Lyonnaise des Eaux, which was founded in 1880 with the sponsorship of the bank, Credit Lyonnais. Both Generale des Eaux and Lyonnaise des Eaux established the tradition of private water delivery in France, benefiting from years of protectionism, and have now emerged as part of a powerful force on the world scene. Taken together, the two water giants – Veolia and Suez – provide service in more than 100 countries with approximately 200 million customers. Only RWE/Thames would come close, benefiting from Margaret Thatcher's privatization of water services in Great Britain in 1989, and more recently from the merger of Thames water with the German electricity leader. In many cases, low margins and limited potential in the European water market have encouraged the multinationals to spread their financial risk to other parts of the world (Barlow and Clarke, 2001: 12–13; Glassman, 1999: E1; Grunsky, 2001: 14; IICJ, 2003; Tolhurst, 1999: 39).

Some American-based corporations attempted to challenge Veolia and Suez – most notably Azurix, which was a subsidiary of the later much-maligned energy-trading company Enron. Enron had hoped to be a major player in the freshwater market on a scale equal to its core businesses in natural gas and electricity. It met with little success, however, because it could not raise sufficient capital to operate effectively in both the water and energy markets (Fleming, 1967: A1; Taylor, 1998: 19; Warsh, 2001: E2).[14] In 1999, American Water Works Company was the leading water company in the United States, serving 16 million customers in twenty-nine states, but its revenue was

---

[12] Included here are Veolia Environment, Suez, Bouygues Saur, RWE-Thames Water and Bechtel-United Utilities, and several other smaller companies.

[13] They also have holdings in other businesses as well.

[14] Enron was interested in water company acquisitions within the US borders and throughout other parts of the world. For example, they attempted to invade markets in Rio de Janeiro, Berlin and Panama, and thus posed a threat to the French companies.

less than 10% of Veolia's. German conglomerate RWE AG purchased American Water Works for US$8.6 billion, which has further taken American water companies outside the leadership of the industry. Through its ownership of US Filter, Veolia is the largest private wastewater firm operating in the United States (*Economist*, 2003: 7; Lenze, 2003: C12).[15]

The water giants have not been without their failures as well as successes. Allegations of corruption and unfair business practices regularly dog them. Ventures in developing companies generally have been less successful than elsewhere. After initial success in conquering new markets Suez retrenched, especially in Asia and Latin America, while Veolia proved more successful focusing on Eastern Europe and North America. In 2006, RWE sold Thames Water to Kemble Water Limited, a consortium led by Macquarie's European Infrastructure Funds, thereby also retreating from the American market. The China market appears promising, but has yet to be effectively penetrated.

A wholesale trend toward privatization of water supplies and water supplies management has yet to occur in the United States. However, in 1999 alone there were US$15 billion in acquisitions in the US water industry. For example, Suez purchased Nalco Chemical Company of Illinois – a water treatment group – for US$4.1 billion, and also acquired Calgon Corporation – the third-largest water-conditioning company, which is based in Pittsburgh – for US$425 million. The 1996 Safe Drinking Water Act and other federal and state laws requiring renovation or improvement of deteriorating water systems place a financial burden on several cities, which are now ready to explore a relationship with a private water company. Also, a 1997 executive order, tax-rule changes by the Internal Revenue Service, and privatization advocates in Congress have opened up the possibility of more shifts from public to private service. Cities such as Indianapolis, Milwaukee and Gary, Indiana, have contracted with private companies to manage their waterworks (Barlow and Clarke, 2002: 12–13; Hairston, 1999: 3B; Helton, 1997: 5C; Iskandar, 1999: 37; Jehl, 2003: 14; *St. Petersburg Times*, 1999: 2E; Taylor, 1998: 19).

In the United States as elsewhere the global water company juggernaut has not always prevailed. Atlanta officials struck a twenty-year operations and maintenance contract with United Water, Inc. (a subsidiary of Suez) in 1998 which paid the company US$21.4 million per year. What had been one of the first large privatization awards in the US, however, was terminated in 2003, ostensibly because of faulty contract provisions, but also because of poor service and the protest (and lawsuits) of environmentalists over the construction of suburban reservoirs (see box below). As one journalist noted, 'The decision, in many ways, takes Atlanta back to square one' (Jehl, 2003: 14). While the action was a setback for privatization of water, and cities such as New Orleans (before Katrina), Louisiana and Stockton, California, were rethinking plans to privatize (or to further privatize in the case of New Orleans), Atlanta's decision was not likely to have long-range implications for water privatization in the United States (Brooks, 2003: C4; Carr, 2003: 8; Powers, 2003: 14–15; Powers and Rubin, 2003: 14–15; Rosta, 2003: 12).[16]

---

[15] *Economist* (2003: 7) and Lenze (2003: C12).
[16] On other debates, see Cook (2002: 20–21), Davis (1999: 4), Hightower (2002: 8) and Russell (2003: 1).

Box 4.2 The Atlanta water conflict

In the summer of 1998, a twenty-year operation and maintenance contract was signed between the head city of Georgia and Suez's subsidiary United Water (UW). However, this contract was terminated less than five years later, after a conflict which recalled those of more than 100 years ago, with the initial formula of concession. Yet, this contract was not a concession (privatization is an inappropriate term in this case), only a management contract, where the company was roughly supposed to invest only in leaks control. What went so wrong that the newly elected mayor Shirley Franklin reversed decisions taken by her predecessor only one year after being in office?

Atlanta accumulated typical problems of rapid demographic expansion (from 1.3 to 4.1 million inhabitants between 1960 and 2000), problems arose concerning low density (27 m pipe length per connection) and ageing infrastructure (frequent breakdowns and repairs, brown water, etc.). On top of this, the sewage collection and treatment system was in a poor state, and the city was being sued by the US EPA for violation of the Clean Water Act (not to mention a conflict over water resources scarcity with cities in other states, such as Alabama). The amount of investment was such that an initial settlement of the dispute had to be re-negotiated, pushing back the deadline to comply with regulations from 2007 to 2014. The PPP signed with UW was meant to alleviate the city's investment burden.

While the city previously spent between US$40 and US$50 million on municipal water operation, United Water was supposed to improve the service and drastically reduce operations and maintenance costs: its yearly fee was fixed at US$21.4 million only, and its task was to reduce unaccounted-for water to less than 15%, push bill recovery above 98.5% (but bills were still paid to the city), while improving staff management.

As its first item of conflict, the city contended that it could save only US$3 million/yr through the contract, while UW estimated the savings at US$18 million. The city indeed had to invest deeply in infrastructure renewal and was highly indebted, but the wide difference between the two figures is related to typical problems with public accountancy. Besides, bill recovery had not reached the requested level of 98.5%, and UW had to admit that this represented a loss of US$9 million to the city.

The second item of conflict concerned fighting leaks on a day-by-day basis. This was shown to be pointless if important investment was not made in infrastructure renewal. While it was obvious that the water system was in disrepair, both parties had accepted to underestimate the level of investment (and water rates increases) to reach an agreement that proved unsustainable: mains and fire hydrant repairs and their cost, borne by UW, were grossly underestimated. Suez thought that a twenty-year long contract could be renegotiated, as is frequently the case in France. They argued that the previous mayor had agreed to pay US$10 million to compensate work done beyond the contract, and to increase the fee by more than US$4 million/yr. But the new mayor refused, and despite significant improvement in water operation, gave the company ninety days to 'cure' complaints beyond what was achieved. A 'scorecard' was created to evaluate the performance, but it was too late. The initial situation of poor data had not been considered in the initial contract, and for this, the American subsidiary of the experienced number two water company in the world was to blame.

As a result, the same type of distrust developed as had occurred in the nineteenth century concessions (and in many contracts in Third World cities today). Each of the parties accused the other of not fulfilling its commitments. In the end, termination of the contract was the only solution, and was agreed upon by United Water. Incidentally, the four-year period before termination of the contract corresponded to a normal management contract, while the agreed upon twenty years corresponded more to a lease contract, where the company should have adopted more responsibility: to invest more in the whole system upgrade and bill customers directly, but be given more time to prove its performance.

Source: Contribution by B. Barraqué.

4.6 CONCLUSION

Many environmental activists have encouraged the public to 'Think globally, act locally'. Multinational water companies have taken up that call, but for very different ends. The historical record on urban water service in the United States has long been viewed from the vantage point of the triumph of public control over private action – a model in many cases unique in the world. A deeper look at that record suggests that a wholesale shift from private means of water delivery to public means is a little too simplistic, since historians are most comfortable demonstrating the shades of gray that makes up much of our lives. Nonetheless, for a very long time municipal control or oversight of water service has been part of the local fabric of cities – a venture that set precedents for many other services to follow. The new century may have something different in store for municipal water supplies and delivery as freshwater becomes an increasingly scarce resource and a more profitable commodity. The line between water as a public good and water as a product has always been blurred, but never to the extent it has in recent years.

ACKNOWLEDGEMENT

Acknowledgements are due to Tom McKinney for his research on privatization for the last section of this chapter.

REFERENCES

American City. 1920. Water-Supply Statistics of Metered Cities. *American City*, (December).

American City. 1921. Water-Supply Statistics of Metered Cities. *American City*, (January).

American City. 1925. Water-Supply Statistics of Metered Cities. *American City*, (February, March, April, May, June, July).

American Public Works Association. 1982. *Proceedings of the National Water Symposium: Changing Directions in Water Management*, November. Washington DC: APWA.

Anderson, L. 1980. The diffusion of technology in the nineteenth century American city (PhD dissertation) Northwestern University.

Armstrong, E., Robinson, M. and Hoy, S. (eds) 1976. *History of Public Works in the United States*. Washington DC: American Public Works Association.

Ausubel, J.H. and Herman, R. (eds) 1988. *Cities and Their Vital Systems: infrastructure past, present, and future*. Washington DC: National Academy Press.

Babbitt, H.E. and Doland, J.J. 1949. *Water Supply Engineering*. New York: McGraw-Hill.

Ball, B.R. 1990. Marysville seeks control of private water system. *Business First-Columbus*, 6 (12 February).

Barlow, M. and Clarke, T. 2002. Who owns water? *Nation*, 275 (2 September).

Besselievre, E.B. 1931. The disposal of industrial chemical waste. *Chemical* Age, 25 (12 December).

Bollens, J.C. and Schmandt, H.J. 1970. *The Metropolis: its people, politics, and economic life* (2nd edn). New York: Harper and Row.

Bolton, C.M. 1959. A metropolitan water works is best. *American City*, 74 (January).

Brooks, R. 2003. Deals and deal makers: a deal all wet. *Wall Street Journal* (31 January).

Carr, M. 2003. Water board hopes to learn from Atlanta: city's privatization venture went south. *New Orleans Times-Picayune* (17 April).

Committee on Municipal Administration. 1898. Evolution of the City. *Municipal Affairs*, 2 (September).

Cook, C.D. 2002. Drilling for water in the Mojave. *Progressive*, 66 (October).

Croes, J.J.R. 1885. *Statistical Tables from the History and Statistics of American Water Works.* New York: Engineering News.

Daniels, R. 1975. Public works in the 1930s: a preliminary reconnaissance. *The Relevancy of Public Works History: 1930s – A Case Study*. Washington DC: Public Works Historical Society.

Davis, C.V. 1933. Water conservation – the key to national development. *Scientific American*, 148 (February).

Davis, J. 1999. Furor over privatization stirs debate over its merits. *Kansas City Business Journal*, 17 (11 June).

Economist. 2003. Savoir faire. *Economist*, 368 (19 July).

Eddy, H.P. 1932. Water purification – a century of progress. *Civil Engineering*, 2 (February).

Elie, L.E. 2003. Privatization argument has its leaks. *New Orleans Times-Picayune* (31 March).

Elkind, S.S. 1998. *Bay Cities and Water Politics: the battle for resources in Boston and Oakland*. Lawrence: University Press of Kansas.

Environment and Natural Resources Policy Division, Congressional Research Service. 1980. *Nonpoint Pollution and the Area-Wide Waste Treatment Management Program Under the Federal Water Pollution Control Act*. Washington DC: GPO.

Everett, C.T. 1996. So is there an infrastructure crisis or what? *Public Works Management and Policy*, 1 (July).

Fair, G.M., Geyer, J.L. and Okun, D.A. 1977. *Elements of Water Supply and Wastewater Disposal* (2nd edn). New York: Wiley.

Finegold, K. 1995. *Experts and Politicians: Reform Challenges to Machine Politics in New York, Cleveland, and Chicago*. Princeton: Princeton University Press.

Fleming, C. 1998. Sofia's choice: water business is hot as more cities decide to tap private sector. *Wall Street Journal* (9 November).

Fleming, R.R. 1967. The big questions… *American City*, 82 (June).

Fougères, D. 2004. *Approvisionnement en eau à Montréal: du privé au public 1796–1865*. Sillery: Septentrion.

Fuller, G.W. 1927. Water-works. *Proceedings of the American Society of Civil Engineers*, 53 (September).

Gaura, M.A. 2002. Water a hot commodity. *San Francisco Chronicle* (1 December).

Gibson, J.M. 1988. The fairmount waterworks. *Bulletin of the Philadelphia Museum of Art*, 84 (Summer).

Glaab, C.N. and Brown, A.T. 1976. *A History of Urban America* (2 edn). New York: Macmillan.

Glassman, J.K. 1999. In Europe, going for the water. *Washington Post* (7 April).

Gleick, P. 2002. The big idea water, water-where? *Boston Globe* (6 January).

Gleick, P., Wolff, G., Chalecki, E.L. and Reyes, R. 2004. *Executive Overview, The New Economy of Water: The Risks and Benefits of Globalization and Privatization of Fresh Water* (Accessed 4 May) www.pacinst.org/reports/new_economy_overview.htm

Griffith, E.S. 1983 [1974]. *A History of American City Government, A Conspicuous Failure, 1870–1900*. Washington DC: University Press of America.

Grigg, N.S. 1986. *Urban Water Infrastructure*. New York: Wiley.

Grunsky, S. 2001. Privatization tidal wave. *Multinational Monitor*, 22 (September).

Haarmeyer, D. 1993. Privatize Seattle water? Study has wrong answer, *Seattle Times* (27 December).

Hanna, Jr., G.P. 1961. Domestic use and reuse of water supply. *Journal of Geography*, 60 (January).

Hairston, J.B. 1999. Treatment plant bidding could be fierce. *Atlanta Constitution* (9 April).

Helton, C. 1997. Atlanta's sewer problems. *Atlanta Constitution* (4 March).

Hightower, J. 2002. The water profiteers. Nation, 275 (2 September).

Hinderlider, M.C. and Meeker, R.I. 1926. Interstate water problems and their solution. *Proceedings of the American Society of Civil Engineers*, 52 (April).

Holtz, D. and Sebastian, S. (eds) 1978. *Municipal Water Systems: the challenge for urban resource management*. Bloomington: Indiana University Press.

Hyman, U. 1992. Wastewater partnerships. *American City & County*, 107 (April).

ICIJ. 2003. Defending the internal water empire. ICIJ. www.icij.org/water

Iskandar, S. 1999. Suez buys Calgon. *Financial Times* (London) (16 June).

Jackson, D.C. 1989. The Fairmount Waterworks, 1812-1911. *Technology and Culture*, 30 (July).

Jacobsen, C.D. 2000. *Ties That Bind: economic and political dilemmas of urban utility networks, 1800–1990*. Pittsburgh: University of Pittsburgh Press.

Jacobson, C.D., Klepper, S. and Tarr, J.A. 1985. Water, electricity, and cable television: a study of contrasting historical patterns of ownership and regulation. *Technology and the Future of Our Cities*, 3 (Fall).

Jacobson, C.D. and Tarr, J.A. 1996. Public or Private? Some Notes from the History of Infrastructure. A Report to the World Bank (Unpublished report).

Jehl, D. 2003. As cities move to privatize water, Atlanta steps back. *New York Times* (10 February).

Knickerbocker, B. 2002. Privatizing water: a glass half empty? *Christian Science Monitor* (24 October).

Krisberg, K. 2003. Privatizing water systems draws mixed reviews. *Nation's Health*, 33 (March).

Lanza, J. 1992. Cities mull privatizing waterworks. *Boston Business Journal*, 12 (11 May).

Larson, T.E. 1968. Deterioration of water quality in distribution systems. *Journal of the American Water Works Association*, 58 (October).

Lee, P. 2001. The wellspring of life, or just a commodity. *Ottawa Citizen* (16 August).

Lenze, A. 2003. Liquid assets. *Pittsburgh Post-Gazette* (16 September).

Lopez-de-Silanes, F., Shleifer, A. and Vishny, R.W. 1997. Privatization in the United States. *RAND Journal of Economics*, 28 (Autumn).

Luken, R.A. and Pechan, E.H. 1977. *Water Pollution Control: assessing the impacts and costs of environmental standards*. New York: Praeger.

McClain, Jr., W.E. (ed.) 1994. *U.S. Environmental Laws*. Washington DC.

McMahon, M. 1988. Makeshift technology: water and politics in 19th-century Philadelphia. *Environmental Review*, 12 (Winter).

Melosi, M.V. 2000. *The Sanitary City: Urban Infrastructure in America from Colonial Times to the Present*. Baltimore: Johns Hopkins University Press.

Micklin, P.P. 1970. Water quality: a question of standards. R.A. Cooley and G. Wandesforde-Smith (eds) *Congress and the Environment*. Seattle: University of Washington Press.

Monger, J.E. 1926. Administrative phases of steam pollution control. *Journal of the American Public Health Association*, 16 (August).

National Council on Public Works Improvement. (1987). *The Nation's Public Works: report on water supply*, 16. Washington DC: National Council on Public Works Improvement.

Nichols, J. 1996. Chance to save lures cities to private sector. *Cleveland Plain Dealer* (22 June).

Powers, M.B. 2003. Atlanta ends privatization deal and faces reservoir lawsuit. *Engineering News-Record*, 250 (3 February).

Powers, M.B. and Rubin, D. 2003. Severed Atlanta water contract was tied to unclear language. *Engineering News-Record*, 250 (10 February).

Public Works Administration. 1939. *America Builds*. Washington DC: US Government Printing Office.

Radford, G. 2003. From municipal socialism to public authorities: institutional factors in the shaping of American public enterprise. *Journal of American History*, 90 (December).

Ross, B.H. and Levine, M.A. 1996. *Urban Politics: power in metropolitan America*. Itasca, IL: F.E. Peacock.

Rosta, P. 2003. Stockton mulls outsourcing after Atlanta changes course. *Engineering News-Record*, 250 (24 February).

Runyan, C. 2003. Privatizing water. *World Watch*, 16 (January–February).

Russell, G. 2003. S&WB adds new duties for privatization bidders. *New Orleans Times-Picayune* (4 April).

Schundler, B. 1997. City chooses private manager for its water utility. *American City & County*, 112 (March).

Scientific American. 1944. Water supplies will be widely extended after the war. *Scientific American*, 171 (July).

Shleifer, A. 1998. State *versus* ownership. *Journal of Economic Perspectives*, 12 (Autumn).

Siems, V.B. 1925. The advantages of metropolitan water-supply districts. *American City*, 32 (June).

Skinner, H.J. 1939. Waste problems in the pulp and paper industry. *Industrial and Engineering Chemistry*, 31 (November).

St. Petersburg Times (Florida). 1999 (29 June).

Swomley, J.M. 2000. When blue becomes gold. *Humanist*, 60 (September/October).

Tarr, J.A. 1984. The evolution of the urban infrastructure in the nineteenth and twentieth centuries. R. Hanson (ed.) *Perspectives on Urban Infrastructure*. Washington DC: National Academy Press.

Tarr, J.A. 1985. Industrial wastes and public health. *American Journal of Public Health*, 75 (September).

Taylor, A. 1998. Enron steps into global water market. *Financial Times* (London) (25 July).

Teaford, J.C. 1984. *The Unheralded Triumph: city government in America, 1870–1900*. Baltimore: Johns Hopkins University Press.

Thompson, E.T. 1958. The worst public-works problem. *Fortune*, 58 (December).

Tolhurst, C. 1999. Drinking at the front of opportunity. *Australian Financial Review* (18 May).

Turneaure, F.E. and Russell, H.L. 1948. *Public Water-Supplies: requirements, resources, and the construction of works* 4th edn). New York: John Wiley.

US Bureau of Census 1961. *Census of Population: 1960*, vol. 1, *Characteristics of the Population*. Washington DC: Department of Commerce.

US Department of Commerce, Bureau of the Census. 1975. *Historical Statistics of the United States* Part 2. Washington DC: Department of Commerce.

Vesilind, P.A. 1981. Hazardous waste: historical and ethical perspectives. J.J. Peirce and P.A. Vesilind (eds) *Hazardous Waste Management*. Ann Arbor, MI: Ann Arbor Sci. Publ.

Warrick, L.F. 1933. Relative importance of industrial wastes in stream pollution, *Civil Engineering*, 3 (September).

Warsh, D. 2001. What Enron got right. *Boston Globe* (9 December).

Water Resources Council. 1968. *The Nation's Water Resources*. Washington DC: GPO.

Waterman, E.L. 1934. *Elements of Water Supply Engineering*. New York: John Wiley.

Chapter 5

Public-private partnership in courts: The rise and fall of concessions to supply drinking water in France (1875–1928)

Christelle Pezon

IRC International Water and Sanitation Centre, The Hague, The Netherlands

In France, back in the mid-nineteenth century, large cities started to develop water networks, mainly under very long concession contracts (ninety-nine years). Fifty years later, other cities organizing their water supply made a different choice: they mainly created in-house services, rather than contracting them out. In 1908, a majority of urban water services were run through local *régies*, or municipal organizations (Burel, 1912). In 1938, out of a population of 40 million, 25 million people were receiving safe tap water, and 18 million were supplied by publicly operated water services (Loosdregt, 1990). By this time, the remaining delegation contracts were no longer concessions – with investment and operating costs under concessionaires' responsibility – but rather *affermage* or lease contracts, where the operator only supported operating costs and the local authority supported the investment costs (Pezon, 2003).

What went wrong with the concession formula in the late nineteenth century – so wrong that French local authorities stopped considering it as an appropriate model for securing a safe water supply? What made the concession unable to meet local authorities strategies?

The jurisprudence of the *Conseil d'Etat* offers a unique opportunity to understand the concession's failure, and analyse the shift in water-supply management that occurred in France in the early twentieth century. The *Conseil d'Etat* (CE) is the Supreme Administrative Court, and regarding water supply, had jurisdiction over any litigation made on the grounds of a delegation contract (concession, lease, etc).[1]

With seventy-eight cases, the 1875–1928 period is the most conflictive period ever seen in the upper court, pitting public against private partners in the water supply sector. It began with the case of the city of Le Havre, which asked the judge to order its concessionaire, *La Compagnie des Eaux du Havre*, to undertake extra-contractual waterworks, and ended with the final decision of the CE to terminate the city of Toulon's concession with the *Compagnie Générale des Eaux*.

---

[1] More generally, the *Conseil d'Etat* is responsible for any litigation regarding water supply, except those made on the ground of private contracts, for example, subscription contracts that users hold from their water services and contracts between private water companies and possible subcontractors. These fall under the jurisdiction of civil courts, the Supreme Court being in this case the *Cour de Cassation*.

The importance of the CE decisions lies first in the enforcement power of the institution as well as in the scope of the decision. The enforcement power of the CE is almost unlimited and the court has never been contested as Supreme Administrative Court. With regard to cases under consideration, the CE's decisions are valid for any similar litigation: the *Conseil d'Etat* effectively 'empties' the case. It gives to any public and private partner committed in similar contracts capital information on their rights and duties, their risk-sharing schemes, and so on. The CE rules cases. It provides parties with details on what they should do according to the laws, on the way it has interpreted the laws (its jurisprudence), and on the particular circumstances of the litigation.

To analyse these seventy-eight concession cases, this chapter first examines the context of the late nineteenth century and explains why these CE decisions were made. It then categorizes the seventy-eight decisions in terms of the nature of the conflicts that opposed local authorities to concessionaires: the extension of concessionaires' duties, the decrease in the water tariff applied to households, and the termination of concession contracts. Third, the chapter demonstrates that while protecting the concessionaires rights, CE decisions disqualified the concession as a contract providing safe access to water for all. In conclusion, it emphasizes that concessions were set aside because they did not fit the public service provision of water.

5.1   CORPUS AND CONTEXT

From 1848 to October 2006, the CE ruled on 486 'water' cases. Of these, 131 opposed local authorities to their private partners. One hundred of these Public-Private

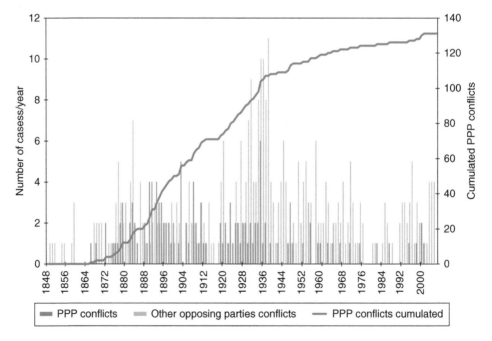

Figure 5.1 PPP cases and other opposed parties cases, 1848–2006

Source: Author.

Partnership water cases took place between 1875 and 1937, sixty-six prior to the First World War and thirty-four between 1919 and 1937. The other cases involved other stakeholders, for example, households or building owners and concerned tariffs or service quality.

From 1875 to 1913, about half of the water cases opposed public and private part-ners (66 vs 143) and, more precisely, local authorities and concessionaires. These 'con-cession' cases dealt with a common issue: conditions concerning renegotiation and/or termination of concession contracts. Such conflicts ceased before the First World War, but twelve post-war cases also dealt with pre-war matters. Seven cases referred to pre-war conflicts were only ruled on by the CE four to thirteen years after the first level administrative courts had made their decision, whereas the average delay between the first and last court decision was three years and three months after 1875 (Figure 5.2). These seven cases will be included within the scope of our analysis.

The five other cases referred to conflicts that followed the decisions of the cities of Lyon in 1888, Nantes in 1895 and Rouen and Toulon in 1911 to terminate their con-cession contracts before the end dates. These decisions entailed eight conflicts prior to 1914 and five more afterwards, ruled by CE between 1924 and 1928. As these five cases also dealt with pre-war issues, they are included in the scope of our analysis, which is then based on seventy-eight CE decisions.

During the 1919–37 period, apart from the five above-mentioned cases, conflicts mainly concerned adjustments of contracts to new macro-economics conditions (social laws, energy prices, inflation). In this second wave of 'PPP water' cases, private

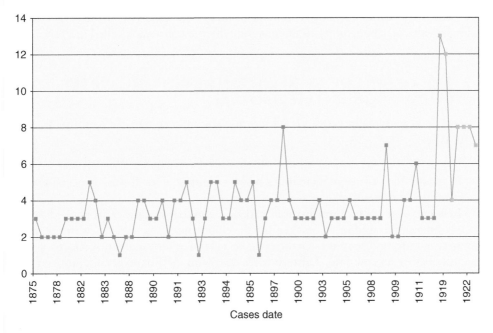

Figure 5.2 Delay between the first administrative court decisions and the Conseil d'Etat decisions (1875–1923)

Source: Author.

partners requested the enforcement of the *Théorie de l'Imprévision*, developed by the CE during the First World War to protect private partners from the consequences of unforeseen events, which would otherwise inflate their costs (Long et al, 1990). The contracts involved were no longer concessions, but either *affermages* – lease contracts – or *régies intéressées* – management contracts (Pezon, 2000).

At the end of the nineteenth century, the concession contract was a favourite option for local authorities working to develop domestic drinking water supply networks. Before 1900, the concession had dominated the French water supply industry. Private companies or entrepreneurs carried out investments in infrastructure and, when stipulated, in filtration plants. As a counterpart to this, they obtained a long-term territorial monopoly on supplying collective water to stand-pipes and private water to households. From the 1850s until the end of the nineteenth century, the concession was almost exclusively the institutional arrangement chosen by the largest local authorities to develop their water services, rather than organizing in-house water service (Copper-Royer, 1896).

However, the situation changed with the turn of the twentieth century. Cities creating their own water services chose to establish in-house water distributions rather than commit to concession contracts. In 1908, more than half of the 500 urban water services were publicly owned and managed (Burel, 1912). Ten years later, two-thirds of cities with more than 5,000 inhabitants managed their own water services (Monsarrat, 1920); and in 1938, there were six times more in-house water services than privately managed ones (Loriferne, 1987).

The cases that opposed public to private partners from 1875 to 1928 referred to concession contracts agreed upon between the 1850s and the 1880s, with a duration of at least fifty years and, more commonly, ninety-nine. Analysis of the seventy-eight cases provides an opportunity to understand why the concession failed in France.

5.2 THE ORIGIN OF CONFLICTS: IS DRINKING WATER A PROFIT-ORIENTED SERVICE?

Among the seventy-eight cases related to pre-war conflicts, one can identify three different types:

1. In twenty-nine cases municipalities make a claim to extend their concessionaires' contractual obligations.
2. In twenty-one cases, municipalities make a claim to lower the tariff applied to supply households and/or to quicken the development of the 'private' water service.
3. In twenty-four cases, municipalities make a claim to terminate the concession contract before its end date. Thirteen cases deal with the financial conditions of concession termination, on which the cities of Nantes, Lyon, Rouen and Toulon are separately and repeatedly opposed to the *Compagnie Générale des Eaux*.

In the first twenty-nine cases, cities are opposed to concessionaires who refuse to make extra-contractual investment either in additional water works or in the improvement of water quality.

In 1877, for example, the city of *Laon* claimed that its concessionaire should serve more streets in order to increase the number of free water delivery points.[2] The city referred to streets that were not mentioned in the contract. Under concession contracts, the private partner was committed to finance, build and operate a water network which delivered a given quantity of water at predefined 'public' points (fountains, municipal buildings). The design of the facility and the corresponding investment costs were agreed upon during the negotiation stage of the contract. However, years after the concession agreement, cities sometimes found that they needed additional water facilities (more delivery points or more water delivered to each point).

With regard to water quality, the city of *Nantes* complained in 1883 about the water supplied by the *Compagnie Générale des Eaux* to fountains and households, and asked for an improved filtration process.[3] In the nineteenth century there were few public health criteria dictating the quality of drinking water. Quality was a contractual agreement: partners might define it in terms of the technologies the concessionaire was committed to invest in and operate properly (a given filtration process as in Nantes), or in terms of the expected result, such as the degree of hydrotrimetry (water hardness) (as in Aix-les-bains).[4] Specifications mentioned in contracts might be updated long before the contracts ended. The concessionaire might comply with its contractual duties, but still be supplying people with undrinkable water, either at fountains (for free) or at home (at their expense).

During the same period, almost as many cities brought cases to lower the rates charged for the private household service and/or to speed up the development of the 'private' service. For example, the cities of Neuilly sur Seine, Clichy, Sceaux and Biarritz requested that users receive a metered subscription at the same price for a fixed quantity of water as the price that the concession contract had defined for the private household service twenty to thirty years before.[5] With the technical improvement in metering since the 1880s, it had become feasible to remove fixed quantity constraints for private users. Users were requesting meters that allowed them to consume as much water as they needed.[6] But this kind of subscription had not been envisaged under the concession contract. Private companies developed extra-contractual rules: they charged for the sale and installation of the meters, and also applied a higher rate for water consumption which exceeded the original quantity foreseen in the concession contract for private service. Cities complained against these conditions and argued that the concessionaires should not charge a different rate for the two types of private service.

Cities also wanted their concessionaire to quicken the development of 'private' services. Concession contracts usually prescribed the annual number of network

---

[2] CE, 9 February 1877, Sieurs Fortin-Hermann et Cie c. Ville de Laon.

[3] CE, 11 May 1883, CGE c. Ville de Nantes.

[4] CE, 19 May 1893, Ville d'Aix-les-Bains c. Compagnie des Travaux Hydrauliques.

[5] CE, 8 August 1888, Commune de Neuilly sur Seine c. CGE, CE, 3 March 1893, Commune de Clichy c. CGE, 13 June 1902, Sinet et commune de Sceaux c. CGE, CE, 20 November 1903, Compagnie des Eaux de Creil.

[6] During the first half of the nineteenth century, the standard for private water use was 20 litres/capita/day. By the end of the century, it had increased to 100 litres/capita/day + the water needed to clean the street and fight fires (e.g. a total of 200 litres/capita/day).

kilometres (in addition to the 'public' network) that the concessionaire was committed to finance, build and operate to develop the 'private service' until the end of the contract, as well as the conditions that households must satisfy to be connected. Concessionaires elected the streets to be serviced according to the number of household connection subscriptions. Households generally had to subscribe for a minimum duration of three years and were offered a fixed and daily quantity of water. Concessionaires might develop the 'private' service faster than the contracts prescribed when, for instance, household demand exceeded the allocated prevision. The profit of concession contracts was basically rooted in the 'private service', development of which served the interests of the concessionaires. But cities such as Rouen in 1912, long disappointed by the quality of the water supplied by the *Compagnie Générale des Eaux*,[7] protested because the Company refused to lay 8 km of pipes per year, as decided by the local council in 1904, and had instead constructed a 2 km/year extension on the grounds of the concession contract.[8]

Some cities asked for and obtained termination of their concession contracts from the first administrative court (the city of Monthléry in 1879[9] and the city of Bayonne in 1905).[10] Others, like the city of Lorient in 1891,[11] unilaterally declared the termination of their contracts. Their concessionaires brought the cases up before the CE. They claimed that their contracts could not be prematurely terminated without financial compensation, as long as they fulfilled their commitment to supply the expected quantity and quality of water (agreed upon *ex ante*). Should they fall short of their obligations, their contracts may be terminated, but in any other cases, the local authorities must either keep the contract or terminate it in compliance with the contractual clause dedicated to the financial compensation that the concessionaires are entitled to in such cases.

5.3   CE POSITION: UNDER A CONCESSION CONTRACT, DRINKING WATER IS A PROFIT-ORIENTED ACTIVITY

Weighing the arguments of the cities and the concessionaires, the CE adopted a literal interpretation of the contracts. The Court implicitly considered that there were no limits to the parties' rationality during the negotiation stage, when terms were agreed concerning the capital investment, the quantity and quality of the water supplied, the 'public' rate as a fraction of the investment made to supply fountains and municipal facilities, the 'private rate' as a fraction of the investment made to supply households, and the contract duration. All these terms were supposed to remain unchanged until the end of the contract, regardless of unforeseen contingencies. CE decisions mandated partners to enforce the contracts they agreed upon decades ago. As the Supreme Court for settling conflicts between any public authority and private body (person or company), the CE was first concerned with preventing private interests from overriding public authority decisions that might interfere with the economic balance of the

---

[7] CGE, Assemblée Générale Ordinaire du 29 May 1905, CE, 27 January 1911, CGE c. Ville de Rouen.
[8] CE, 22 November 1912, Ville de Rouen.
[9] CE, 24 March 1882, Sieurs Dalifol, Huet et autres c. commune de Montlhéry.
[10] CE, 26 December 1919, Compagnie des Eaux de Bayonne c. ville de Bayonne.
[11] CE, 6 April 1895, Sieur Deshayes c. ville de Lorient.

original agreements, and curb contractual duties at the expense of the concessionaires. The CE's jurisprudence then covered three major issues:

5.3.1 Amendments should be negotiated by the parties on the grounds of the initial contract's status quo

The *Conseil d'Etat* never allocated any residual control rights to any party that would extend the scope of the initial contract.[12] If municipalities became unhappy with the contracts they had agreed upon, even long ago, they could only invite their private partners to renegotiate. The concessionaires might or might not be interested in a new negotiation stage. Moreover, municipalities could not force their private partners to accept an extension to the contract, which would reduce its profitability. Concession contracts usually granted concessionaires an annual fee, paid by the public partner, of 4% or 5% of the investment made on the 'public' network (municipal water bill), over twenty to twenty-five years. Then municipalities, which claimed for extra-contractual investments in the public network or in water quality improvement, were invited by the CE to provide their concessionaires with a compensation scheme in line with the initial scheme, implicitly denying any economies of scale or of scope.

5.3.2 Local authorities had no right to renegotiate access to private service

The same rule applied to the development of the 'private' water network. Municipalities could not redefine access to this service or ask the operator, as Rouen did, to serve streets four times faster than stipulated in the contract, regardless of the size of connection request.

Moreover, municipalities had no right to interfere with the commercial relationship between concessionaire and households. As long as the concessionaires offered the private service prescribed in concession contracts (a fixed quantity of daily water for an annual bill equal to 10% of the waterworks), they might also sell additional services to meet household demand. These extra-contractual private service arrangements were not the business of the municipalities. They had no right to discuss these arrangements or to undertake legal actions on behalf of private water users in order to obtain more favourable conditions.[13] Concessionaires were afforded the right to offer meter-type subscription, which included the sale and a fee to maintain the meter, and a higher rate for extra water consumption.[14]

To some extent, the CE reminded municipalities that the concessionaires' profit chiefly originated in 'private' service revenue. Concessionaires had the exclusive right to develop this activity on a monopoly base, as a compensation for their commitment to supply municipalities with 'public' water. For municipalities, 'public' water came first when negotiating initial contracts. This included the service of municipal buildings (administration but also hospitals and schools) and stand-pipes. Concession contracts allowed municipalities to be supplied without raising the capital, by paying

---

[12] Under the Economics of Contract, an entity which has the residual right of control can take decisions when neither the contract nor the law has expressly designated the decision-maker.

[13] CE, 8 August 1888, Commune de Neuilly S/Seine c. CGE.

[14] CE, 3 March 1893, Commune de Clichy c. CGE.

an annual water bill on their general budget. Concessionaires expected to recover their investment in the 'public' network after twenty to twenty-five years – even longer if costs overran during the construction stage. Thereafter municipal water bills were no longer due, and concession profitability relied entirely on private service revenue. This meant that private service should cover the cost of capital (interest and/or dividend) during the first twenty to twenty-five years of concession contracts, and both the capital and the capital cost afterwards, not to mention the operating costs.

5.3.3 The CE restrictions on contract termination

It should be noted that a municipality was not entitled to terminate the concession contract by itself, even in cases where it had become obvious that the concessionaire had lost the ability to operate the contract. Only a judge could formally take such a decision, based on due consideration and evidence that the concessionaire had stopped fulfilling its contractual duties and, more precisely, had ceased operating the water service, either public or private. This unique situation necessitated a systematic contract termination without any guarantee in order for the concessionaire to recover the amount invested, even in part. When a concessionaire had ceased operations, a municipality might decide:

● to build a new facility, in which case, the concessionaire lost its initial investment
● to buy the existing facility back, in which case the concessionaire received the proceeds of the sale, after bargaining with the municipality
● to auction the facility, in which case he concessionaire received the proceeds of the sale.

In all other cases, a municipality could not terminate a concession contract without substantial financial loss, as defined contractually. According to the eight cases dealing with the conflicting termination of the contracts of Lyon, Nantes, Rouen and Toulon, municipalities owed their former partner the capital not yet depreciated, as well as the profit the *Compagnie Générale des Eaux* would have made until the end of the contracts, based on the ten last years average profit made on each concession.

5.4 THE CONSEQUENCE OF CE DECISIONS ON WATER SUPPLY MANAGEMENT

In the end, the CE decisions contributed to disqualifying the concession as a form of contract under which municipalities could develop water networks and provide safe water to all. At the beginning of the twentieth century, dozens of middle and small-size cities, which progressively considered the supply of networked water as the norm, became convinced that a concession contract could not secure them against unsafe and short water distribution, at a reasonable price.

In one to two generations, there were tremendous scientific and technical innovations. Hygienists made it clear that water supply was a top public health issue (Murard and Zylberman, 1996), and engineers started to successfully implement treatment processes both on spring and surface water (Pezon, 2000). Network extension and safe tap water supply became technically feasible and strongly supported by health professionals. This contributed to changes in the ways politicians envisioned water services. Improving health implied generalized access to the former 'private' service rather than free collective water. Local authorities stopped considering 'private' water as a profit-oriented service, but rather as a health-first-oriented service. Public water supply management symbolized the vocation of water public service to provide safe water for all.

The 1884 Municipal Law granted local authorities administrative authority over their territory.[15] This law also made the 36,000 municipalities – inherited from the *Ancien Régime* – the first political authority in France to be elected on a 'one man, one vote' basis (Morgand, 1952). In a situation where few people received 'private' access to water, the political benefit of promoting widespread access to water explained why local politicians wanted to increase connection rates without delay. In 1892, tap water was a reality for less than 130,000 people in France, while the cities supplied with 'public' water, most of which were committed to a concession contract, totalled about 4.5 million inhabitants (Goubert, 1987). In Lyon, for example, the *Compagnie Générale des Eaux* had only 16,000 clients after thirty years of concession (Villard, 1885). In Paris, water carriers had almost vanished, swept away by the facility of tap water, but who, apart from those who could buy water from vendors, could afford tap water?

Unlike the developments in other networked services (gas and electricity), local authorities were able to get involved in water supply even before legal changes enabled them to directly manage industrial and risky activities.[16] Local authorities were always acknowledged as having the right to organize their own water supply themselves (Duroy, 1996). Even the strongest opponents to *municipal socialism* considered that water supply was not concerned by restrictions on public undertakings to protect *'la liberté du commerce et de l'industrie'* (Mimin, 1911). The CE itself, which always held to a restrictive interpretation of the 1884 law regarding the involvement of municipalities in industrial matters,[17] never set limitations on municipal water supply; central government shared the same position (Duroy, 1996). These elements underline the point that the shift in water supply management in the early twentieth century was rooted in a political change regarding what local authorities considered as a public water service. A modern public service must provide each household with safe and unrestricted water. Local authorities financed the development of water networks on their general budget and facilitated the access of households to water (frequently, no-connection fee policies).

But what about local authorities that were still committed to concession contracts in the early twentieth century with perhaps decades to run? A few local authorities bought back the concessionaires' operating rights through very lengthy (thirteen to twenty-five years) and conflicting processes. The position of the *Conseil d'Etat* concerning the financial settlement, clearly in favour of concessionaires, prevented most municipalities from engaging in buyback actions.

Aside from the short list of cities that terminated concession contracts before the end date, others, whose concession contracts were more than twenty to twenty-five years

---

[15] Law of 5 April 1884 on municipal organization, Art. 61.

[16] Décret du 8 October 1917, which enabled municipalities to supply electricity to households. In 1915, municipalities were given the right to organize urban transport.

[17] Opinion of the Conseil d'Etat of 2 August 1894.

old, converted their former annual bill into a subsidy to improve water quality and/or to develop 'private' service, including this *de facto* within the scope of public water service. The increasing contribution of municipalities to water supply investment slowly turned concession contracts into *affermage* or lease contracts (Pezon, 2000).

5.5 CONCLUSION

Under the concession formula, both partners shared the vision that private connections and water came as a compensation for the provision of public and collective water. From that point of view, a concessionaire's profit was subjected to the satisfaction of primary needs, as defined by local authorities to comply with their public health and fire protection duties. The so-called 'public' service was free for users (paid for by local tax-payers), whereas the 'private' service was entirely financed by users (network extension, connection, meter). Through this arrangement, local authorities cross-subsidized those who could not afford a private service by those who could (previously satisfied by water carriers).

This smart contractual arrangement collapsed in the early twentieth century when local authorities tried to transform the former 'private' service into the new core of their water policy. The jurisprudence of the *Conseil d'Etat* emphasized just how hard local authorities tried to implement new policies within the concession framework, and why such a contract was not adapted to making access to water available to all.

Local authorities and concessionaires finally stopped engaging in contracts. In 1894, the *Compagnie Générale des Eaux* reported that it had ceased to enter into contracts until a solution was found to the water quality issue.[18] It would wait until 1913. Meanwhile, the biggest French water company added the option of meter subscriptions for all its local private services. The revenue from metered services provided a profit of 5% a year – at least in its biggest concession, the Parisian suburbs.[19] In 1898, the Company stated that the revenue from the extra water consumption was the major driver for its overall revenue.[20] Without improving the overall connection rate overmuch, the concessionaire made more money by increasing the unit profit rate on each household connected to private service (Pezon, 2007). Water concession was stuck in a vicious circle whereby operators' practices were increasingly unreconcilable with local authorities' perception of what a water service should be.

Public partners finally disentangled themselves from concession contracts either by buying back the concessionaires' contractual rights or by defining a new partnership that only left to concessionaires the responsibility for operation of the redefined public service. Local authorities would then largely impose a water tariff system, which was accepted by the population.

To conclude, two lessons can be drawn from the French experience in water concessions.

Compared with other developed countries, where private water companies collapsed in the early twentieth century, France has always combined public and private management for local water services. Water companies survived the concession

---

[18] CGE, Assemblée Générale Ordinaire of 28 May 1894.
[19] CGE, Assemblée Générale Ordinaire of 27 May 1889.
[20] CGE, Assemblée Générale Ordinaire of 16 May 1898.

debacle, to remain operators or sub-contractors for public water services. They would gain momentum soon after the State implemented a new regulatory framework in the early 1950s, and promoted *affermage* contracts (public financing and private operation and billing).

The development of domestic water supply in France, or other developed countries, never took place on the basis of the full cost recovery principle. Concession contracts were no exception to this. Today, developing countries and the international community acknowledge that water services cannot be charged investment costs up front, and limit its financial requirements to the recovery of operating costs. More research needs to be undertaken in order to adjust financing mechanisms towards a development process for improved access to water.

REFERENCES

Burel, J. 1912. *La régie directe considérée du point de vue de l'hygiène dans les villes, la question à Lyon*. Paris: A. Rousseau éditeur.

Copper-Royer. 1896. *Des sociétés de distribution d'eau*. Paris: A. Pedone éditeur.

Duroy, S. 1996. *La distribution d'eau potable en France*. Contribution à l'étude d'un service public local. Paris: Librairie Générale de Droit et de Jurisprudence.

Goubert, J-P. 1987. *La conquête de l'eau*. Paris: Hachette.

Long, M., Weil, P., Braibant, G., Dévolvé, P. and Genevois, B. 1990. *Les grands arrêts de la jurisprudence administrative*, 9th edn. Paris: Sirey.

Loosdregt, H.B. 1990. *Services publics locaux: l'exemple de l'eau*. Actualité Juridique Droit Administratif, Vol. 11, 20 November.

Loriferne, H. (ed.) 1987. *40 ans de politique de l'eau en France*. Paris : Economica.

Mimin, P. 1911. *Le socialisme municipal devant le Conseil d'Etat*. Critique juridique et politique des régies communales. Pithiviers: Librairie de la Société du recueil Sirey.

Monsarrat, G. 1920. *Contrats et concessions des communes et des établissements communaux de bienfaisance*. Paris: Bibliothèque municipale et rurale.

Morgand, L. 1952. *La loi municipale: commentaire de la loi du 5 avril 1884. Vols 1 and 2*. Paris: Ed. Berger-Levrault.

Murard, L. and Zylberman, P. 1996. *L'hygiène dans la République. La santé publique en France ou l'utopie contrariée 1870-1918*. Paris: Fayard.

Pezon, C. 2000. *Le service d'eau potable en France de 1850 à 1995*. Paris: Cnam.

Pezon, C. 2003. *Water Supply Regulation in France From 1848 to 2001: a Jurisprudence Based Analysis*. Communication to the Annual Conference of the International Society for New Institutional Economics, 11–13 September, Budapest, Hungary.

Pezon, C. 2007. The role Of 'users' cases in drinking water services development and regulation in France: a historical perspective, *Utilities Policy*, April.

Villard, G. 1885. Etude d'un service d'eau pour la ville de Lyon. Principes généraux d'alimentation des villes en eau potable. Etude d'un tarif rationnel des eaux ménagères. Bibliothèque Nationale, Paris, (self-published).

Chapter 6

In search of (hidden) Portuguese urban water conflicts: The Lisbon water story (1856–2006)

Luísa Schmidt, Tiago Saraiva and João Pato
Institute of Social Sciences, University of Lisbon, Portugal

6.1 A CENTURY OF PORTUGUESE WATER SERVICES: EVOLUTION, ACCOMPLISHMENTS AND FAILURES

The history of urban water services in Portugal can be explored through the permanent tension between central government and municipalities. The creation of the hydraulic services in the last quarter of the nineteenth century[1] set the scene for contemporary water public policies in Portugal: most waters are public, the state bears the responsibility for its administration, and its private use is regulated under two different regimes – license and concession contracts. This policy framework was also valid for water supply, drainage and treatment systems in urban areas: municipalities assumed administrative and management responsibilities, and the central government would provide financial and technical support through the hydraulic services and its regional branches. The former would decide whether to run the systems directly, to create administrative services or public companies, or even to grant concession contracts to private companies. The latter would supervise public works developed by municipalities and the development of water infrastructure. This model was applied in all urban areas with the exception of the city of Lisbon where central government intervened directly and decided itself the concession conditions with a private company.

As is usually the case in long alliances, the relationship was not an easy one. Water indicators, when available,[2] show that the results were far from satisfactory – moreover, if we assume its main purpose was first to guarantee universal water supply and drainage and later wastewater treatment (see Table 6.1). In the 1980s only half the

---

[1] The law published on the 6 March 1884 approved the Organization Plan for the Hydraulic Services, an administrative branch of the Ministry of Public Works that lasted until 1987. The revision of the legislation that defined the frontier between public and private waters was launched eight years later (Decree no. 8, 1 December 1892 – Organization of the Hydraulic Services and Respective Staff).

[2] These numbers were not regularly updated, no distinction was made between rural and urban services, and the evaluation procedures lacked methodological consistency. Even today, distinct governmental bureaus produce different numbers with respect to water services in Portugal. Consequently, these numbers should be seen as indirect indicators, not as exact values.

Table 6.1 **Total population served with water services**

Year	(source)	Water Supply	Water drainage	Wastewater treatment
1941	(MOPC, 1941)	26.32%	–	–
1970	(Lencastre, 2003)	37%	17%	–
1980	(DGSB, 1981)	39.4%		
1990	(IRAR, 2004)	80%	61.80%	31%
2000	(MAOT, 2000)	90%	75%	55%
2007	(ERSAR, 2008)	92.00%	80.00%	72.00%

Source: Various sources were used to collect information on water services in Portugal over the period. See the references below.

1941 – MOPC. 1941. *Anuário dos Serviços Hidráulicos*, Lisbon: Imprensa Nacional.
1970 – Lencastre. A. 2003. Hidráulica urbana e industrial, *Memórias Técnicas*, Vol. II. Lisbon: LNEC.
1980 – DGSB. 1981. *Plano director de saneamento básico para o decénio de 1981–1990*. Lisbon: MHOP.
1990 – IRAR. 2004: *Relatório Anual do Sector de Águas e Resíduos em Portugal*, Vol. I, Lisbon: IRAR.
2000 – MAOT. 2000. *Plano Estratégico de Abastecimento de Água e de Saneamento de Águas Residuais 2000–2006*. Lisbon: MAOT.
2007 – ERSAR. 2008. *Relatório Anual do Sector de Águas e Resíduos em Portugal*, Vol. I, Lisbon: ERSAR.

population was served with water supply systems, not to mention sewage systems, with less than a third of Portugal inhabitants having their wastewater treated at the beginning of the 1990s (Table 6.1). Technical supervision and financial support weren't sufficient for an effective state water policy and most solutions tested by municipalities, including the concession of services to private companies, proved inefficient.[3]

Is it correct to assume an enduring institutional conflict in urban water services between the two levels of government from the end of the nineteenth century till the 1980s? Probably yes, but it never took the formal character of a judicial case. Nevertheless, it is clear that municipalities resented the lack of interest of central government in the well being of their inhabitants as revealed by the small amounts of investment. However, central government wasn't eager to delegate the responsibility for large infrastructure works to local authorities, knowing their limited technical capacities and proverbial mismanagement of public funds. In any case, the population was not in a position to express discontent during the long period of authoritarian government. The most obvious conclusion from urban water services data in Portugal through the twentieth century is their slow progress, lagging behind development compared to other European countries – the recurrent benchmark of Portuguese policy-makers.

This old obsession with Portuguese delay was the main rationale for numerous initiatives. In 1932, the sanitary problems caused by water epidemics (cholera surges and others), typical of a country still stuck in the nineteenth century, constituted a good opportunity for the new dictatorial regime – which had emerged from the 1926 coup

---

[3] See the case of water supply to the city of Porto: a concession to Compagnie Générale dês Eaux pour l'Étranger was signed in 1887, but deficiencies in systems operations were sufficient motives for the municipality to cancel the contract and install municipalized services (Cordeiro, 1993: 11–34).

d'état, and started an authoritarian conservative rule that would last until 1974[4] – to prompt a new relation between central and local power. Local demands to expand water supply and sewage infrastructure were a strong political argument for closer institutional and technical control by central government, and the creation of the Water Sanitary Council (1933) was thought to be enough to accomplish such a purpose.

The Ministry of Public Works, probably the most active branch of government with regard to materializing the visions of the authoritarian New State, adopted an even tighter position by creating the Bureau of Urbanization Services in 1944 (Direcção Geral dos Seviços Urbanos). The Bureau surveyed the needs in urban water infrastructure, designed the projects, and forced municipalities to implement them. The state financed 50% of the building costs of water supply and sewer networks, and 75% in the case of distribution through public fountains for small villages (MOP, 1954). The dictator Oliveira Salazar, in his typical vindication of traditionalist values, feared that good old Portuguese habits would disappear if people ceased to gather round water fountains (Freitas do Amaral, 1995). It is also useful to remind ourselves that by 1940 only 20% of the population, out of a total of 8 million people, was living in urban areas. The truth is that a substantial increase in municipal funding had occurred[5] as well as technical orientation and support, but progress, as already stated, was just too slow. Although the number of interventions subsidized by the central government increased steadily in the 1940s and 1950s (from 200 in 1946 to 800 in 1960), the number of interventions concluded by the municipalities remained almost steady at a much lower level (less than ten in 1946 to around seventy in 1960).[6] By the end of the 1960s, the government itself recognized that the policy framework was inadequate to cover urban areas with necessary infrastructures, but no substantive change took place before the end of the dictatorship.

The April revolution in 1974 and the two-year period that followed, dominated by radical left-wing politics, were times of sudden and substantial political change in Portugal. This affected not only on social and political values, but also created high expectations for the creation of an effective welfare state that would finally reach the entire population.[7] Water policies were no exception, and a new model for water services was designed, dividing the country into sanitation regions, geographically corresponding to districts. Public companies, under the control of the central government, would be established in each region promoting scale economies and technical cooperation.[8]

---

[4] There is a long discussion among historians regarding the proper classification of the Portuguese New State that came out of the 1926 coup with 'conservative authoritarianism' being the most consensual typology. Nevertheless, it is hard to miss the fascist nature of many of the state institutions, supporting the thesis of Manuel Lucena that there was no other country like Portugal which took the institutionalization of fascism so far, making Oliveira Salazar's New State 'a fascism without fascist movement'. See Manuel Lucena (1976), António Costa Pinto (1992) and Fernando Rosas (1992).

[5] At least 50% of the total costs of infrastructure would be supported by the central government and special credit conditions would be made available for the remaining investment.

[6] Data collected from annual reports of the Bureau of Urbanization Services (1944–1960).

[7] International comparisons between the structure of the Portuguese budget and other western European countries reveal the low percentage of resources allocated to welfare policies by the Portuguese state prior to the revolution (José da Silva Lopes, 2005: 265–304).

[8] Council of Ministers Resolution, 23 January 1976.

Such drastic reductions in the influence of municipalities was clearly against the political spirit of a revolution that promised to promote democratic institutions at the local level. This is probably the most significant cause for the failure of the regional model. Some of the public companies were even created in the years that followed, but commenced operations. Nevertheless, the growing responsibilities of municipalities under the new regime were no solution either for better water supply or for the universality of sewer systems. During the 1980s, local power was facing a double challenge: to construct basic infrastructure for water supply (only 50% of the population was covered at this time) and wastewater treatment, in tune with the new environmental paradigm that had emerged since the 1970s.[9]

Only an exogenous stimulus such as entry into the European Community in 1986 was able to dramatically change the Portuguese water policy model. The abundance of structural funds for environmental intervention, as well as the consideration that innovative environmental policies were necessary for this purpose, demanded profound changes to the water institutional framework and water services model. The first step was undoubtedly taken in 1987 with the extinction of the Hydraulic Services, ending 103 years of explicit hydraulic tradition, and the creation of the *Instituto da Água* (Water Institute) in 1993, which was intended to consolidate the new environmental paradigm in water policies.

In the mid-1990s new laws regulating water services defined new terms for the relationship between municipalities and central government. The division of water services into upper (wholesale) and lower systems (retail), as well as the creation of both a state-owned water company – AdP-Águas de Portugal (Portuguese Waters)[10] and a regulatory agency,[11] were the chief innovations. The central government would intervene in water services through this public holding, whose main objective would be to create, manage and finance water companies (upper systems) that would serve multiple municipalities in water supply, drainage and treatment. The majority of these companies' stock would be state-owned with municipalities always in a minority position. The latter would, in turn, be encouraged to create second-level water companies that would serve citizens with water services (lower systems), but could also grant the concession of these services to public or private companies, create municipal services, or even municipally owned companies. European funds would be channeled to water systems through the public holding 'AdP – Águas de Portugal', on the one hand, and through the municipalities on lower-level systems, on the other.

The creation of the public holding 'AdP' was an incentive for scale economies and technical integration. The spirit of the reform was embedded with the need to achieve efficiency, as well as managerial and technical competencies in water services, both in upper and lower systems. Nevertheless, even nowadays, most water services run by municipalities still lack the sufficient know-how. The regulatory agency reveals in its

---

[9] On the different water paradigms see Chapter 1 and Chapter 2 in this book.

[10] Created in 1993, the company AdP – Águas de Portugal would only start operating effectively in 1995.

[11] In 1995 the Observatory for Multimunicipal Systems was created as a first regulatory body, with no practical results. It would be replaced in 1997 by the IRAR (Instituto Regulador de Águas e Resíduos), that only in 2003 started its operations. In November 2009 a new law renamed the institution as ERSAR (Entidade Reguladora dos Serviços de Águas e Resíduos) and reinforced its power by expanding its regulatory powers to all operators and enhancing its independence concerning political and economical pressures.

annual reports[12] that the inexistence of regular business accounting in some municipal services, which is essential to produce a water service cost structure, has been one of the main causes for the present undervalued price of water. Most municipalities do not charge the real cost of water services, disregarding the principles recommended in the European Water Framework Directive.

The model defined in the 1990s produced significant changes and channeled major investments into the water sector. In 2004 there were nineteen concessions controlled by the central government (Águas de Portugal, SGPS) and municipalities on the upper-level systems, as well as twenty-three concessions (private and public) on the lower systems.[13] The remaining systems (lower level), which serve the large majority of the population (potentially 8.7 million), are directly run by the municipalities.[14] Although this remains the most representative share of the water services market, no financial information is available from the regulatory agency. Thus, not only do municipalities lack accounting and technical know-how, but they are also unaccountable to the regulatory agency. But there is a clear lack of a multi-level policy specifically dedicated to small units. Moreover, most investment has been channeled to water supply: water drainage and wastewater treatment still have significant deficiencies in quantity and quality terms.[15]

The water price can also be considered as a sign that the reform is still far from concluding. Analysing the two systems described (upper and lower systems), a high dispersion of water prices can be noted for the upper-level systems, with a price differential of more than 100%.[16] Moreover, the maximum price paid by Portuguese consumers for water supply is less than half of the mean price paid by their European counterparts. Such values reveal that either the prices do not reflect real costs, or that the necessary investments are not being made. Furthermore, it is widely accepted that it is impossible to increase water prices too fast.

Despite considerable evolution, there is still a long way to go. Water supply and treatment systems have been prioritized over the last thirty years, with over two-thirds of the total funding for environmental sector investment from the first and second European Community Support Frameworks (CSF) (1986–1992; 1993–1999). The irony of the situation is that both the third CSF (2000–2006) and the recently

---

[12] See RASARP since 2004 (www.ersar.pt).

[13] The upper-level systems are a natural monopoly with no participation by private companies, and no private sector investment allowed. For the lower systems concessions are distributed as follows. The public company Aquapor-Luságua (owned by the public holding Águas de Portugal) is the most important, having thirteen concessions in the market. The biggest private company is controlled by the Somague Ambiente holding (an international private group), with ten concessions. Indáqua (a Portuguese private company) participates in three concessions, and Compagnie Générale des Eaux Portugal represents the interests of this group in four concessions.

[14] Distribution of the services by type: municipal companies (11), municipal services (223) and municipalized services (32) (see RASARP, 2004 for this data).

[15] In March 2001, Margot Wallstrom, the European Environmental Commissioner, considered the condition of Portuguese wastewater services to be unacceptable after the enormous amount of money invested with contributions from EU funds. For example, half of the 115 wastewater treatment plants of the Tagus basin did not respect the minimum standards set for their operation. For data on wastewater treatment, see Ministério do Ambiente (1999).

[16] In 2009 the national medium price is €0.49/m^3, but values vary from the highest (Ä0.58/m^3) in Águas do Douro e Paiva to the lowest (€0.32/m^3) in Águas do Minho e Lima.

approved National Strategic Reference Framework (2007) still define the water sector as an investment priority. In fact, the problem doesn't seem to be exclusively of a financial nature. In spite of all the EU funds, the lack of local technical know-how and an inspection structure meant that much of the money was simply thrown away. A report published in 1995 by the National Laboratory of Civil Engineering revealed not only the lack of skills to properly operate the new sewage water treatment plants, but also the obsolete technological solutions adopted (Melo Baptista and Matos, 1995).

To sum up, although a new policy framework is gradually taking effect, its results do not yet validate its choice as the most adequate solution. The relationship between central government and municipalities is increasingly centralized, and the tendency to create large regional water services is underlined by the recent technical integration of systems. In fact, such a framework converts what was an exception – the case of Lisbon – into the general pattern of water policy across the country. Indeed, both the regional scale and the direct role of the state differentiated the Lisbon case from other urban centres. But this is not just a matter of formal similarities; the relation goes much deeper, for the creation of the public holding *Águas de Portugal – AdP* allowed the transferrence of the existing technical and management expertise of the state-owned Lisbon Company, EPAL, to the new local companies. The dissemination of many 'EPALs' throughout the territory (at least on the upper-level system) was taken as the best solution to properly managing the boom in water infrastructure, which resulted from European funds. It thus makes sense to examine the Lisbon water story in detail. The Lisbon area now has some 2.5 million inhabitants, and constitutes the only real metropolis in the country where the urban dimension of water conflicts is a relevant issue. What could be seen as a legitimate methodological option of describing the best documented case, a typical 'case study' methodology, then becomes an inquiry into the main laboratory of water policies for the entire country.

6.2 THE LISBON WATER STORY

6.2.1 Liberal waters (1858–1926)

The monumental character of the *Águas Livres* aqueduct, built in the eighteenth century, might suggest that the problem of water supply to Lisbon was solved many years ago.[17] But the truth is that the cyclical outbreaks of epidemics in the nineteenth century reveal the poor sanitary conditions of a capital, repeatedly facing water scarcity and sewage troubles. Foreign engineers visiting the city denounced the uselessness of the impressive masonry of the aqueduct, 'a landmark of ignominy to the Portuguese people', which delivered 'no more than 4 litres per head' in the 1850s (Valle, 1856: 133–36, 145–47).[18] It was quite obvious that the large majority of the population didn't use the aqueduct waters, instead resorting to the numerous wells and sources, many of them private, within the city perimeter.[19] The cholera and yellow fever epidemics of

---

[17] On the history of the eighteenth-century aqueduct see Moita (1990); Caseiro, Pena and Vital (1999); Gentil Berger (1994) and Oliveira Caetano (1994: 293–312).

[18] The waters from the Aguas Livres aqueduct came from Belas 15 km northwest from Lisbon.

[19] For a detailed description of the water sources inside the city perimeter in the middle of the nineteenth century, see Veloso de Andrade (1851).

the years 1856 and 1857, with a death toll of some 9,000 Lisbon inhabitants out of a total population of 160,000, were the direct cause of central government involvement in a subject that was previously exclusively municipal.[20]

In 1858, the state signed a concession contract with Empresa de Águas de Lisboa (EAL), the first Lisbon water company, formed with the capital of sixty-four Portuguese shareholders, that promised to deliver 93.75 litres per day for every Lisbon inhabitant. The company hired the French engineer Charles Louis Mary, *ponts et chaussées* inspector of the Seine department, who designed a project comprising several new lateral aqueducts to be connected to the existing eighteenth-century aqueduct. Just as important as finding new water sources was the design of a distribution network, which relied on four new reservoirs to conduct water to every Lisbon building. This network was a significant innovation when compared to all previous municipal projects, which only offered public fountains distributed along city streets and squares. For the first time the entire city was subsumed to an engineer's rationale, with the aim of transforming it into an efficient organism.

Pushing to make every Lisbon inhabitant its client, EAL transformed water into a commodity. In the following years the municipality would become its first opponent. The local administration never accepted the loss of control over water supply issues and its members repeatedly denounced the dangers of a private monopoly as a menace to the public interest. Despite all criticisms, the granting of concessions was the solution the 'regenerationist' government found to expand infrastructure over the territory. This government took power in 1851 with a policy of material improvements that would bring to an end the tumultuous first decades of the liberal regime.[21] But it is also true that the municipality had its reasons, for Lisbon's first water company was never able to deliver enough water and never demonstrated enough financial capacity to supply the lacking infrastructure.

In 1863, under the pressure of municipality disapproval and public protests during a drier than usual summer, the government formed a commission of inquiry which revealed that the company only supplied 8% of the agreed water.[22] However, local power did not recover control of Lisbon's waters; instead the Ministry of Public Works took charge in the name of the capital's hygiene. The former argued that water supply and sewerage (which was already a local responsibility) should be taken as a single service, thus claiming control of both. But central government maintained that the new infrastructure designed by ministry engineers was just too expensive and complex to be managed at local level. Searching for new sources abundant enough, state engineers had presented a project for a new aqueduct bringing water from more than 100 km away from Lisbon. This long Alviela Canal put an end to the municipality's demands, and in 1867 the state signed a new contract with a private company, the

---

[20] On the tight relation between water infrastructure and epidemics in the nineteenth century, see Saraiva (2005) and Ferreira da Silva (2006). On epidemics and the evolution of Lisbon population see Rodrigues (1995).
[21] Regenerationism promised to end civil disputes by building roads, railways and ports, replacing politics by technology, in an ideology that resonates with Saint Simon utopia. It is no coincidence that many of its leaders, namely Fontes Pereira de Melo, were engineers. On regenerationism, see de Fátima Bonifácio (1999).
[22] On this polemic, see Saraiva (2005) op cit: 124–34.

Companhia das Águas de Lisboa (CAL) which hired the engineers of the Ministry of Public Works to lead its technical section.

The building of the canal suffered long delays as is often the case with big public works. To expropriate the estates crossed by the lengthy canal line was a difficult task. However, reaching an agreement between the company, the state, the municipality and Lisbon proprietors regarding canalization regulation proved even more difficult. The first agreement insisted upon implementation of the approved regulation such that every proprietor was obliged to build a connection to the company's distribution network (Alves, 1940: 5–23; Leite Pinto, 1989). Only after approval of the regulation did the company guarantee its financial viability to cover the infrastructure building expenses. In 1880 the Alviela canal was finally inaugurated.

The arrival of the Alviela water in 1880 was celebrated in a mass event staged to welcome the new age of Lisbon sanitation. Together with the railway lines connecting the capital to all the country's provinces, the works of the big port and the new large avenue that 'Haussmanized' Lisbon, the water works were presented as proof of the ability of the liberal state to put Portugal on the path of progress. Lisbon was hailed as the capital of a modern country, with engineers designing networks that controlled the fluxes of people, goods and water.[23] The new daily 30,000 m^3 were taken by the press as a kind of magic solution that converted Lisbon from a dry North African city into a green European Capital, freed from the epidemics so much feared by urbanites.

But to eradicate epidemics it is not merely sufficient to have abundant potable water. Sewerage plays an equally fundamental role. As we saw, the claim by local authorities to take control of supply was justified with a hygienist rhetoric tightly connecting distribution and sewerage. However, it was engineers of the central government that first envisaged, in 1874, a waste carriage system using water as a draining and cleansing agent of the sewage pipes. Conversely, the municipality favoured cesspools and privy vaults with pipes admitting exclusively rain runoff and water used in sinks and bathtubs. It was quite a paradox that a typical argument for municipalization of water supply all over Europe – the need of copious quantities of water for the proper work of the water-carried sewer system – was thus absent from the Lisbon debate on control of urban waters. Only in 1880, the year when Alviela water arrived, did a municipal commission involving government engineers start to design a sewer system relying on the water-carriage model that would be approved four years later (Ferreira da Silva, 2006).

The new sewer network would nevertheless grow at a very slow pace. It was impossible to universalize sewage collection when supply itself was covering only about 50% of the city's households in 1890, and some 60% ten years later. There was nothing more dangerous than building an extended network of sewage pipes without enough water to clean them, for they could become the main locus for epidemics, as had been the case in the previous cholera outbreak. The main consumers of the company waters were still by far the municipality and the central state, accounting for 76% of total consumption for the years 1900–09. It is no surprise, then, that the main concern of the company was to obtain an assurance that the government and the municipality would pay their growing water debts, a dispute that had to be solved by the Administrative Court. Water in Lisbon was thus facing a double bottleneck: the

---

[23] On the mass event and its relation to the renewal of the image's capital, see Saraiva (2005) op cit: 137–42.

Figure 6.1 **Thirsty day in Lisbon**
Source: Joshua Benoliel (1912).

sewer system could not grow because there was not enough water supply coverage for private households; private consumption did not grow sufficiently because the sewage system (one of the main water consumers) was not complete. This was the price to pay for separating water and wastewater.

What was obvious at the turn of the century was that water infrastructure development was not keeping pace with the city's population growth. In 1900, Lisbon already had some 350,000 inhabitants, a huge upsurge when compared to the 200,000 people living in the city in 1864. In the 1920s, the numbers would climb to almost 600,000. In the first decade of the twentieth century, the water company was already facing a problem of scarcity of available water. For the years 1900–1909 the total daily consumption per capita was down to 74.1 litres, with private consumption limited to 17.7 litres. In 1905, only twenty-five years after the grand-opening of the Alviela Canal, the company started looking for other sources, namely the surface water of the Tagus River (Leite Pinto, 1989; Ferreira da Silva, 2006; Alves, 1940).

Although the new project to bring water from the Tagus was praised as the most complete and detailed engineering project ever produced by Portuguese technicians, the company was never able to launch it in the early decades of the twentieth century when Lisbon's population was exploding. The expensive 80 km Tagus Canal, designed to supply 108,000 m^3, was constantly postponed among complaints of the debts of the government and municipality towards the company, or of the low water price paid by private consumers. The resistance against a company that saw its share price rise by four times from 1870 to 1909 was reason enough for bad press among the Lisbon population. The case became even worse in 1913, when it became clear that the Alviela Canal wasn't enough to meet growing demand during the dry season. In July the residents of the more elevated areas of the city, where supply was interrupted for several days, broke the fire hydrants (*Ilustração Portuguesa*, 1913). The fighting of fires inside the urban perimeter became itself a critical issue, with the press denouncing how small fires reached catastrophic dimensions due to lack of available water (*Ilustração Portuguesa*, 1917). The constant water shortages during the summers led the company, in accordance with central government and supported by the national guard, to set a plan for water rationing, distributing it through sixty improvised public fountains and by aid of water tankers circulating through the city streets (Leite Pinto, 1989: 269–79). After having converted the majority of the Lisbon population into its clients, the company with police support now controlled public access to water. The newspapers, exhorted by the municipality's complaints, excited public opinion against a company whose headquarters suffered a bombing assault in 1924 (Leite Pinto, 1989: 278). That same year water rationing was discussed in Parliament with members of the government themselves denouncing the rationing measures taken up by the water company.

During the first three decades of the century, scarce water volume was not the only reason for public distrust towards the company. Since the typhus outbreak of 1912 water quality had also become a subject of public concern (*O Século*, 1940). One of the company reservoirs was identified by the hygienist doctors as the source of the epidemics affecting 2,615 people and killing 254. Only after 1918 was water chlorinated. Consumers repeatedly denounced its bad taste, which was reason enough to interrupt the water treatment in 1926, with chlorination returning only in 1931 (Pinho, 1942: 37–45). By then the number of deaths caused by typhoid fever had receded to thirteen cases per 100,000 inhabitants,[24] but such figures were still ten times bigger than those of Berlin or London. The main reason for such a gap was of course the need of Lisbon's population to make up for the company's rationing measures during the dry season, through the use of non-reliable private water sources and wells still existing in the city area and its suburbs.[25]

6.2.2 Authoritarian waters (1926–1974)

The big tensions around water supply in Lisbon were of course unbearable to the dictatorial regime that emerged from the military coup of 1926. The constant quarrels between the company, the municipality, the central government and the population, were taken as paradigmatic of the powerlessness of the Republican regime (1910–1926) to solve social disputes and properly manage public affairs (Alves, 1947: 142–47). In 1932, the very same year he was nominated Ministry of Public Works of the first cabinet formed by Oliveira Salazar, Duarte Pacheco[26] launched the new basis for Lisbon water

---

[24] Major progress when compared to the thirty-five cases per 100,000 inhabitants for the years 1916–20.

[25] In 1945 an inventory by the government engineers counted some 2,512 wells inside the city perimeter, with only ten belonging to the municipality (Pinho, 1945: 37–49).

[26] Duarte Pacheco (1900–43) was the Ministry of Public Works of the authoritarian regime led by Oliveira Salazar from 1932 to 1936 and from 1938 to 1943, the year of his death. He was also the Mayor of Lisbon from 1938 till 1943. Duarte Pacheco, an engineer by training, represents the technocratic character of the regime, a soft version of Albert Speer, launching a vast programme of public works to change Portugal's image, and in particular to convert Lisbon into the metropolis of the new empire. He died in 1943 in a car accident that symbolized in a tragic way the modernism of the most dynamic leader of the Portuguese New State.

supply. The solution couldn't be simpler: the water price was abruptly increased by some 40% with the company obliged to immediately start the construction of the Tagus Canal, planned back in 1908. The company was now also responsible for water quality, having to install chlorination plants at its own expense. If the company didn't accept the conditions set by the government, the state would directly take over Lisbon water supply and the company would be dismantled. Even though the solution was presented without any negotiation, the company learning of the terms of the agreement in the morning newspapers, the new contract was quickly signed (Alves, 1940: op cit).

Once again the main loser was the municipality: despite its ambition of taking over a company that didn't fulfill its duties, it never had sufficient political power or financial capacity to take care of the city's water supply.[27] The only dialogue now took place between the company and central government, with the creation within the Ministry of Public Works of a Commission for the Inspection of Lisbon Waters (CFAL). Beyond controlling the company's activities and inspecting every new work, this government agency could also present new projects to be undertaken by the company. The solution of keeping a private monopoly together with a strong interference capacity by the state was an approach typical of Salazar's corporate New State. In the following years such state interventions in the economy would also be applied to hydroelectricity production or steel manufacture. If the first water concession of the nineteenth century was paradigmatic of the way liberal governments extended infrastructure over the territory, Lisbon waters set the tone for the many years to come under the dictatorship. The best proof of such a claim is the ironic comment made by the head of CFAL that the people of the capital didn't protest against the over 40% increase in price imposed by government (Alves, 1947: 143). Public conflicts had now been officially banished, urban water conflicts included, and would remain so for a long time, as further confirmed by media analysis.

In 1933–34 under the pressure from CFAL, the company had already doubled its supply capacity. During the following years, CFAL engineers replaced the project to supply Lisbon with surface water by the much cheaper solution of extracting groundwater from the Tagus alluviums in an area closer to Lisbon, shortening by some 20 km the huge Tagus canal with a daily capacity of $257,500 \, m^3$. In 1940, the CFAL could already be proud of a daily $137,000 \, m^3$ summer supply (for a maximum consumption of $80,000 \, m^3$), with the previous infrastructure guaranteeing only $37,000 \, m^3$. CFAL engineers could not accept that such a grand improvement would remain unwitnessed by the city's inhabitants, much of the works lying beneath their feet or far from the urban centre; the Ministry of Public Works thus commanded a monumental fountain to celebrate the end of water scarcity in Lisbon.[28]

---

[27] During the authoritarian regime of the New State, the municipal powers were directly nominated by the central government. It is sometimes hard to distinguish what is local and what is central, especially in the case of Lisbon, where the ministry of Public Works, Duarte Pacheco, was nominated Mayor of Lisbon in 1938.

[28] The monument was placed facing the impressive new buildings of the Superior Technical Institute, the institution breeding the technocrats of the authoritarian New State, CFAL engineers included. The fountain and the Engineering School were built following the architectural canons of the regime, established by the Ministry of Public Works, both standing as urban landmarks of the new fascist capital. On the evolution of Lisbon under the authoritarian New State, see Matias Ferreira (1986) and Acciaiuoli (1998).

But the capital city of the *Estado Novo* (New State) dreamt by Duarte Pacheco was much bigger than the old city limits. It included the planning of a prestigious tourist area to the west connected to the city centre by a scenic highway offering quick access to its sandy beaches. A copious water supply was of course needed to sustain such expansion and a new large conduit was built feeding the Sun Coast (as it was named) with Lisbon water (Alves Costa, (1940: 15–21). In this case, the company limited itself to supplying the water, with the local municipalities along the coast being responsible for its distribution. This was also the case of the conduit connecting the city centre to Sintra, another favorite tourist attraction and currently listed as a UNESCO World Heritage site. The curious thing is that even if the primary reason for guaranteeing supply to such places over 20km away from Lisbon was satisfying growing tourism activity, the conduits more or less followed existing railway lines, thus providing the infrastructure for future migration from the countryside to the city.

Actually, those two axes – along the coast to Cascais to the west and to Sintra to the north-west – were to become two of the main axes of development in the Lisbon metropolitan area during the second half of the century. Further expansion followed the water conduits to the east, from where the water was entering Lisbon. In all these areas served by the company's water, the local authorities were responsible for its distribution. It is thus hard, not to say impossible, to understand the spatial patterns of the expansion of the Lisbon metropolitan area without taking into account the layout of the waterpipes.[29] The extension of water supply infrastructure by the company to the Sun Coast, to Sintra and along the canals that brought water into the city from the east, was the first manifestation of what would become the Lisbon metropolitan area. In fact, the company and CFAL were the only entities that operated at the scale of the metropolis in the 1940s and for many years to come. For the first time it seemed that water was leading urban expansion and not the other way round.

The dramatic increase of population served by the company's water meant an upsurge of the total consumption of 50% between 1942 and 1947, with private consumption now accounting for the main share (57% of the total). And so, in 1949, Lisbon was facing shortages again, with water being rationed by the company during summer nights.[30] This time the newspapers were not used as a tribune for discussion of the infrastructure problems of the capital; instead they were used to publicize restrictions on consumption with the population, according to the company's account, willing to aid, namely by promptly reporting any leak.[31] As previously stated, all

---

[29] It may be argued that we thus leave aside all the southern urban expansion across the river where company pipes never arrived, but one of the main features responsible for the very dispersed southern settlement pattern is exactly the lack of infrastructure (see Portas, Domingues and Cabral, 2004) namely water supply, with the population relying, till the 1980s, mainly on private wells (River Basin Plan). The lack of an extended sewage system made things worse with the aquifers used by the population being polluted by the increasing population.

[30] See 'Nota referente às restrições de abastecimento de água em Lisboa, no Verão de 1949', *Boletim da Comissão de Fiscalização das Águas de Lisboa*, 30: 81–84.

[31] Also, and in spite of the extension of the distribution network inside the city limits, 33% of Lisbon households were still not connected by 1943. The numbers gathered by CFAL's initiative also revealed, as expected, that the main proportion of people unconnected comprised the poorer social segments.

urban water conflicts had been banished;[32] this remained the case even now that water was once more scarce and summer rationing had been reinstated. News related to water supply had a distinguished positive tone when compared with the repeated criticisms prior to the establishment of authoritarian rule (Schmidt, 2003).

In the following years investments in distribution and the introduction of new surface water for the Tejo Canal (1963) would progressively improve records. Following the unreliable official figures used by authorities for the big development plans of the next decades (Comissão de Planeamento da Região de Lisboa, 1973), the proportion of the population supplied in Lisbon finally reached 100% at the beginning of the 1970s. What such numbers were hiding was that much of the demographical expansion of the Lisbon area during the 1960s, fed by a massive rural exodus, was being absorbed by the uncontrolled proliferation of unplanned suburbs and even slums where the water pipes of the company didn't reach. Between 1960 and 1970 the population of the metropolitan area would grow roughly from 1.5 million to 1.8 million inhabitants. In 1981 the total population for the area was already close to 2.5 million (Ferrão, 1996). In twenty years 1 million people settled but the urban core remained almost unchanged with little more than 800,000 inhabitants, while the poorly served peripheries boomed.

Under the new leadership of Marcello Caetano (1968–74), the dictatorial regime began demonstrating a wider awareness of the social problems of development;[33] meanwhile, the newspapers began publishing stories about the drama of Lisbon's peripheral neighbourhoods. In 1969, a ten-storey building collapsed in the clandestine neighbourhood of Brandoa, home to some 18,000 to 20,000 people, in what would become a symbol of the chaotic expansion of the Lisbon metropolis. In their descriptions of the many Brandoas around Lisbon, journalists talked about barefoot children inhabiting the unplanned areas of former farms that surrounded Lisbon, where lack of piped water was denounced as the central problem (*O Século Ilustrado*, 1970). The problem, once again, was not just the incapacity of the distribution network to follow the frenzied rhythm of metropolitan territorial expansion, but also the shortage of total water at the disposal of the water company to supply the unforseen growth in population. In fact, water service interruptions were scarce at the urban core where distribution was guaranteed by the company. But in the suburbs underfinanced municipalities like Sintra or Cascais, which also received water from the company, weren't able to cope with the demography boom with their limited infrastructure of reservoirs and old narrow pipes.

6.2.3 Democratic waters (1974–2006)

The magnitude of the problem would become clear with the return of cholera to Lisbon in 1971. Epidemics broke out in the slums, which sheltered some 150,000

---

[32] The only exceptions were the conflicts between the company and the landowners of estates where new sources for city supply were being established. Complaints also reached the National Assembly on the excessive zeal placed on isolating water conduits and sources from their surroundings, as denounced by farmers and shepherds.

[33] Marcello Caetano (1906–80) assumed control of the authoritarian regime in 1968 softening its repressive character in what become known in the historiography as the Marcelist Spring. Nevertheless his strategy of 'keep the course' in the colonial wars in Angola, Mozambique and Guiné Bissau eventually led to his overthrow by the 1974 revolution.

people. Mass vaccination of slum dwellers and health service campaigns stressing the importance of boiling water taken from non-treated sources, were enough to limit the death toll to ten people among the hundreds of choleric patients. Nevertheless, repeating the gesture of nineteenth-century hygienists, doctors started to survey the sanitary conditions of the new urbanites and denounced the lack of access to company water for the majority of the population outside the city core (*O Médico*, 1971: 605–09). With few exceptions, the newcomers were relying mainly on isolated wells and springs. The bacteriological analysis of such sources revealed that 70% of their water was contaminated and non-potable. The picture concerning sewage wasn't much better, with only one rudimentary plant for sewage treatment in the eastern outskirts designed to receive the wastewater of no more than 50,000 people. Non-treated effluent polluted the beaches of the Sun Coast, but of greater concern was the proliferation of individual cesspits in the new expansion areas, far from central Lisbon. These were contaminating the same water sources people were using. It is important to remember that the large majority of this population comprised first-generation urbanites, who brought with them a rural culture of water. Rather than expect, or even request the company to bring its pipes to their homes, they just improvised their own supply. In April 1974, as the Carnation Revolution brought to an end forty-eight years of dictatorial rule, a new cholera outbreak started to spread in southern Portugal, reaching Lisbon during the summer and killing thirteen out of 600 identified cholera cases (Figueiredo, 1974).

With water shortages and cholera outbreaks occuring in a revolutionary context, the position of the old water company became untenable.[34] On 21 June, 1974, company workers occupied the facilities and demanded the firing of the board of directors. Three days later the Government took direct control of Lisbon's water supply with the members of the former CFAL assuming the leading role. In October, the Empresa Pública de Águas de Lisboa (EPAL) was formed: the first nationalized company of the revolutionary period. Once again, Lisbon waters were an indicator of future government intervention in the economy, no longer following the corporative state model of the 1930s.

In the following years the public company was generously funded to put an end to the scandalous lack of water in the Lisbon region.[35] Besides doubling the volume of surface water taken from the River Tagus at Valada, a new system was designed to bring water from the large reservoir at the Castelo do Bode dam on the Zêzere river, inaugurated only in 1988. The public company was finally able to deliver enough water to the metropolitan area, although shortages were still common in areas where the peripheral municipalities were responsible for distribution of water supplied by EPAL. It was only in the 1990s and with EU funding, that a new main system was built, following the external Lisbon ring road, which enabled water to bypass the central Lisbon reservoirs and be distributed directly to the different peripheries. Water supply infrastructure was finally losing its centralized character and was keeping pace with the spatial expansion of the Lisbon metropolitan area. Unlike the 1930s, waters were now following the urban sprawl and not the other way round.

---

[34] By coincidence the contract ended the concession that same year in October. See Empresa Pública de Águas de Lisboa (1975).

[35] See the annual reports of EPAL (1974–80).

Since the 1970s, this hydraulic approach of offering progressively more and more water to the population had to be complemented with environmental concerns about sewage. The problem of water pollution in Lisbon was magnified by the coastal tourist region, west of the city, with newspapers in the 1980s repeatedly denouncing the polluted water of the beaches. The rapid expansion of the western suburbs, with Sintra becoming over the next decade the foremost urban agglomeration, leaving Lisbon behind,[36] meant a massive increase in wastewaters arriving to beaches with no treatment. The construction of a large sewage collector along the coast proved a complicated project, beginning in the 1970s and ending only in the late 1990s. Once again, EU funds provided the financial resources to construct the necessary infrastructure along the beaches of the Sun Coast beaches. Nevertheless, newspapers still reported overspills from the interceptor under heavy rain conditions (combined sewer). This was not the only problem, as many houses were still not connected to the drainage system, and kept throwing sewage directly into the river basins flowing into the sea (Schmidt, 2007). Only with the passing of the Municipal Urban Plans of 1995, which tried to bring some order to the chaotic urban expansion, did the large collector start to receive wastewater from all new households.

If the EU was fundamental to improvements in water quality, it now requires quality levels that according to SANEST, the company responsible for sewage along the Sun Coast, are too high for wastewaters deposited into the sea. Required investment in secondary and tertiary treatments will surely bring about an increase in operational costs. The example of SANEST, which together with SIMTEJO, the company operating Lisbon wastewater, still leaves one-third of Lisbon inhabitants without wastewater treatment,[37] confirms the difficulties of the Portuguese context, with problems of lack of infrastructure overlapping environmental problems typical of late modernity (Schmidt, 2009).

The polemics around the water quality of the Castelo do Bode reservoir, from where EPAL takes most of its water to supply 2.6 million people of the Lisbon region, illustrate such difficulties. If water quality is guaranteed by the treatment station of Asseisseira, one must still emphasize the negative effects of construction on the reservoir shores, with water indexes strongly deteriorating.[38] In spite of the major significance of this water reservoir for the future of the metropolitan area, the public discussion that took place in 2003 on the development plan for the dam area didn't include representatives from the Lisbon area. The inference was that the 100km distance implied a geographical independence between the two areas: out of sight, out of mind. Having addressed the difficult step of adding the sanitary approach to the hydraulic, water experts now had to start considering territorial approaches, where sustainable environmental practices play a major role. In addition to hydraulic and sanitary engineers, the complexities of urban waters now require that companies engage environmental experts and landscape architects.

The Lisbon water story constitutes an important part of Lisbon's story over the last 150 years. The urban expansion of the nineteenth century; the fascist design of a new

---

[36] For numbers on Lisbon urban growth go to www.ecoline.ics.ul.pt.

[37] The works to solve the lasting issue of Lisbon sewage were reinitiated in 2009, and they are scheduled to end by 2011.

[38] The water indexes deteriorated from A1 to A3. See Almeida Vieira (2003).

metropolis; the hidden slums of the authoritarian regime; and the difficult 'Europeization' of Lisbon in the last decades of the twentieth century, may only be understood by including water in the narrative. We saw how engineers repeatedly promised to bring an end to water scarcity and epidemics resulting from poor sanitation, and how the built water infrastructure was once and again overstretched by urban dynamics. The best example is perhaps that of the fascist regime's technocratic solution for Lisbon water supply in the beginning of the 1930s, which helped to legitimate the New State, and its subsequent inability to expand infrastructure in the 1960s when all popular protest was suppressed. But it is also significant that the revolutionary solution to the cholera epidemics of 1974 was found once again through government engineers responsible for the inspection of Lisbon waters. The EU funds of the 1980s didn't bring alternatives to this technocratic water culture with Lisbon urbanites relying on the state for cheap and abundant water, caring little about the situation where their water was captured. Water still remains a business for experts, not for politics.

6.3 URBAN WATER CONFLICTS: FROM THE UNFINISHED WELFARE STATE TO THE NEW REGULATORY STATE

Throughout the twentieth century, results from successive policy frameworks for water services in Portugal were far from satisfactory. An exogenous impulse was necessary in order for a long-due structural change to see the light of day. The Lisbon story is illustrative, as we have seen. The impact of European funds on the water sector, as well as the policy reform operated in the 1990s, clearly resulted in considerable quantitative improvements, but financial and technical problems, as well as infrastructural ones, still prevail. An inquiry conducted in 2004 revealed that 70% of Portuguese mayors still identified water and sanitation as their most urgent environmental problem (52% referred to sewage systems and 18% to water supply), and 33% identified sanitation as the most important problem affecting municipalities, over social or economical problems (Schmidt, Nave and Guerra, 2005). This clearly highlights water as a recurrent issue, and one that accumulates first, second and third-generation environmental problems.

Despite all the problems, conflicts over urban waters today do not assume explicit and direct consequences, as was also the case throughout the twentieth century. In fact, most water conflicts in Portugal were reported in rural areas, and occured due to competitive uses of water between farmers and landowners, where no state regulation was available or enforced. An exhaustive research conducted on the archives of the Administrative Supreme Court rulings from 1890 to 2005 confirmed a lack of significative judicial cases concerning urban water between the state and private institutions or citizens. The tension could be felt on successive policy changes, as well as on the preambles of most important pieces of legislation regulating urban waters, where several deficiencies and contradictions where referred; but somehow conflicts – at least explicit conflicts – never emerged.

The nature of urban water conflicts, predominantly latent, is also confirmed with extensive analysis of weekly newspapers in Portugal over the same approximate time period (1900–2005): 113 news items reported latent conflicts concerning urban waters, and only eighty-five directly referred to explicit conflicts. The results of this analysis show that besides some cases concerning water shortages in Lisbon, at the beginning of the century, total civic inertia occurred for most of the period (1926–1974): the dictatorial regime imposed a social and political context where conflicts were diluted, or politically repressed, especially in cases where the state itself was not offering quality services.

At the beginning of the 1970s, and even before the revolution, some news media reported the pollution of Portuguese rivers as one of the most significant environmental problems – experts were given the possibility to express their views on the media – along with water supply and sewage deficiencies that resulted from the exponential growth of Lisbon's suburban areas. Nevertheless, and once again, no conflicts were reported. After the revolution, contrary to what could be expected, conflicts over water sanitation seemed to be totally submerged under more general social and political conflicts, related to a country living a kind of revolutionary euphoria and making its first steps towards a democratic regime. Social policies were finally becoming generalized – healthcare, education, housing – and even though sanitation was part of the political agenda, investments in the sector and effectively implemented practical solutions, were very slow: the state didn't seem to have enough resources to fulfill all social demands (Figure 6.2).

Nevertheless, no organized protests were registered, even for the Lisbon area, where the population had to cope with an incipient water supply service and inexistent water treatment. As we saw, citizens would try in many cases to solve their own problems directly, especially in suburban areas, by managing their own water sources. The low incidence of civic protests in Portugal should not be attributed to the nature of its inhabitants, although the phenomenon is independent of social-economic factors

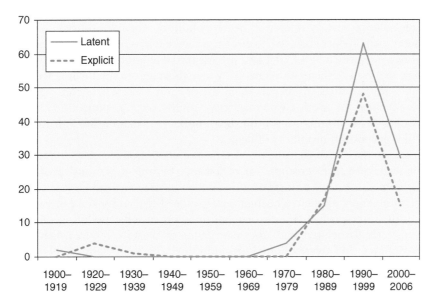

Figure 6.2 Latent and explicit conflicts from 1900 to 2006

Source: The graph is based on an extensive analysis of the main weekly Portuguese newspapers from 1900–2006. The information is available at http://ecoline.ics.ul.pt

Table 6.2 **Water conflicts by subject (1900–2006)**

	Supply	Scarcity	Pollution of underground waters	Beach pollution	River pollution	Price	Water quality	Sewage	Others
1900–1919	2	0	0	0	0	0	0	0	0
1920–1929	3	0	0	0	0	0	1	0	0
1930–1939	1	0	0	0	0	0	0	0	0
1940–1949	0	0	0	0	0	0	0	0	0
1950–1959	0	0	0	0	0	0	0	0	0
1960–1969	0	0	0	0	0	0	0	0	0
1970–1979	2	0	0	0	3	0	0	0	0
1980–1989	9	0	0	5	17	0	2	2	0
1990–1999	12	27	9	4	35	3	10	14	3
2000–2006	6	7	0	4	11	6	8	5	1

Source: www.ecoline.ics.ul.pt (1900–99) and *Jornal Expresso* (2000–2006).

pervading the entire society. As Manuel Villaverde Cabral has shown, the persistent larger power distance index perceived by Portuguese citizens when compared to other European countries, as revealed by European Social Surveys (ESS), derives from a long history of restricted literacy policies and the bureaucratic despotism of the modern Portuguese state (Villaverde Cabral, 2003: 31–60). The positive evolution of economic and social development indexes doesn't automatically bring a better distribution of symbolic resources, fundamental for access to social and political power.

This lack of significant protests among the population is reflected in the news. Only since the 1980s have changes been detected with newspapers adopting a new critical tone. This fact was clearly related to beach pollution, when sea tourism became part of common social habits and sewage discharges were felt as a huge nuisance, especially in the above-mentioned Sun Coast and the Algarve (Schmidt, 2003). Some latent conflicts were reported in the news related to water shortages in the Lisbon Metropolitan Area, but water conflicts still occurred mostly in rural areas where eucalyptus plantations were blamed for causing water scarcity. Conflicts also assumed the form of election boycotts with populations protesting the lack of sanitation infrastructure, a typical first-generation problem (Table 6.2).

But it was in the 1990s that the degradation of water quality, as well as its implications for water use (for consumption or leisure), became a 'serious' problem. Since then, the number of news reports on this matters have grown significantly, with conflicts becoming more explicit. The media were indeed crucial for the general acknowledgement of water problems and their multiple dimensions: pollution of rivers and beaches directly associated with the inexistence of wastewater treatment facilities; strongly polluted and overexploited groundwater; and inefficient water supply and bad water quality. For the first time, in 1993, an official report on water quality status was publicly discussed, informing public opinion about the bad condition of most water bodies, as well the low quality of water served to urban populations (Schmidt, 2000). These reports would become regularly publicized, and showed that still in 2003 around 200,000 Portuguese inhabitants remained without drinking water in contravention of both European and national requisites, while 350,000 lacked water supply (mainly in the north-west of the country). Furthermore, it is now clear that water quality analyses are not being undertaken, disregarding this stipulated in the law.

It was also from the 1990s onwards that the Spanish Hydrological Plan became an issue for the Portuguese public: newspapers regularly reported water captures in Spain, reducing water volumes in shared rivers like the Douro or Tagus, but mainly in the Guadiana. This river saw its water flow diminish by 60% from 1970 to 1990, leading to new conflicts directly related with water scarcity (Schmidt, 2000: 121–31). It was Spanish ambitions on Iberian waters and old mistrust of a neighbour's intentions that finally sparked a debate on Portuguese waters. The conflict around the sharing of the Guadiana waters had the positive effect of awakening the Portuguese government to the issue of proper water management. Until then, no river basin management authorities had been established for the country, with legislation regulating water planning programmes only published in 1994. Six years were needed to prepare River Basin Plans and the National Water Plan (2000), which would scandalously reveal that 70% of Portuguese water courses were highly polluted.[39]

The pollution of rivers thus became one of the most mediatic themes of the 1990s: first, the media reported upon the annihilation of Portuguese rivers, affecting biodiversity and human use (commercial or leisure); then, just as regular scientific data on their condition became available, municipalities and civic groups began demanding protection measures for rivers and water bodies, reaching the mainstream newspapers. Some protests were aimed against existing industrial facilities, or projected ones, while more recently, some have even reached the European courts; however, the results from these are as yet unknown.[40]

Since 2000, the privatization of water services has become a clear issue in the press, mainly through the voice of well-known opinion-makers who underlined the potential risks. Public announcements from successive Ministers of the Environment from 2000 to 2005 – which suggested the possibility of privatization, become a significant political fact, open to speculation. Today, water appears in the pink pages of the 'Economy' section. It is also significant that, for the first time, conflicts associated with water prices (future raises) have been recorded.

6.4 CONCLUDING REMARKS: HIDDEN CONFLICTS OR POTENTIAL CONFLICTS?

Taking together the detailed history of the Lisbon water case, the bird's eye view of sanitary infrastructure evolution in Portugal throughout the twentieth century, and the media analysis on water conflicts, we are now in a privileged position to undertake a prospective exercise on water conflicts. Let us try to present a typology of potential future conflicts: institutions, prices, uses, scarcity and climate change. Of course the types of conflicts are deeply interrelated but we'll isolate them for analytical proposals and for the sake of the argument.

---

[39] In 2009 a new Water National Plan was launched. Following the data from the Water Institute (Instituto da Água), some 40% of the rivers are heavily polluted and 25% of them have a reasonable status.

[40] Two national surveys conducted in 1997 and 2000, showed that the main problem pointed out by the Portuguese population was river pollution (Schmidt et al, 2004).

The institutional conflict is perhaps the most obvious. If Portugal does not comply with European Water Framework Directive standards for both quantity and quality of water services, the European Commission will probably heavily fine detected faults. The EU does have a strict policy for water and the fact that it has been substantially funding Portuguese water services for the last thirteen years gives it a very strong political and moral argument. This issue has already emerged with the EU penalizing SAN-EST (see above) for discharging wastewater with only primary treatment into the Atlantic Ocean until 2007. Another institutional conflict opposes the different regional branches of the state holding Águas de Portugal AdP and many municipalities. The former is becoming less flexible with the fast-growing debts of the latter, defending the 'political price' of water against the internalization of costs.

This last point leads us directly to the issue of prices. From the above sections we may easily conclude that the water culture in Portugal passed directly from individual self-sufficiency, with people themselves digging wells and improvising cesspits, to a total reliance on state ability to deliver cheap and abundant water as well as free wastewater treatment. The 'deterritorialization' of water issues, with people ignoring the complex and expensive infrastructure that brings water to their taps and takes it to wastewater plants, leads to stronger resistance by consumers against abandoning the present political price of water. However, as the EU requests stricter standards from the Portuguese supply and sanitation infrastructure, prices will have to rise to fund growing operational costs. In fact, since in 2007 the Ministry of the Environment started to announce compulsory increases in water prices. In the Council of Almada, for instance, a populated Lisbon suburb in the south riverbank of the Tagus, the municipality recently denounced the inequalities arising from the new water tariffs, with larger families being more penalized than smaller ones. Moreover, the eventual privatization of water services could bring significant price rises, as happened in other countries, promoting the transition from latent forms of conflicts to explicit ones. Private or not, the times of cheap water are coming to an end, and consumers, supported by municipalities, are starting to show resistance.[41]

We previously underlined the emergence of a territorial paradigm with landscape planning assuming a major role. From the news analysis presented, we learned of the growing complaints of populations worried about bad river water quality. Here, we're not only dealing with potable water consumers, but also urbanites that assume the good health of rivers as part of their living quality standards. Water has an aesthetic value that goes beyond its drinkable nature. Urbanites, imposing their leisure values on the rest of the territory, will most likely become more sensitive to the future of Portuguese rivers and thermal waters. The space of previous rural water conflicts may now become the scene of urbanites fighting for their right to use water landscapes.

Lastly, conflicts can emerge where water scarcity and competitive uses occur, especially in areas of the country where the resource is not abundant, or where uses are seasonal. Tremendous tourist pressure in the Algarve (in the south of the country) is an explicit case, where water shortages are constantly reported and the significant growth of high water-consuming infrastructures (golf courses), as well as seasonal consuming

---

[41] In 2007, a curious inversion of roles occurred with the government forbidding water suppliers, among them many municipalities, from charging consumers for the use of water counters, as is current practice in Portugal.

rhythms, are causing growing tensions between citizens, municipalities and private companies. The predictions of climate change impacts for much of the country, but particularly for its southern regions, support the relevance of scarcity problems in the near future (Schmidt and Prista, 2010). Most reliable climate scenarios for the twenty-first century, presented by the various models developed by the SIAM research project, indicate a 'small increase of annual precipitation for the northern region of Portugal and a decrease for the central and southern regions'. These models 'also estimate an increase of the precipitation seasonal asymmetry, with relevant decreases in summer precipitation', and with a consequent 'progressive reduction in the annual river runoff and aquifer recharge' (Santos and Miranda, 2006: 118). It is important to bear in mind that the Intergovernmental Panel for Climate Change (IPCC) report recently identified Portugal as one of the countries of the Mediterranean basin most vulnerable to climate change effects.[42]

If the new geographical distribution of water resources is to become the main source of conflict around water in the years to come, the stories in this chapter suggest that the mobilization of engineers and financial funds alone, will not prove adequate to finding a resolution. Although technology and money are central to any possible solution, democratic societies should be equally committed to developing mechanisms for the sustainable governance of water. After all, water is much more than H_2O, it is also a common good, and as such, it challenges our capacity to imagine appropriate politics and policies for its sustainable management.

REFERENCES

Acciaiuoli, M. 1998. *Exposições do Estado Novo: 1934–1940*. Lisboa, Livros Horizonte.

Alves, J.C. 1940. *Abastecimento de Água a Lisboa. Boletim da Comissão de Fiscalização das Obras de Abastecimento de Água à Cidade de Lisboa*, 16.

Alves, J.C. 1947. *Águas de Lisboa. 15 anos de Obras Públicas (1932–1947)*. Lisbon: Comissão Executiva da Exposição de Obras Públicas.

Alves Costa, M. 1941. O abastecimento da Costa do Sol com água de Lisboa. *Boletim da Comissão de Fiscalização das Águas de Lisboa*, 19.

Bonifácio, M. de F. 1999. *Apologia da História Política. Estudos Sobre o Século XIX Português*. Lisbon: Quetzal.

Caetano, J.O. 1994. O aqueduto das Águas Livres. I. Moita (ed.) *O Livro de Lisboa*. Lisbon: Livros Horizonte.

Caseiro, C., Pena, A. and Vital, R. 1999. *Histórias e Outras Memórias do Aqueduto das Águas Livres*. Lisbon: EPAL.

Comissão de Planeamento da Região de Lisboa. 1973. *Elementos sobre redes de abastecimento de água, electricidade, saneamento, recolha e tratamento de lixos*. Lisbon: Casa da Moeda.

Cordeiro, J. 1993. Um serviço centenário: o abastecimento de água à cidade do Porto (1887–1987). *Arqueologia Industrial*, II Série I (1–2).

Costa Pinto, A. 1992. *O Salazarismo e o Fascismo Europeu*. Lisbon: Estampa.

Empresa Pública de Águas de Lisboa. 1975. *Relatório e Contas*, 1974. Lisbon: EPAL.

Ferrão, J. 1996. Três décadas de consolidação do Portugal demográfico moderno. A. Barreto (ed.) *A Situação social em Portugal, 1960–1995*. Lisbon: ICS-UL.

Ferreira da Silva, A. 2006. Sanitary revolution and technology in nineteenth century Lisbon. M. Merger (ed.) *Transferts de Technologies en Méditerranée*. Paris: PUPS.

---

[42] See the IPCC website, http://www.ipcc.ch/

Figueiredo, L. de 1974. Portugal na 7ª pandemia de cólera. *O Médico*, LXXIII (no. 1214).
Freitas do Amaral, D. 1995. *O Antigo Regime e a Revolução – Memórias Políticas (1941–1975)*. Lisbon: Edn. Círculo de Leitores.
Gentil Berger, F.G. 1994. *Lisboa e os Arquitectos de D. João V. Manuel da Costa Negreiros no Estudo Sistemático do Barroco Joanino na Região de Lisboa*. Lisbon: Cosmos.
GPC (1977). *Plano de Médio Prazo: 77 – 80*. Lisbon: INCM
GPEEF. 1983. *Memória Descritiva e Justificativa*. Lisbon: MOP
Ilustração Portuguesa. 1913. A falta d'agua em Lisboa. *Ilustração Portuguesa* (7 July).
Ilustração Portuguesa. 1917. Um pavoroso incêndio. *Ilustração Portuguesa* (1 October).
IRAR. 2004. *Relatório Anual do Sector de Águas e Resíduos em Portugal, Vol. I.*, Lisbon: IRAR
Leite Pinto, L. 1989. *História do Abastecimento de Água à Região de Lisboa*. Lisbon: INCM/EPAL.
Lucena, M. 1976. *A Evolução do Sistema Corporativo Português, Vol. 1, O Salazarismo*. Lisbon: Perspectivas e Realidades.
MAOT. 2000. *Plano Estratégico de Abastecimento de Água e de Saneamento de Águas Residuais (2000–2006)*. Lisbon: MAOT
MAOTDR. 2007. *Plano Estratégico de Abastecimento de Água e de Saneamento de Águas Residuais II (2007–2013)*. Lisbon: MAOTDR
Matias Ferreira, V. 1986. A cidade de Lisboa. De capital do Império a centro da Metrópole. PhD thesis UNL/ISCTe.
Melo Baptista, J. and Matos, M.R. 1995. *Investigação e Desenvolvimento: instrumentos de apoio a uma política de desenvolvimento sustentável em saneamento básico*, 16 Vols. Lisbon: LNEC.
Ministério do Ambiente. 1999. *Plano de Bacia Hidrográfica do Tejo, Análise e diagnóstico da situação de referência, Vol. III*. Lisbon: Ministério do Ambiente.
Moita, I. (ed.) 1990. *D. João V e o Abastecimento de Água a Lisboa*. Lisbon: Câmara Municipal de Lisboa.
MOP. 1954. *Boletim da Direcção Geral dos Serviços de Urbanização, Vol. I*. Lisbon: MOP.
MOPC. 1941. *Anuário dos Serviços Hidráulicos*. Lisbon: Imprensa Nacional.
O Médico. 1971. A Cólera em Portugal – Entrevista com o Dr. Arnaldo Sampaio. *O Médico*, LXI (no. 1056).
O Século. 1940. A água que Lisboa bebe. *O Século* (16 March).
O Século Ilustrado. 1970. O cerco à cidade – Brandoas há muitas. *O Século Ilustrado* (31 October).
Pato, J. (2008). O valor da água como bem público. Tese doutoral. Lisbon: Instituto de Ciências Sociais da Universidade de Lisboa.
Portas, N., Domingues, A. and Cabral, J. 2004. *Políticas Urbanas*. Lisbon: Fundação Calouste Gulbenkian.
Pinho, B. de. 1942. A purificação bacteriológica das águas de Lisboa. *Boletim da Comissão de Fiscalização das Águas de Lisboa*, 21.
Pinho, B. de. 1945. Inventário de minas, poços, furos e cisternas da área da Cidade de Lisboa. *Boletim da Comissão de Fiscalização das Águas de Lisboa*, 25.
Presidência do Conselho. 1973. *Elementos sobre redes de abastecimento de água*. Lisbon: INCM.
RASARP – Relatórios Anuais do Sector das Águas e Resíduos em Portugal. 2004. *Controlo da Qualidade da Água para Consumo Humano, 2004 Vol. 4*. Available at: http://www.ersar.pt/website/ViewContent.aspx?SubFolderPath=&Section=MenuPrincipal&FolderPath=%5cRoot%5cContents%5cSitio%5cRASARP&FinalPath=RASARP
Rodrigues, T. 1995. *Nascer e Morrer na Lisboa Oitocentista: migrações, mortalidade e desenvolvimento*. Lisbon: Cosmos.

Rosas, F. 1992. Portugal e o Estado Novo (1930–1960), vol. XII of J. Serrão and A.H. de Oliveira Marques (eds) *Nova História de Portugal*. Lisbon: Presença.

Santos, F.D. and Miranda, P. (eds) 2006. *SIAM II – Climatic Change in Portugal. Scenarios, Impacts and Adaptation Measures*. Lisbon: Gradiva.

Saraiva, T. 2005. *Ciencia y Ciudad: Madrid y Lisboa (1851–1900)*. Madrid: Ayuntamiento de Madrid.

Schmidt, L. 2000. *Portugal Ambiental. Casos & Causas*. Lisbon: Círculo de Leitores.

Schmidt, L. 2003. *Ambiente no Ecrã. Emissões e Demissões no Serviço Público Televisivo*. Lisbon: Imprensa de Ciências Sociais, ICS.

Schmidt, L. 2007. *País (In)Sustetável – Ambiente e Qualidade Devida em Portugal*. Lisbon: Ed Esfera do Caos.

Schmidt, L. 2009. 'Civic action and media in the water issues' in C. Buchanan, P. Vicente and E. Vlachos (eds) *Making the Passage through the 21st century – Water as a Catalyst for Change*. Lisbon: Luso-American Foundation (FLAD).

Schmidt, L., Truninger, M., Valente, S. 2004. 'Problemas Ambientais, prioridades e quadro de vida' in Almeida, J.F. (eds). *Os Portugueses e o Ambiente*. Lisboa: CELTA.

Schmidt, L., Gil Nave, J. and Guerra, J. 2005. *Autarquias e Desenvolvimento Sustentável – Agenda 21 Local e Novas Estratégias Ambientais*. Porto: Fronteira do Caos.

Schmidt, L. and Prista, P. 2010. 'Portugal – changement climatique, eau et société' in *Grande Europe*, 19/Avril 2010, Dossier: Face aux désordres climatiques. La Documentation Française. Available at : http://www.ladocumentationfrancaise.fr/revues/grande-europe/dossiers/.

Silva Lopes, J. da,. 2005. Finanças Públicas. P. Lains and Á. Ferreira da Silva (eds) *História Económica de Portugal, 1700–2000, Vol III, O Século XX*. Lisbon: Imprensa de Ciências Sociais.

Valle, L. del. 1856. Abastecimiento de aguas en Lisboa. *Revista de Obras públicas*, IV (12, 13).

Veloso de Andrade, J.S. 1851. *Memória Sobre Chafarizes, Bicas, Fontes e Poços Públicos de Lisboa*. Lisbon: Imprensa Silviana.

Vieira, P.A. 2003. *O Estrago da Nação*. Lisbon: Dom Quixote.

Villaverde Cabral, M. 2003. O exercício da cidadania política em perspectiva histórica (Portugal e Brasil). *Revista Brasileira de Ciências Sociais*, 18(51).

Chapter 7

Water supply services in the cities of Brazil: Conflicts, challenges and new opportunities in regulation

Ana Lucia de Paiva Britto[1] and Ricardo Toledo Silva[2]

[1]Post-Graduate Programme on Urban Studies (PROURB-UFRJ), Federal University of Rio de Janeiro, Brazil
[2]School of Architecture and Urban Studies (FAU-USP), University of São Paulo, Brazil

7.1 INTRODUCTION

Noticeable in studies dealing with water supply and sanitation policies in Brazil is the insistence with which most of them refer to the former National Plan of Water Supply and Sewerage (Planasa), either in favour or against it. The importance attributed to this reference point is curious, since the Plan, and the whole public finance strategy underpinning it, lasted less than fifteen years, from the early 1970s to the mid-1980s. The history of water and sanitation services in Brazil is much older, beginning in the 1850s in main cities like Rio de Janeiro, Sao Paulo, Salvador and Recife. The time that has passed since the formal ending of Planasa, in 1986, now surpasses the entire duration of the Plan. Still, this programme remains emblematic of an unprecedented national concern for services, hitherto seen as a predominantly local matter, having mobilized very large amounts of financial resources on the expansion of services coverage.

The main feature of this model was the centralization of water supply and sanitation services through state-wide platforms, each operated by a state water company created under the aegis of a powerful financing scheme. These were controlled at federal level, based on a mix of fiscal and non-fiscal sources. The model is still inertial as regards present strategies because most of the existing services are still supplied by state water companies, tied to state-wide technical and managerial premises, and drawn from the unprecedented expansion of infrastructure and service coverage experienced at that time. The main strategic argument for the state-wide supply model was cross-subsidization between profitable and non-profitable service areas, in which areas requiring subsidy would be supported by surplus-generating areas. This was an idealized formula to promote social coverage with no direct subsidization from tax revenues, instead keeping distributive measures within the limits of the services' economic self-sustainability.

Between the early 1970s and the mid-1980s the concentration of investments under the Planasa model resulted in a rise in national coverage of water supply services from nearly 54% of urban households to about 90%, while sewerage urban connections expanded from about 22% to nearly 40%. Considering that the urban rate of population growth for this period was still very high, the absolute figures of coverage expansion were indeed outstanding. However, the remaining uncovered segments of urban demand comprised mostly the poorer social strata settled in the outskirts of the cities, where services expansion would involve higher marginal costs, important environmental constraints, and strong interference from other sectors of public investment facing similar difficulties with expansion, such as housing and public transport.

During the almost two decades after Planasa, investments were barely sufficient to maintain the average figures of urban access, with some progress on access to sewage collection, and less progress on proper treatment. With 80% of the entire population living in cities, according to the Brazilian Census of 2000, urban environmental problems have emerged in this decade as one of the most challenging concerns of public policies at all levels of government. Generalized access to water and sanitation, as part of this broad set of problems, remains an open question.

The structural economic adjustment following the monetary orthodoxy that prevailed among most Latin American indebted countries during the 1990s, has dramatically reduced the margins for public investment. Social policies, generally, and expansion of urban services to poor areas, in particular, were severely affected by constraints on public investments. This scarcity of public investment, against the backdrop of a fiscal crisis, implied the aggravation of social vulnerability in general, made visible in the cities by the rapid increase in precarious forms of housing.

From the standpoint of the water and sanitation industry, the scarcity of public investment called for private funds as a possible alternative to the capital needs of water and sanitation services – not necessarily under a model of private operation. For example, in the case of the Sao Paulo State water company, loans were taken out on the international money market, and part of the capital opened for stock exchange markets, nationally and internationally. Both movements, although not implying private control of the company, pushed managing strategies towards a private logic. The capital costs had to be paid according to private interest rates, and the almost 50% of capital stocks negotiated in the market required conformity to the rules of wide open accountability.

Since these first moves towards privatization alternatives in the mid-1990s. Private sector participation progressed in a limited manner. According to 2007 data, there were 63 private concessions for water and sewerage services in the country; most of these contracts started between 1995 and 2002. Today, the private sector operates in 10 states. Tocantins is the state with the largest number of cities with private water and sanitation services (124 municipalities), because the state company was privatized. Next come the states of Mato Grosso, São Paulo, and Rio de Janeiro. The largest number of private concessions is in Mato Grosso, due to the extinction of CESB and subsequent municipalization. Municipalities that were unable to organize direct provision of services, then opted for concession to private companies. In São Paulo and Rio de Janeiro states, municipalities with greater economic dynamism tend to focalize the activity of major private companies. Approximately 30 private companies are involved in water and sanitation concessions in Brazil. Most of them are companies with national capital. Before 2000, large international water companies invested in Brazil: French companies Veolia and Suez-Ondeo, the Spanish AGBAR (Aguas de Barcelona), the English company Water International, the Portuguese Aguas de Portugal and the North American company Enron). But most of these companies withdrew.

Table 7.1 Brazil: consolidated investments on water supply and sanitation in 2005

	Direct investments R$ million	Loans R$ million	Non-refundable transfers R$ million	Other R$ million	Total R$ million
North	40.6	9.4	24.7	6.0	80.7
North-east	153.6	64.9	343.9	40.1	602.6
South-east	930.6	506.3	158.8	173.5	1 769.3
South	350.6	199.9	7.7	58.8	617.0
Mid-west	344.0	65.3	35.9	30.7	475.9
Brazil	1 819.4	845.8	571.0	309.0	3 545.5

Exchange rate (December 2005): R$ 2.20 per US$.
Source: SNIS (2005).

In terms of total investment, the present figures are very modest when compared to the performance of the former Planasa model. In 2005, investments for the expansion of water and sanitation services were estimated, nationally, at around US$1.6 billion, according to official statistics. This national consolidated investment comes to about 0.18% in terms of percentage of GNP. In 1981, the Planasa investments by means of loans to state companies reached an absolute amount of nearly US$960 million (at the exchange rate of 1981), or about 0.37% of GNP. If considered against dollar inflation from 1981 to 2005, and absolute urban growth during the same period, it is clear that the present consolidated level of investments, all sources considered, is dramatically lower than the levels practised by the national financing system alone, at the time of the Planasa.

In Table 7.1, the composition of consolidated investments shows that the direct investments made by service companies, extracted from their operational revenues, is significant. They account for more than 50% of total investments, nationally. For the poorer regions in the north and north-east, non-refundable governmental transfers are still significant, accounting for, respectively, 30.61 and 57.07% of their investment needs. However, loans constitute a relatively low portion of the investment budgets of northern and north-eastern services (around 11%), while they account for around 30% of investments at mid-western and south-eastern services.

This presents a general picture of diversified investment profiles, in which the restraints on public investments have actually enabled wider participation by private capital in water and sanitation services in Brazil. However, this has not evolved into a general movement towards privatization, in the sense of direct private operation of services, as seems to have happened in other countries. In Brazil, public operation of services still prevails, either by state companies created under Planasa rules or by municipal autonomous entities. However, most of these have to conform, to a considerable extent, to private-like procedures in terms of capitalization and generating sufficient surpluses to cover investment needs in a context of diminished public funding.

7.2 THE INSTITUTIONAL CONFLICTS AND CHALLENGES

Public jurisdiction over water supply and sanitation services has long been a major issue in Brazil. The established structure of the country recognizes three autonomous levels of political power: the federal union, the federate states and the municipalities. The Brazilian constitutional power structure has always recognized municipal jurisdiction over local services, by and large comprising water distribution and sewage collection since the first decades of the 1900s. The exception was capital cities in most states and the Federal District (Rio de Janeiro up to 1960 and Brasilia, from 1960 to now), whose local services used to be provided and/or regulated by the state or the district powers themselves.

At the beginning of the 1970s, a huge federal system for financing and regulating water supply and sanitation was launched, by means of a nationwide plan known as 'National Plan of Basic Sanitation' (Planasa).[1] Under this system municipal jurisdiction was practically overruled by a state-wide executive organization. At that time, the Brazilian Government was an authoritarian regime, whose exceptional powers were used to force municipalities to give up their constitutional jurisdiction over local water services. By imposing constraints on access to public financing, the federal government induced municipalities to adhere to state systems of integrated supply, carried out by the newly created water and sewerage companies (CESB) in each federated state. These entities were funded by a mix of federal and state financing, and organized under a hybrid model of public administration that combined public ownership with private-oriented management. This form of state executive structure prevailed in Brazil at the time following the administrative reform of 1967, which aimed primarily to promote the economic sustainability of state executive actions.

For water supply and sanitation, the strong need for subsidization to cover social demand would render unfeasible the economic sustainability of the services, should they be kept as single municipal jurisdictions. Since the specific conditions of provision and the capacity to pay for services were very heterogeneous, the strategy of the national plan was based upon the prospect of inter-municipal cross subsidization. This was the main alleged rationale for the centralized reshaping of the institutional structure, facilitated by the authoritarian character of the political system, under which municipal prerogatives could be reduced in favour of a regional supply strategy. Formally, the municipal jurisdiction was preserved, since a municipal law was required for each municipality as a precondition to joining the state system. But in practice, denial of access to federal and state funding for autonomous municipal services, virtually enforced adhesion processes.

By the second half of the 1980s, the financial system behind the CESB model was seriously jeopardized by limitations in its repayment formulae, linked to the monetary indexes applied to housing loans. In a context of growing inflation, the financial assets of the rotation funds for water and sanitation melted under unrealistically low monetary indexation. The normalization of the political process in Brazil in the early 1990s saw the centralized organization of public services supply come under fire. Democratization of public policies was itself understood as quasi-synonymous to decentralization. Decentralized management of public services was seen by most activist groups as more democratic, efficient and committed with a general concept of social well-being, than the former regionally based planning logic.

---

[1] The literal translation of the Brazilian expression 'basic sanitation' stands for the combined urban services of water supply and sewerage. It contrasts with the notion of 'environmental sanitation', which in Brazil comprises solid waste management and flood control.

In practice, the state supply model and the CESB themselves, as main executive branches of that model, were questioned both by the leftist organization of municipal autonomous services – the ASSEMAE – and by the special programme created by the federal government to reshape the water sector in a more competitive environment. This programme, entitled the 'Water Sector Modernization Programme' (PMSS), partially funded by the World Bank, aimed towards greater flexibility in the existing management structures, which were still operating under the Planasa standards, despite being deprived of centralized sources of subsidized financing. Recognizing the enormous differences between different regions of the country, the PMSS advocated the co-existence of several management models. It would be up to the municipalities and the states to choose the appropriate management model from a number of institutional possibilities, ranging from service concessions to private operators to the modernization of existing state companies.

The PMSS, despite being launched by a federal government institution, has not resumed an interventionist approach. Since the second half of the 1980s, there has been no national plan signalling priorities or incentives for particular lines of action, as was the case under Planasa. However, actions were taken without coordination between the different governmental agencies involved, and no standard procedures existed to evaluate the use of financial resources or the social effectiveness of the investments made.

The programme was committed to a general objective of universal access to services, to be fulfilled by means of two independent but complementary lines of action. The first aimed at the institutional restructuring of the sector through the development of regulatory measures, new financing mechanisms and possible institutional models for the management of services. The second line of action aimed at the financial, technical and organizational modernization of certain CESBs, looking to create a sort of demonstration effect for other CESBs. In this second line, investments in CESBs were made with the assistance of the World Bank and local financing.

Nevertheless, the concession of services to CESBs remains the dominant option: nearly 70% of the drinking water systems of Brazilian municipalities are still managed by state companies, a number of them under expired concession contracts.[3] For sanitation, the percentage drops to about 15%. Municipalities partially operated by CESBs, that is, water services contracted with no relevant sanitation concession, may have sanitation services directly operated by themselves, mostly in precarious conditions.

Territorial conflicts associated with state-centralized management emerged as a result of inadequate standards of urban land use and occupation, both under municipal jurisdiction. The institutional and technical organization of water and sanitation services do not follow the guidelines of urban policy, defined at the municipal level. The supply-side logic of water and sanitation services is sector-based and not necessarily related to urban or territorial planning. The urban process and, therefore, urban planning and management cannot be dissociated from the territory, even when the

---

[3] The legal provisions brought by the national water and sanitation law of January 2007 oblige either the complete write-off or the reshaping of expired concession contracts in the form of programme contracts. Under this format, the main objects and targets of the services must be explicitly agreed upon between the municipality and the state company, over a service platform mutually recognized as sufficient to attain these targets.

most elementary ideas of functionality are an issue. However, a number of strategic decisions concerning water supply expansion do not consider the specifics of particular urban growth processes. This sector-based logic has evolved under the aegis of asymmetric concession contracts, the terms of which did not take into account municipal priorities concerning the services. The asymmetry of power between municipalities and state water companies was reinforced by those contracts, the latter being entitled to determine their investment plans, regardless of the priorities defined by the municipalities.

Despite the general principles established for the decentralization of public services by the Brazilian Constitution of 1988, the CESB still accounts for a considerable proportion of active water and sanitation services. No significant claims for municipal take-overs have occured, even when most of the original concession contracts had almost expired. The lack of a stable regulatory framework for water and sanitation services meant that the situation concerning the rights and liabilities of both the municipalities and concession holders remained unclear, as regards still active contracts and the prospects for new arrangements. One main divergence has been the value of non-amortized assets, whose depreciation criteria were not clearly set by the original concession agreements. Municipalities tended to be held hostage to state companies because they could not afford the remaining investment costs to do away with concessions, whereas state companies, facing serious financial difficulties ever since the decline of federal funding schemes, would not agree to write-off the remaining contractual obligations of municipalities, notwithstanding the alleged illegitimacy of these contracts. For the case of larger and profitable services in big cities, the dispute focused less on unpaid investments in the service areas themselves, and more on their strategic role in cross-subsidizing the supply for smaller and poorer areas.

Aside from the disputes involving service concessions and their potential to generate surpluses necessary to cross-subsidization, the metropolitan service areas present peculiar problems regarding the functional integration of service phases. Despite a general recognition of municipal power over local services, metropolitan networks are particularly complex with regard to central capacities and the main vectors of inter-municipal distribution. Conflicts between municipally managed services and state companies still persist, wherever clear agreements regarding bulk supplies of treated water and the final disposal of wastewaters are lacking. Regarding water supply, disputes generally focus on the price of the dispatched water, in the face of rising marginal costs for water taking, treatment and transport, equalized across the metropolitan supply system. Regarding sewage primary collection, treatment and disposal, disputes focus both on the costs involved and acceptable standards of effluent disposal, since the costs of pollution become tangible in the downstream intakes for urban supply.

The lack of a regulatory framework has played a crucial role in the creation of conflicts between state companies and municipalities. None of the relationships and disputes mentioned previously have been subject to a stable set of specific rules, but rather to particular interpretations of general principles regarding the administrative responsibilities of state and municipal governments, and their respective executive entities. Since the ending of Planasa rule in 1986, the federal government, under varying political orientations, has tried to establish a specific regulatory framework to fulfill this need, despite polarized debates on the matter. The definition of this framework and of a national sanitation policy were also considered essential concerns for the new

Ministry of Cities, established during the first government term of the Workers Party, in 2003. Between the end of its first term and the beginning of its second in late 2006, the federal government finally succeeded in approving a new regulatory framework at the National Congress, sanctioned in January 2007.

The new act requires the establishment of regulatory provisions for drinking water supply, sanitation, solid waste management and urban flood control services.[4] The first significant innovation brought about by the national act has been a comprehensive overview of services, with an interdependent approach. This is in strong contrast to the Planasa concept of basic sanitation, limited to water supply and sewerage. This comprehensive view is particularly important in areas of new urban expansion on the outskirts of larger urban concentrations, which frequently interfere with vulnerable environmental systems. In the Metropolitan Area of Sao Paulo, some of the largest informal housing settlements are located in areas of permanent conservation around metropolitan water reservoirs, where cross-contamination between point and non-point source pollution makes water supply, sewage, solid waste and stormwater disposal complementary components in the overall control of sanitary and environmental degradation.

Another innovative dimension of the national law is the need for autonomous regulation of services, necessarily exercised by a public institution independent from the service supplier. During the centralized rule of finance, regulation was subsumed as part of the financing package, under control of the national authorities and their state branches, the state companies. In practice, although not formally invested with public authority to exercise planning and regulation of the services, state companies to a large extent performed this role. They did so because they had economic power and the requisite technical skills, while the municipalities more of less acquiesed, accepting the intrusion.[5] Under the new law, independent regulation is a precondition to any managerial arrangement for service provision, regardless of the public or private character of the provider. The new act also establishes a whole set of users' rights and provisions concerning public accountability, such as the publication of all reports, studies, decisions and instruments related to regulation and inspection, as well as full access to information pertaining to services, including periodic reports on quality. The traditionally asymmetric concession contracts that used to establish the rights and obligations of state companies and municipalities regarding water supply and sanitation, have now been replaced by programme contracts, necessarily based on municipal plans and autonomous regulatory bodies.

7.3 ECONOMIC CONFLICTS AND CHALLENGES

The main source of long-term financing for water and sanitation services in Brazil has, for many years, been the national fund for the compensation of labour severance (FGTS). Created in 1966 to support the national housing policy, the fund later diversified, in the early 1970s under the Planasa rule, into financing investments for water

---

[4] Law 11445, January 2007.

[5] For the sake of this discussion the very motive of this acceptance, should it result from authoritarian pressure or consent, is meaningless. The important fact is that the legal framework was vulnerable and insufficient to ensure a local voice on local interests.

and sanitation infrastructure. While Planasa was the sector plan, the financing scheme above it was the powerful Sanitation Financing System (SFS) operated by the National Housing Bank. The SFS comprised a mix of FGTS resources and foreign loans, kept by the federal government, aside from some minor fiscal transfers. The system was designed to operate mostly as a long-term loan provider, whose debts would be assumed by the state water companies (CESB) with guarantees issued by the states' treasuries. Apart from the rationale of state-wide cross subsidization, this guarantee system, reliant on the state treasuries, constituted a strong reason for the centralized nature of this national policy.

To access federal financing, each state created a water and sanitation fund (FAE). Each fund comprised 75% loans for direct investment on the expansion of water and sanitation infrastructure, to be repaid via the operational revenue of the state companies, and 25% seed money to launch the first necessary investments for expansion, assumed to be non-profitable. This latter part should have come from state fiscal revenues, but in practice this only occured in a few cases of richer states. By and large, most states did not have sufficient funds to establish the initial investments alone, and demanded a special line of federal credit to cover their part. In practice, the direct and basic investments for many states were almost entirely financed by the federal system. But this was not free. Many of the state administrators at that time did not realize that they were actually assuming a huge long-term debt with the federal system, based on repayable sources, and not on ordinary fiscal transfer, as had been the case in the past.

By the mid-1980s, both the federal financial schemes headed by the National Housing Bank, for housing and water and sanitation, were broke. A large part of their failure was due to insufficient repayment, as a consequence of undervalued monetary indexing in a context of uncontrolled inflation. The National Housing Bank closed in 1986, and the remaining assets and obligations of the sector's financial schemes were transferred to the National Bank of Savings (CEF). The Sanitation Financing System retained more or less the same procedures regarding loans for direct investments in water and sanitation infrastructure, but no more funding for basic investment in state companies would be available. Foreign loans were discontinued, and the system became almost exclusively dependent on the labour severance compensation fund (FGTS).

The fund is still operating, but the constraints on public debt and the high cost of opportunity imposed by the macroeconomic policy in Brazil after the early 1990s, made access to it in practice very difficult, both for state companies and municipal services. With the growing restrictions on monetary policy, the services that have done best are those that adopted an independent finance strategy, fundamentally based on user rates revenue.

In most cases, the investment capacity of the state companies has been reduced because of their limited scope and high indebtedness at elevated interest rates. Abicalil (2004) has shown that no more than 19.8% of the CESB incomes, on average, were destined for investment, while 23.6% were used to pay the financial costs of contracted debts. With regard to the application of user rates and water consumption, in an increasing block rate structure, the situation is equally problematic: between 1994 and 2003, water consumption in the top consumption bracket has dropped by 18%, whereas the number of users in the lowest bracket, less profitable, rose by 30%. Revenue per customer has consequently fallen by 28.3%.

Despite the new institutional prospects brought about by a more liberal approach towards municipal borrowers and, very recently, by the regulatory framework of 2007, important challenges remain to financing universal access to services, particularly in the face of economic exhaustion and the legal limitations of inter-social cross-subsidization. Financing poor users with the surpluses generated by wealthier users was the chief strategy of the CESB/Planasa model. This rationale was based on the idea that in a large country with significant social inequalities, the larger territorial jurisdiction of a service area could better accommodate the juxtaposition of economic surpluses and losses. In each state, the areas where costs could not be directly recovered by user charges would be subsidized by surpluses generated elsewhere. This was not a process of direct transfer from rich to poor areas, but a complex of compensating measures between areas whose operational conditions involved different levels of technical difficulty and marginal costs.

The technical difficulties for water extraction, treatment and distribution, as well as those related to sewage collection, treatment and disposal, vary dramatically from case to case, depending on raw water availability, topography, urban morphology and network topology. These technical differences are combined with huge socio-economic discrepancies, giving rise to a wide range of difficulties for service provision. The more favourable mixes of technical and socio-economic conditions make these natural candidates to be the first service areas to operate, since they are potential generators of significant economic surpluses to subsidize the less favoured areas. Should a purely economic logic be applied to service expansion, networks would clearly evolve following the gradient of growing physical and socio-economic combined difficulties.

In some cases, physical and socio-economic constraints do not necessarily evolve together: the wealthier clusters may be located in areas relatively less convenient from the standpoint of physical servicing – for example, elite neighbourhoods located on top of hills, where water supplies are overpressurized – while some poor settlements may be favoured by being close to natural or infrastructural facilities that render access to services relatively cheap. But this is not the rule: systematic research on Brazilian metropolitan areas show that the poorer settlements tend to be located in the least accessible areas regarding most infrastructure modes, and are subject to strong environmental constraints.

During the first stages of network expansion, a wide range of cross-subsidizing opportunities tend to emerge, as there are a considerable number of profitable fields able to subsidize service expansion in socially vulnerable areas whose technological difficulties are relatively moderate. The more service coverage evolves, the more margins for cross-subsidization tend to narrow. The next profitable area for expansion will not generate as much surplus as its predecessor, and the next subsidy-demanding coverage will be more vulnerable and costly. This is a logical tendency of the cross-subsidization process, up to the limit where profitable expansion is no longer a prospect, and there remains no margin to raise prices within an increasing block rate structure, regardless of the absolute needs of growing social demand in uncovered urban outskirts. At this point, cross-subsidization is saturated.

This is a general trend, not equally applicable to all cases. For the most significant Brazilian concentrations, there are examples in which the break-even point between profitable and subsidy-demanding expansions is manageable, thus allowing an almost universal expansion with no need for external subsidy. This is the case of the Metropolitan

Table 7.2 **Selected indicators of operational efficiency**

	Produced volumes Mm³/year	Macro-metered volumes %	Micro-metered connections %	Unaccounted for water %	Number of active units per employee Units/employee
COMPESA – PE	529656	50.23	62.13	59.90	317.15
CEDAE – RJ	1787910	87.64	66.18	51.69	339.73
SABESP SP	2826309	100.00	99.97	34.19	572.72
Campinas – SP	102732	100.00	99.92	22.02	378.64
Petrópolis – RJ	12755	100.00	97.25	20.14	420.59

Source: SNIS (2005).

Table 7.3 **Selected indicators of social effectiveness**

	Urban population connected to water network Inhabitants	Urban households connected to water network %	Average consumption per unit m³/month. unit	Average per capita consumption l/inhab. day	Sewerage connection per water connection %
COMPESA – PE	5823182	93.15	9.84	81.13	18.16
CEDAE – RJ	10162450	85.23	22.20	213.09	40.65
SABESP SP	22570120	97.17	13.76	160.27	73.33
Campinas – SP	995010	96.77	16.46	208.11	83.73
Petrópolis – RJ	232166	80.32	12.70	115.55	62.03

Source: SNIS (2005).

Area of São Paulo, where most water services and bulk supplies are operated by the state company Sabesp. When this happens, there is an almost direct causality between service efficiency and social effectiveness. But when the margin for cross-subsidization is saturated prior to service expansion to the poorer outskirts, no self-sustaining solution is possible.

Tables 7.2 and 7.3 depict some selected indicators of efficiency and effectiveness for three state companies – Pernambuco (PE), Rio de Janeiro (RJ) and São Paulo (SP) – and two municipal services, one operated by a public company controlled by the municipality (Campinas, in the state of São Paulo) and another by a private concession (Petropolis, in the state of Rio de Janeiro).

Produced volumes, in millions of cubic meters per year, are depicted in order to give an idea of the magnitudes involved. The services provided by the state companies do not cover all the municipalities in each state. In the case of São Paulo, Sabesp was responsible, in 2005, for the water services in 367 of the 645 municipalities in the state. In Rio de Janeiro, Cedae covered sixty-five of its ninety-two municipalities. Meanwhile, in Pernambuco, Compesa covered 169, out of 185 municipalities. The proportion of unaccounted water includes both physical and non-physical losses, in a rough proportion between the total volume produced and the effectively billed water. It is not a precise indicator for service performance on physical water usage, but it provides a proxy on the level of control in overall operation. Metering, both for the bulk volumes dispatched from the treatment station and for each user unit, is shown to be a requirement crucial to operational control. The cases of Compesa and Cedae, whose total losses on unaccounted for water reach more than 50%, show a clear correlation between poor metering and loss.

For the case of Compesa, where macro-metering is lower than micro-metering, the prospect of control is still bleaker than in any other of the examples, since nearly 50% of the estimated volumes are uncertain following bulk dispatch, regardless of the accuracy of micro-metering at consumption. This explains why total losses reach about 60% in a service where more than 60% of the consuming units are metered.

The number of active units per employee includes – according to SNIS 2005 criteria – both the number of directly hired employees and those contracted by service providers. The lower the number of service connections (active units) per employee, the less productive the service in terms of labour productivity. This indicator shall not be taken alone as a definitive yardstick for service efficiency, but poor performance of labour productivity is clearly and directly detrimental as regards margins for cross-subsidization. Positive results, if any, tend to be mostly absorbed by the internal labour costs of the service.

Some selected indicators of social effectiveness for the same services, depicted in Table 7.2, seem to support the broad hypothesis of causality with respect to operational efficiency.

The less efficient services – Compesa and Cedae according to the indicators in Table 7.1 – are those that present poorer performance in terms of effectiveness, both for water supply and sewage collection coverage. In the case of Compesa, a relatively high percentage of households connected to the water distribution network are shown to be ineffective in terms of final results, since the substandard consumption per unit (less than $10\,m^3/month$) reveals a strong constraint on dispatch and distribution. Conversely, Cedae in Rio de Janeiro demonstrates unitary consumption higher than the standard Brazilian level. This higher consumption level, together with the important total losses shown in Table 7.1, forms a picture of widespread waste. In both cases the coverage of sewerage connections remains very low in comparison to the more efficient services.

The small number of cases and selected indicators depicted in Tables 7.1 and 7.2 do not lead to any definitive conclusion on prospects of economic sustainability for services under state, municipal or private provision. But the basic relationships raised are sufficient to define an analytical path to evaluate the remaining margins for potential cross-subsidization on existing services. In any case, the theoretical limits of this approach seem to imply a paradox: the more inefficient services are those in which any move towards technological and managerial improvement tends to be positively drawn to relevant increases in terms of social effectiveness, whereas for those who have already achieved more developed standards of efficiency, the remaining margin for cross-subsidization, within the water service itself, tends to be insufficient to cover still-needed social response.

Considering the concrete difficulties experienced by state and municipal services in accessing funding from the FGTS, as explained at the beginning of this section, and the severe restrictions to fiscal debt at all levels of government, direct subsidization to social supply is almost impracticable. As internal cross-subsidization has, in most cases, achieved saturation, a possible way-out is to treat it from a broader cross-sectoral perspective. Since the late 1990s, the institutional framework of water resources management – based on a rationale of multi-purpose use of resources and a broad participatory decision-making process – has opened perspectives for intersectoral subsidization. One of the baseline understandings in water resources planning and policy has been the recognition of priority water use for human demand, even if possibly subsidized by other uses. The new federal law on environmental sanitation services, when establishing integrated approaches to water supply, sewerage, solid wastes and urban flood control, actually launched the basis of a broader prospect of cross-service subsidization, and a common agenda with water resources priorities, at basin level.

7.4 SOCIO-ENVIRONMENTAL CHALLENGES AND CONFLICTS: SOCIAL INEQUALITY AND ENVIRONMENTAL DEGRADATION

The general coverage of water and sanitation services for urban households in Brazil is, at present, not as bad as the conflicts and challenges described above might suggest. Despite the problems on the regulatory and public finance fronts, practically all urban households from north-east to south have access to water supplies, while sewage collection and disposal have a more heterogeneous distribution.

Despite the clear insufficiency of access to sewerage, this service has also evolved in terms of coverage since the times of Planasa. The aforementioned programme openly prioritized the extension of water supply networks, and never significantly invested in sewage collection and treatment during its almost fifteen years of existence. In spite of this, a number of remarkable projects on metropolitan sewerage were undertaken, such as the Great São Paulo Sanitation System (SANEGRAN), in the late 1970s, whose main interceptors, designed to conduct all the metropolitan collected sewage to a central treatment station, had transport capacities as large as 16.5 cubic meters per second. However, the grandiosity of this project, and its remarkable failure to effectively treat and properly dispose of metropolitan wastewaters, explains many of the contradictions of pollution control and social vulnerability in Brazilian cities.

Table 7.4 Brazil: macro-regional coverage of urban water supply and sewerage services (2005)

Region	Water supply % urban households	Sewage collection % urban households	Sewage treatment % of collected volume
North	68.5	6.7	10.0
North-east	98.6	26.7	36.1
South-east	96.8	69.4	32.6
South	100.0	33.7	25.3
Mid-west	100.0	45.4	39.7
Brazil	96.3	47.9	31.7

Source: SNIS (2005).

As with most of the interventions designed for urban sewerage during the Planasa rule, this project took for granted the allocation of investment principally to structural works on bulk collection and treatment. In fact there were important challenges to be faced at the macro scale, since there were clear signs of an escalation in environmental pollution, with significant impacts on urban water availability. The bulk/structural approach had worked adequately for water supply, as the active character of the distribution networks for these systems, which depend on dispatched capacities upstream of usage points, established a logical hierarchy for capacity expansion. Still, in the case of water supply, distribution is pressurized and there are no significant physical thresholds for capillary distribution (Silva, 2004). This renders it relatively cheap in comparison with bulk investments on central capacity and primary distribution mains. The problem with sewage collection networks is their intense interaction with the urban tissue and topography, since they work basically under the influence of gravity. The technocratic approach of the Planasa projects on sewerage have, in most cases, underestimated the costs and difficulties of developing comprehensive collection networks, especially in explosively growing cities where outskirt settlements evolved on unfavourable topography and environmentally vulnerable plots.

This is key to understanding the main reasons for the gap in coverage levels for sewerage and water supply in most Brazilian cities, up to now. Apart from the asymmetry in investment, there was also a problem relating to technological approaches, which had been distorted by a vision of subsidiary importance initially attributed to collection. By the mid-1980s, most sewerage system projects had already taken on the expansion of capillary collection networks as a public health priority, regardless of the possibility of immediate connection to treatment facilities.

When discussing these types of physical and technological interaction, the links with equality, in terms of social access, might not be immediately clear. To establish these it is necessary to bear in mind the prevailing morphology of most Brazilian cities regarding urban expansion and the location of the poor. With the exception of *favelas*, which are geographically inserted in central areas, most poor settlements in Brazilian cities are located amid the peripheral remains of open land and dense vegetation. These areas are environmentally vulnerable and generally essential for water supply. The extension of conventional networks to these areas is generally more difficult and costly than in central areas, because of the various physical obstacles creating complex network topologies. Sewage collection in these areas is particularly challenging because of its dependency on gravity. Topographical obstacles, landslides, irregular urban tissue and flooding can all make separate sewage collection unfeasible. Cross-connections with stormwaters suggest that integrative solutions should be pursued, by means of partially unitary networks of collection and ancillary components.

Regardless of these remarks, general figures of access related to the income of the family head do not, at first glance, reflect a more significant concentration than has been observed regionally. However, unlike regional distribution – which may be justified by the presence of alternative forms of individual sewerage – income distribution of access is a direct function of inequality.

Distribution of access in relation to income is less unfair for water supply than for sewage collection. While the range of variation for water supply varies from 85.16% to 96.43%, directly proportional to income range, access to sewage collection ranges from 34.80% to 77.76%. Therefore, the necessary expansion of the collection

Table 7.5 **Access to urban water supply and sewage collection related to income range (2005)**

Income range (minimum wages)*	Population Number of urban inhabitants	Urban water supply % access	Sewage collection % access
Up to 1	13 006 249	85.16	34.80
1–2	28 431 964	88.58	42.13
2–3	24 744 805	91.19	48.63
3–5	33 027 396	93.05	56.40
5–10	30 053 760	94.91	64.06
10–20	12 904 311	95.87	70.52
>20	5 859 541	96.43	77.76
No income	1 224 280	90.78	47.94
Brazil (urban)	152 013 993	91.97	54.27

Legal minimum wage in Brazil: approximately US$210/month (December 2007).
Source: PNAD (IBGE)/AESBE (2007).

system – which tends to be more costly at the irregular and vulnerable urban outskirts than in the formal sectors of the cities – is clearly concentrated on the poorer segments of demand, who are unable to pay for the full marginal costs of this expansion. This is the reason why the institutional and economic challenges referred to above are inextricably linked to the question of equality. Moreover, this is why conventional mechanisms of cross-subsidization, limited to sector-oriented management of services, have been shown to be exhausted since the Planasa financing model came to an end.

There are a number of possible ways out that can be explored under a broader perspective of cross-subsidization, in which different services may benefit from common structures and non-structural measures. A wider conception of environmental services as a result of the new regulatory framework, defines the grounds for integration between services, as well as for effective links with water resources planning and management.

7.5 NEW OPPORTUNITIES IN WATER AND SERVICES MANAGEMENT: REGULATION AND CONFLICT-RESOLUTION

The specificity of technical and managerial problems in metropolitan areas, compared to other urban and territorial organizations of environmental sanitation services, is self-evident from most studies and researches carried out in Brazil. Table 7.6 highlights the bottle-necks in sewage treatment that indirectly restrain water availability in all metropolitan areas.

When considering water availability as a double function of quantity and quality, and the complexity of densely urbanized areas in terms of the different locations of water intakes and frequent downstream discharges, sewage interferences are crucial. These are more than a problem of public health; they are main determinants of water availability for urban use.

To form an objective idea of the problems associated with sewerage actually affecting water availability in the main Brazilian metropolises, it is necessary to distinguish different situations regarding the topology of the collection system. At the most precarious

Table 7.6 Selected metropolitan areas: access to urban water supply and sewage collection in 2005

Metropolitan area	Geographic region	Sewage collection (1992) % urban households	Sewage collection (2006) % urban households
Belo Horizonte	South-east	68.91	83.58
São Paulo	South-east	74.9	78.64
Salvador	North-east	33.74	78.42
Rio de Janeiro	South-east	52.65	62.28
Curitiba	South	33.27	59.32
Fortaleza	North-east	11.5	43.81
Recife	North-east	25.04	38.97
Porto Alegre	South	19.55	10.01
Belém	North	5.41	9.27

Source: PNAD (IBGE)/FGV (2007).

levels are municipalities and districts lacking sewage collection. In such cases, local collection is a public health must, a priority placed before the environmental dimensions of sanitation. Table 7.6 depicts the situation for certain densely urbanized metropolitan areas with huge concentrations of people, presenting figures on sewage collection coverage far below the respective regional averages. This in itself is a warning sign regarding the specificity and priority of this particular form of territorial organization, concerning cases of smaller and less dense urban agglomerations.

Another dimension to be explored from the figures in Table 7.6, combined with Tables 7.4 and 7.5, is the insertion of these metropolitan areas into the respective regional indicators for sewage treatment and the structure of service access regarding income. Lack of coverage for sewage collection is determinant as regards the prospect of treatment. As mentioned above, point-source pollution following concentrated sewage discharges may give raise to ambitious projects on pollution control, employing advanced technological resources. However, the bottle-necks on collection, clearly shown in Table 7.6, are a definitive warning against unrealistic targets for pollution control. Environmental problems cannot be resolved independent of the equity dimensions of service coverage. This is why metropolitan areas may constitute a prime laboratory for integrating regulatory frameworks for water resources and environmental sanitation services.

A number of scholars and institutions, notably in Latin America, have recently labelled a new comprehensive approach to water and urban processes 'Total Urban Water Management' (TUWM). This approach has been explored at the State University of São Paulo by a group comprising members of the Department of Hydraulics and Sanitary Engineering at the Polytechnic School and the Centre of Urban Information Studies at the School of Architecture and Urban Studies (Braga et al, 2006). One of the key issues of this approach, largely based on the group's concrete experiences in developing an integrated plan for the *Alto Tietê* hydrologic unit (FUSP, 2002) in the Metropolitan Area of São Paulo, is the identification of specific conflicts and challenges associated with dense urban occupation.

Water planning and management is a key issue for economic production and social reproduction everywhere. The main branches of economic activity have always disputed over water uses for irrigation, hydropower, urban supply, flood control, navigation and recreation, despite the existence of any arbitrating body formally empowered to solve these disputes. In Brazil, the majority of electricity generation has been hydro-based since the beginning of the twentieth century, in a context where extensive irrigation is required for agriculture. Since the 1940s, multiple resources planning has been the preferred approach to establish regulatory and operational frameworks for national and regional policies, with hydropower logic possessing a clear hegemony.

This hegemony became increasingly questioned as other uses evolved in the urban economy, namely urban water supply, industrial, flood control and sewage disposal. Socio-environmental organizations exerted significant pressure to accord urban uses absolute priority under the new institutional framework organizing Brazilian water resources policy since the mid-1990s. Under this new institutional framework, integrated water resources planning and management evolved, giving rise to broader planning boundaries, concerning not only water resources themselves, but also the entire complex of land and water uses that determine the quantitative and qualitative conditions of water uses. In this context, the combination of structural and non-structural measures for most water uses became an all-pervading feature of water resources planning and management, including the land uses linked to them.

Total urban water management developed as a specific approach to densely urbanized basins, under this broader framework of integrated water resources planning. The Metropolitan Region of São Paulo (MRSP), with an area of about 8,000 square kilometers, houses 19.5 million inhabitants in thirty-nine municipalities. It is located within a single hydrographic unit in the upstream section of the Tietê river. The plan for this region was developed for the São Paulo State Water Council by the university group (mentioned above), from 1999 to 2002. The experience of developing that plan, and a further cooperative process in interdisciplinary research in water resources planning and urban studies, gave raise to the integrated TUWM approach.

The application of TUWM principles in the development of *Alto Tietê Plan* involved four types of interaction between different sectors and jurisdictions with an active role for the sustainable use of metropolitan water resources (Braga et al, 2006):

- Integration between systems/activities directly related to water use in the river basin area, in particular, water supply, wastewater treatment, flood control, irrigation, industrial use, energy use or other systems with a direct impact on the sources, such as solid waste, taking into account joint management of quality and quantity.
- Territorial/jurisdictional integration with institutions in charge of urban planning and management – municipalities and the metropolitan planning system – including better control upon the urbanization process, so as to avoid excessive demands on quantities and qualities of existing resources, and to reduce the impacts of stormwater floods.
- Regulating communication with sectorial systems that are not direct users of the water resources, such as housing and urban transport, taking into account the creation of real alternatives to the occupation of source protection areas and floodplains, and allowing urban development patterns that, as a whole, will not worsen the impervious conditions of the urban land and pollution in the basin.

- Communication with neighbouring basins, in order to establish stable agreements on the current and future conditions of flows and the export of water used in the basin. This type of communication will tend to become decisive and extremely complex in the future, given that all options to increase water offers to the MRSP, involve importing flows from other neighbouring basins.

All these are complex challenges and cannot be fully met within the specific competencies of the water resources management system. The three former challenges require strong institutional communication with the environmental systems and metropolitan planning, as well as the development of relationships with the municipal governments involved. The latter challenge involves communication with the management systems of neighbouring basins. A broader view of conservation and rational water use programmes and actions has also emerged as an essential element for an integrated concept of basin management systems and metropolitan planning.

These strategic principles and lines of action were proposed prior to the coming into effect of the new regulatory framework for urban environmental services, sanctioned in January 2007 (Brazil, Law 11445/2007). The wide range of services included within the concept of environmental sanitation and the regulatory flexibility with regard to different alternatives for territorial and functional aggregation between services – aside from the provisions for active user participation in service planning – now approximate the organizational structures of environmental sanitation services and water management. This is particularly the case in metropolitan areas, where conflicts have arisen regarding concessionary power over common services, such as bulk supplies and sewage treatment. With the new regulatory framework a new services management structure has become possible. The states' companies retain an essential role in the production of drinking water (catchment and treatment). The municipalities are therefore able to choose between the following management options: (i) managing services via a direct or autonomous management structure (managing by means of a public department or municipal corporation); (ii) managing in association with other municipalities through the development of intermunicipal cooperation, in accordance with the new law on intermunicipal associations (*consorcios*); or (iii) managing by delegating management either to state companies, such as the CESB, or to private corporations. According to the new law, public bidding is not mandatory for municipal delegation to the CESB, in contrast with the procedures applicable for delegation to private organizations, generally.

Another opportunity established by the new law is the possible combination of provision from the existing water resources framework and the new regulatory framework for environmental sanitation services, under which a promising institutional space may be created to solve jurisdictional conflicts. The combined rationales of integrated water management and interconnected sector services may underpin more precise bases for the definition of territorial and functional levels of integration and regulatory competences.

REFERENCES

Law 11.445 of 5 January 2007. Establishment of national guidelines for basic health, modifying Laws 6.766 of 19 December 1979, 8.036, of 11 May 1990, 8.666, of 21 June 1993,

8.987, of 13 February 1995; and revoking Law no. 6.528, of 11 May 1978; and other powers.

Abicalil, M.T. 2004. Financiamento e Partcipação do Setor Publico no Saneamento. Paper presented at *I Colóquio Jurídico-Econômico de Saneamento Básico: definindo o marco régulatório*, organized by Escola de Direito de São Paulo da Fundação Getulio Vargas (DIREITO GV), from 16 to 17 November (in Portuguese).

Braga, B.F., Porto, M.F.A. and Silva, R.T. 2006. Water management in metropolitan São Paulo. *Water Resources Management*, 22(2) June: 337–52.

FUSP (Fundação Universidade de São Paulo). 2002. *Plano de Bacia do Alto-Tietê, Relatório final, versão 2.0.*, São Paulo, Comitê da Bacia Hidrográfica do Alto-Tietê.

IBGE, 2007. Pesquisa Nacional por Amostra de Domicílios, PNAD: síntese de indicadores.

PNAD, Programa de Modernização do setor de Saneamento. 2005. Sistema Nacional de Informações em Saneamento, 2004. Basilia MCIDADES.SNSA. Available at: http://www.snis.gov.br.

Silva, R.T. 2004. Infra-estrutura urbana, necessidades sociais e regulação pública: avanços institucionais e metodológicos a partir da gestão integrada de bacias. L.C. de Queiroz Ribeiro (ed.) *Metrópoles. Entre a coesão, a fragmentação, a coesão e o conflito*. Rio de Janeiro: Fundação Perseu Abramo, pp. 365–94 (in Portuguese).

Vargas, M.C. 2002. Desafios da Transição para o Mercado Regulado no Setor de Saneamento. *Anuario 2002 GEDIM: Serviços Urbanos, Cidade e Cidadania*. Rio de Janeiro: Editora Lumen Juris, pp. 113–58 (in Portuguese).

Chapter 8

Urban water conflicts in Buenos Aires: Voices questioning the sustainability of the water and sewerage concession

Sarah Botton[1] and Gabriela Merlinsky[2]

[1]Agence Française de Développement (AFD), Marseille, France
[2]Research Institute 'Gino Germani', Faculty of Social Sciences, Universidad de Buenos Aires, Argentina

8.1 INTRODUCTION

The city of Buenos Aires lies along the banks of the Río de la Plata, an inexhaustible source of freshwater. It also sits on the largest reserve of groundwater in the world. However, a significant portion of the inhabitants of the Buenos Aires Metropolitan Region (BAMR) do not have access to a supply of quality water. Some of them, mainly in the southern part of the concession, also suffer from serious environmental externalities. This observation merits research into the conditions under which the resource is managed and distributed.

At the beginning of the 1990s, Carlos Menem's 'justicialist' government launched a series of wide-ranging reforms in the face of pressure from international organizations.[1] These reforms enabled the International Monetary Fund (IMF) and the World Bank (WB) to undertake and assess the results of 'on-site' tests based upon neo-liberal policies they advocated.[2] An analysis of the various developments in the Aguas Argentinas Sociedad Anónima (AASA) Buenos Aires water company concession agreement makes it possible to retrace the gradual whittling away of the neo-liberal economic model. This was based on the idea that private operators, under public authority control, would be able to inject large investments[3] into the water sector to enable the operator to make up for its lost time. Such investment was to be recouped via a suitably adapted pricing mechanism accompanied by effective commercial management. The ultimate objective of the agreement was a full quality service (serving 100% of users) in the largest concession area in the world (Alcazar et al, 2002). The

---

[1] The introduction of Argentine peso/US dollar parity and privatization of all key sectors of the economy led Argentina to break permanently with an approach to industrial development based on a centralized state in favour of the complete opening up of the Argentine economy to foreign investment.

[2] This took place without any regard for the constant and significant deterioration in the principal social welfare indicators (unemployment, poverty, healthcare, education, etc.) and the increasing burden of external debt (which had rapidly become the major item of public expenditure). The image of Argentina as one of the international financial organizations' 'star pupils' resulted in even greater surprise on the part of the international community when confronted with the extent of the social catastrophe triggered by the crisis of December 2001.

[3] In order to maintain and expand the network.

AASA's concession agreement and regulatory framework stipulated a certain number of principles which formed the basis for organizing the concession. In this study, we will seek to analyse how these principles may be perceived as incorporating the seeds of urban water conflicts[4] and how they gradually changed with the development of the conflicts,[5] in order to draw conclusions with respect to the dynamics at work in the concession area.

8.1.1 The social urban context

The city of Buenos Aires, capital of Argentina, encompasses both the characteristics of a modern Latin American city which has benefited from particularly strong industrial development during the second half of the twentieth century, and those of a highly fragmented territory in social and economic terms. As a result, the city and its peri-urban ring are marked by strong contrasts in terms of poverty: from the slum *Villa de Emergencia 31*, located behind the train station, to the extreme precariousness and (for the *porteño*)[6] almost unimaginably deteriorated and unsanitary conditions of the neighbourhoods on the outskirts of the city. Moreover, poverty in Buenos Aires has another facet: the pauperization of the middle classes resulting from the national crisis of the past few years, occured on top of the structural poverty – *Necesidades Básicas Insatisfechas* (NBI)[7] – of the outskirts and small-ring neighbourhoods of the federal district (Prévot-Schapira, 2002). According to data collected by the company *Aguas Argentinas* S.A. (AASA), there are 593 poor neighbourhoods (comprising 2.5 million inhabitants) in the concession area, of which 445 (1.1 million inhabitants) are within the area served by the network. For the sake of clarity, let us recall that of the 12 million inhabitants living in the city of Buenos Aires in October 2002, 54.3%[8] were living below the poverty line, that is, on less than 700 pesos – US$240 – per month and per adult equivalent); in other words, 21.2% of the population of the federal district and 64% of the population of the outskirts, more than half of whom were living below the extreme poverty line.

8.1.2 Water services before privatization

In Argentina, the state-owned company for water and sewerage, *Obras Sanitarias de la Nación* (OSN), aimed from its inception in 1912 to present itself as a 'model' public utility by virtue of its triple ambition: public hygiene, income redistribution and land organization (De Gouvello, 2001). The water tariffs were highly representative of

---

[4] In our approach, the definition of the urban conflict implies consideration of the city or the urban environment as a whole set of material products and services dedicated to the satisfaction of collective or individual needs. In this sense, the city is a public object upon which private interests do exist and, in this sense, constitute an object of social dispute. This social dispute materializes particularly in processes of social integration/exclusion that produce territorial segregation through the configuration of socially homogeneous spaces within heterogeneous cities (Pirez, 1994).

[5] The urban conflicts we study in this paper are 'open' conflicts. According to Albert Hirschman's grid, they would be analysed as 'voices' (neither 'exit' nor 'loyalty'). For further details, *see* Hirschman 1970.

[6] A Porteño is an inhabitant of Buenos Aires.

[7] Unsatisfied Basic Demands.

[8] Data provided by INDEC (Instituto Nacional de Estadísticas y Censos) 27 December, 2002.

Table 8.1 Water and sewerage coverage before privatization

	Water	Sewerage
Federal district (city of Buenos Aires)	99%	99%
Outskirts	55%	36%
Concession total	70%	58%
Number of connections (millions)	1.2	0.7

Source: AASA concession contract (1993).

this ambition: the Río de la Plata enabled the provision of water to the entire city of Buenos Aires in large quantities (the aim of OSN was to provide 700 litres per day per inhabitant, the highest volume in the world). The issues of supply and rational utilization of the resource were therefore not of primary concern, nor was the issue of cost recovery because the *infrastructure* rationale predominated over that of efficient service provision. Hence, the tariff was defined not as a function of the quantity of water consumed – as the system provided for *canilla libre* (all you can use) – but as a function of a calculation of indexes. This was a platform similar to that of a tax system based on the rental value of a dwelling (surface of the land, of the constructed area, type and age of construction, zone coefficient, etc.), so as to allow for a more 'equitable' income distribution. However, this universal-access project encountered a major hurdle. On the one hand, the lowest revenue groups were scattered in an eccentric manner, which meant that expansion integrated an increasing population that contributed less and less towards the service. On the other hand, the public utility, because of the heavy structural financial losses of the system, stopped investing in infrastructure early on, leaving the peripheral areas of the federal district awaiting connection to the network. The 'OSN model' was not in a position to finalize the project because of its unbounded ambition. This ultimately led to the opposite result: a good service in the federal district and fringe peripheral areas, but others awaiting connections that the state-owned company was not in a position to provide. Moreover, even in areas not covered by the network, water was considered an 'OSN question': users, political actors, mayors and local administrators were simply excluded from having any voice in the system (Schneier-Madanes, 2005).

8.1.3 The private sector operating the largest water concession in the world

In 1993, after the Dublin conference stated that water was an 'economic and social good', a number of privatizations of water and sewerage state-owned companies occured throughout the world. In this context, the Argentine government launched a call for tender for the Buenos Aires concession in order to continue and improve the activities of *Obras Sanitarias de la Nación* – a mainly loss-making public company in need of hefty investment into its seriously deteriorated infrastructure. The Suez group won the tender. The concession contract, based upon the 'universal service' notion (Arza, 2002), stated that over the long run (thirty years) almost all the population of the concession (comprising the federal district and the greater Buenos Aires area) had to be connected to both services – water and sewerage – whenever the urban

Table 8.2 **Population connected and to be connected to water and sanitation services per type of neighbourhood (1993)**

1993	Population connected (million inhabitants)		Population to be connected (million inhabitants)		
	Water	Sewerage	Water	Sewerage	Total
Standard neighbourhoods	5.6	4.7	1.4	2.3	7
Poor neighbourhoods	0.4	0.2	2.1	2.3	2.5
Total	6	4.9	3.5	4.6	9.5

Source: AASA data (1998).

configuration so allowed. Every five years, the company provides the water regulatory agency *Ente Tripartito de Obras y Servicios Sanitarios* (ETOSS) with a plan encompassing all the expansion works to be carried out over the next five years, as well as the corresponding tariff adjustments. The five-year plan must be accepted by ETOSS and represents a firm commitment from the company, which will be fined by the regulatory agency in the event of non-compliance. The technical and commercial elements at stake for Aguas Argentinas are found in the expansion goals, most of which target the poorest neighbourhoods and those located furthest from the concession (mainly precarious neighbourhoods). The expansion goal at the time of takeover (1993) was to integrate 3.5 million customers, of whom 65% lived in poor neighbourhoods.[9] The challenge was and remains huge.

Technical access to the network: the technical approach looked for in the contract was simple: a surface network with water coming from the Río de la Plata. In spite of the presence of good quality water tables in most areas, programme organizers did not originally envisage the drilling of wells by poor neighbourhoods to gain access to water. As of late, the idea of a single supply technique has been partially revisited with the launching of projects.[10] Indeed, the contract actually specifies the obligation that households are to be connected to the network once expansion works are finished. The sole connection envisaged is individual and for houses; there is no possibility of public taps. Except for the possibility of obtaining water from a well (permitted before the works of the company reach a particular area), AASA envisaged no alternative linkage to the public water network.

Economic access to the service: the cross-subsidy tariff system that existed when the company was state-owned was still valid at the time or writing. It is used on the one

---

[9] AASA data

[10] Agua + Trabajo (Water + Work) programme, municipality of La Matanza (2004). This takes into account the construction of local networks using water from the water tables (via wells), which themselves should be connected to the primary network within a few years (once expansion works are complete). This programme, which was initiated at the direct request of the Argentine president in 2003, aims to connect 178 poor neighbourhoods of the La Matanza municipality (on the second ring of the outskirts) to the network. It envisions the participation of neighbourhood cooperatives benefiting from a fund of more than 35 million pesos (around 13 million US dollars). It is an ambitious project, which should eventually connect more than 400,000 people.

hand to finance the operation (as a redistribution tariff), and on the other hand (since 1997) for financing network expansion.[11]

8.1.4 Development of the chapter

In this chapter we analyse the water conflicts from the perspective of 'urban sustainability' with a heuristic grid composed of three main axes: economic, social and environmental issues, which lead or have led to water urban conflicts in metropolitan Buenos Aires.

8.2 ECONOMIC SUSTAINABILITY ISSUES LEADING TO POLITICAL CONFLICT AND CONFLICT AMONG USERS

We study here the conflicts linked to the economic sustainability of the concession on the basis of two emblematic examples: firstly, *user conflicts*, resulting from the scheme for financing expansion of the network and, thus, to territorial solidarity issues at stake within the concession; and secondly, the conflicts due to the devaluation of the Argentinean peso (in January 2002) which could be termed *political conflicts*, and which concern financial balances and monetary risks in the concession contract.

8.2.1 Financing the expansion of the network: the SUMA conflict

In line with the 'consumer-payer' approach adopted in the concession agreement, the first extensions of the network concerned predominantly economically profitable areas (particularly in the north of the concession area), which were in a position to pay for the major costs involved in developing the infrastructure. This triggered a series of splintering phenomena, mainly between the inhabitants of the capital, already being served, who benefited from a higher quality service[12] (at no extra cost), and the residents of peripheral zones who had to bear the enormous cost of expanding the network.[13]

This situation led to two kinds of user conflicts:

- The first, **concerted demonstrations by non-connected people** unable to pay for the new connection fee, led to the reform of the expansion financing scheme.
- The second comprised **organized protests by already-connected people** not wanting to have their bill raised as a consequence of the new financing system.

The AASA concession agreement initially stated that network expansion would be financed only by newly connected users (via an 'Infrastructure and Connection' charge). The charge ranged from US$400 to US$600 for water and $1,000 for sanitation. Most low-income customers could not afford this charge and, therefore, were unable to pay their bill. Moreover, some began what they called a 'bill strike' (there were 80,000 unpaid bills in 1996).

---

[11] We analyse the 1997 reform in section 1.1.
[12] Repair and maintenance of the capital's networks and treatment plants.
[13] Secondary and domiciliary networks.

Table 8.3 Comparison of average costs per bi-monthly bill before/after the renegotiation of 1997

		Before	After
Average cost for users already connected	Water and sanitation services	30.00	30.00
	Regulatory charges	0.80	0.80
	SUMA Tax	–	6.00
	VAT	5.46	7.72
	Total	**37.26**	**44.52**
Average costs for new users (water only)	Water and sanitation services	6.00	6.00
	Regulatory charges	0.16	0.16
	SUMA tax	–	3.00
	CIS charge	–	4.00
	Infrastructure charge	44.00	–
	VAT	10.53	2.76
	Total	**60.69**	**15.92**

Source: Data taken from *La Nación*, dated 24/02/98, used by Alcazar, Abdala and Shirley (2002).

The first neighbourhood protests had begun in 1995 in the western part of the concession (La Matanza, Lomas de Zamora) when local residents realized that the connection fee (*carga inicial de conexión* (CIC)) was more expensive than contract work by third parties – the *obras públicas contractadas a terceros* (OPCT) rates. Neighbourhood associations began to organize action to prevent *Aguas Argentinas* from working on the network expansion in the area: a human barrier of 300 people prevented engineering work on a project; lawyers were engaged to negotiate a change to a specific project and its financing; formal complaints were made to a regulatory agency; people held street demonstrations and staged sits-in in front of the AASA local headquarters, and issued denunciations on TV and so on (Schneier-Madanes, 2005).

In 1997 the methods for financing the concession were renegotiated at the company's request, following blatant economic distortions and concerted action by those excluded from the network as a result of their inability to pay. Since then, network expansion has been financed via the participation of all network users (based on two new concepts: the service incorporation charge – *carga de incorporación al servicio* (CIS) – which replaced the CIC, and a new 'universal service and environment tax' – *Servicio Universal y Medio Ambiente* (SUMA).[14] The result was a slight increase for users who were already connected and a significant fall in the cost of connection for new users.[15]

In spite of a strategy focused on renewed solidarity at the territorial level, the creation of SUMA was the cause of numerous debates, widely covered in the media, between local political representatives, as well as numerous legal actions. The concept of SUMA altered the splintering *capital/periphery* effect for many users as it challenged the 'new consumer as payer principle', which was one of the basic precepts of the international financial institutions (*full cost recovery*). The introduction of the concept of SUMA was denounced by the Counsel for the Argentine people (*el Ombudsman*), with respect to its SU (universal service) component. This declaration halted all expansion work in

[14] The SUMA comprised two aspects: SU for universal service and MA (medio ambiente) for 'environment'.
[15] Calculations by pro- and anti-privatization analysts highlight the same trends.

progress for over a year pending resolution of the issue. The municipalities concerned by the stoppage viewed the declaration as defending special interests of the City of Buenos Aires (due to the considerable increase in water charges involved for users already receiving the service) and banded together to take legal action against this intervention. This concerted action also provided an opportunity to establish a forum comprising all the Greater Buenos Aires municipalities to initiate a debate on all themes related to water and sanitation services. The creation of this institutionalized forum via the intermediation of the Ministry of Public Works for the Province of Buenos Aires, constituted an essential step in the shift of power from the national sphere to the local level. As such, we should bear in mind that the concession agreement had been signed between the Argentine State (grantor of concession) and AASA (concession holder). Although the local public authorities were directly concerned by decisions relating to the management of water services, they were not involved in such decisions due to the decision-making structure on which the concession agreement was based.

8.2.2 Devaluation of the Argentinean peso: renegotiation of the concession contract

The neo-liberal model rested on convertibility between the Argentinean peso and the US dollar, and the government held this option at arm's length until the complete exhaustion of its possible application. The crisis of December 2001 and the ensuing social movements led to the fall of the government and total rejection of the on-going economic model. The peso-dollar parity, which could not be supported any further by the national economy, was rapidly abandoned. In January 2002, the temporary government announced[16] the devaluation of the national money and instituted *pesification* (conversion to peso) of public services' tariffs (the tariffs stipulated in the concession contracts had been defined in US dollars), thus putting an end to the concession contract terms. The end of the monetary model based on convertibility had a huge impact on operators' economic and financial balances with renegotiations of the concession contract for all public utilities taking place in the following years.

According to the new law, the renegotiation of contracts had to take into account

> the impact of the tariffs on the competitiveness of the national economy, on the income distribution, on the services quality level, on the investment plans – as soon as they were contractually defined – on the users' interests, on the accessibility to services, on the technical systems security and on the operators' profitability.

The purpose of the law was to confirm the clear will of the public authorities to widen the revisions of the contracts to include 'not only the interests of the private companies but also the level of completion of the contracts and, above all, the population capacity to pay for the services, because the majority had experienced an important fall in their income level' (Thwaites Rey and Lopez, 2003). Article 10 in the law prevented the companies from 'stopping or altering the completion of their obligations', while Article 13 authorized the government to 'regulate in a temporary way

---

[16] Law no. 25.561 (6 January, 2002) called the 'law of economic emergency and reform of the monetary regime'.

the incomes and prices of the goods and services considered as "critical" so as to protect users' and consumers' rights from a hypothetical distortion of the markets and of actions of a monopolistic or oligopolistic nature'.

In many aspects, this law constitutes a complete reversal in services management as a result of a new economic context, much less favourable to the companies.[17] Beyond the water sector, the renegotiations context was the framework for many events on the Argentinean political and economic scene. It opened the way to a new kind of urban water conflicts characterized by confrontation between the government and water (and other sector's) companies, including: pressure for and against the raise of public service tariffs, companies taking cases to international courts[18] to denounce the 'pesi-fication' of contracts, 'crisis' events in services (electrical cut-offs, etc.), and the deci-sion by some international companies to leave the country.[19] The first analyses of this complete reversal were highly critical of the private companies (Aspiazu and Schorr, 2003), explicitly denouncing the unfair balance of negotiations (in favour of the com-panies); nevertheless, they also admitted the importance of the economic and social elements at stake in the negotiations.

In actual fact, the Argentine economic crisis of December 2001, which had been on the cards since the mid-1990s, clearly highlighted the limits of this doctrine. However, the impacts within the Buenos Aires Metropolitan Region (BAMR) were in clear contrast to this. The crisis also demonstrated that the southern areas are subject to global economic and financial circumstances largely beyond their control. The Enron crisis, which trig-gered the hasty departure of *Azurix*, plunged an entire zone of the southern agglomera-tion into uncertainty with respect to the future of its water services. On the other hand, the changes in political orientation in such contexts also contribute to jeopardizing long-term PPP agreements,[20] as illustrated by the creation of Argentinean presidential decree no. 303/06[21] in March 2006. This led to the renationalization of the water and sewerage operator (i.e. Suez' departure from Buenos Aires) and the creation of a new public oper-ator: *Agua Y Saneamiento Argentinos* (AySA).[22] This decision appears to be the direct consequence of a new political line, very critical towards the neo-liberal policy of the

---

[17] The majority of the services contracts rested on the US dollar reference. These contractual definitions were very profitable to the operators and had led to significant exterior debts for expansion investments (in US dollars). In a context where the peso was worth a dollar, the exterior debts were largely covered by local incomes but once 'pesification' was instituted, the exterior debts became a real burden for the com-panies, which could not compensate them with incomes converted to pesos (divided by three on average) during a global recession.

[18] The CIRDI (World Bank arbitration institution) within the framework of bilateral treaties for the promo-tion and protection of foreign investments, signed by Argentina. This 'offensive' by the companies rapidly drew a response from the Argentinean public authorities: the Ministry of the Economy emitted a decree (no. 308/02) to exclude companies that had been to court from the renegotiation framework. Many com-panies decided to give up their claims in other sectors (telecommunications, electricity, etc.), but the water companies are all still claiming compensation from the Argentinean State. The outcome of the Tucuman case (Veolia) should be decided shortly.

[19] France Télécom sold its participation in Telecom Argentina in September 2003; EDF lowered its partici-pation in Edenor to 25% in September 2005; Suez announced its will to disengage its participation in Aguas Argentinas in September 2005.

[20] On this question, also see the Bolivian Case (Box 8.1).

[21] Decreto de necesidad y urgencia no. 303, 21 March, 2006.

[22] This new public company is held by the State (90%) and the employees (10%) and is headed by the leader of the trade union (SGBATOS).

Box 8.1 The Bolivian case

The La Paz – el Alto[23] concession (1.8 million inhabitants) formed part of the 'second wave' of Suez' major PPP agreements abroad (along with Casablanca, Manila and Djakarta in 1997). It was defined along similar lines to the Argentinean contract (call for tender, a thirty-year concession aimed at universal access to services, five-year plans for network extension goals, etc.). During the first extension plan, *Aguas del Illimani* largely exceeded the contractual goals in terms of water and sewerage servicing (100% in water and 88% in sewerage in La Paz, 100% in water and 54% in sewerage in El Alto), clearly favouring intensification of connections in already-served areas over extension of networks.

The negotiation of the second five-year plan led to contractual redefinitions: the tariff remained the same but the cost of connections was substantially increased. Furthermore, these redefinitions coincided with a period of budgetary austerity, social agitation at the national level (following social movements against water privatization in Cochabamba), and political instability (the resignation of President Sanchez de Lozada, replaced by Carlos Mesa, after violent popular demonstrations in February and October 2003). In spite of some 'pro-poor' initiatives (condominium projects for sewerage, 'district 7' for water network extensions in extra-contractual suburbs), the relationship between the company and Bolivian society continued to deteriorate. In December 2005, President Evo Morales, former trade unionist and the first indigenous president, was elected. He created the first Ministry of Water in Bolivia, thus making water policy a political priority. He issued a presidential decree 'firing' the private company and his government has since attempted to define a model for a 'social water company'. While awaiting the outcome of these discussions, Suez is preparing its departure.

In both the cases of La Paz and Buenos Aires, beyond the differences in socio-economical, political and cultural contexts, the State brutally ended the water concession agreements. The rejection of the private operator by the Argentinean and Bolivian States had its origins in very different contexts: if the popular demonstrations in La Paz showed a clear rejection of the company by the poor communities, in Buenos Aires poor populations, on the contrary, tended to approve the programmes developed by the operator. Ultimately, the great project of a private sector that would participate so as to 'rescue' developing countries 'unable' to organize universal basic services has clearly demonstrated its limitations. This acknowledgement raises serious questions.

In this respect, it is astonishing to note the lack of analysis concerning the genuine dynamics of these failures. Recent topicality could be an explanation. However, there are various processes obstructing a thorough and open reading of the history of PPP in Latin America. Even if the many militant points of view were expressed, and analyses were to be published concerning the main outcomes of the concessions (number of connections, financial results, tariff and institutional evolutions, regulation schemes), there remains a need for reflection on essential themes such as: modifications in the relationship between States and citizens or between citizens themselves (popular/middle class), the emergent need for citizenship (since the return to democracy at the beginning of the 1980s), and the translation of State-centered social solidarity *à la française* into foreign contexts (indigenous issues in Bolivia). The private company, caught within historical and cultural new contexts, poses problems to the political model since it contributes to creating strong expectations from populations.

---

[23] The city of El Alto, geographically attached to the capital-city La Paz, is mainly inhabited by immigrants, most of them very poor. The water concession agreement offers the operator exclusive rights on the service in this territory but only partial obligations for the network extensions.

1990s, with the new government announcing the end of its 'benevolent attitude towards multinational firms'. This decree was adopted after more than four years of unfruitful negotiations between the government and the water company (in spite of the fact that Suez appeared to wish to stand by its 'showcase' concession, in spite of the economic crisis, until at least September 2005). This decision was based on the (questionable) arguments that the operator had contributed to worsening the quality of the drinking water distributed in the networks (citing high nitrates rates), and more generally, had not fully fulfilled its contractual commitments. The political arm-wrestling between the public authorities and the private company during the negotiations finally degenerated into a conflict, brought to arbitration, which is still underway.

8.3 SOCIAL SUSTAINABILITY ISSUES: BRINGING WATER SERVICES TO THE POOR

In this section we study the question of the social sustainability of the concession through the examples of two measures adopted and applied to deal with the collective claims of poor people in Buenos Aires' suburban zones. First, we examine the 'participative management model' as an answer to the demands of poor neighbourhoods for connection to the drinking water network (technical access); and second, we examine the 'social tariff' as an answer to the situation of poor users no longer able to pay water bills in the new recessive economic context (economic access).

8.3.1 Bringing water and sewerage networks to poor neighbourhoods

In her work, Cristina Cravino analyses in detail the different modes of mobilization of poor people in the peri-urban neighbourhoods of Buenos Aires (in particular, the *asentamientos* at the beginning of the 1980s) to obtain land property titles. She insists upon the importance of *claiming* public services (water, electricity, telephone) in the dynamics of the situation (Cravino, 2001). In terms of access to citizenship, recognition of land property and access to public services are presented as two related processes, each impacting the other.

At the beginning of the 1990s, the privatization of public services deeply transformed the nature of the water operator's response to this form of claim. Since this time, the population has not addressed its claim for connection to the public authorities directly, but rather to a private company acting according to a concession contract, whose terms are not clear regarding this specific question.

The contract actually excludes slums as it assumes only that networks will only be expanded to urbanized areas, as well as the internal networks of the *barrios armados* – large groups of dwellings under the responsibility of the municipalities. This means that there is no contractual obligation whatsoever for the provision of services to these two types of neighbourhoods which, in terms of population, represent more than 25% of the poor neighbourhoods within AASA's concession area.[24]

---

[24] Of the total population of the concession area's poor neighbourhoods (more than 2 million people), around 15% live in slums, 10% in barrios armados and 75% in precarious neighbourhoods (data from the IIED-LA-UADE report) Participation of the private sector in drinking water and sewerage in Buenos Aires, balancing the economic, environmental and social goals, July 1999.

Table 8.4 **Population connected and to be connected to water and sanitation services per type of neighbourhood (1998)**

1998	Population connected (million inhabitants)		Population to be connected (million inhabitants)		
	Water	Sewerage	Water	Sewerage	Total
Standard neighbourhoods	6.8	5.5	0.2	1.5	7
Poor neighbourhoods	0.8	0.3	1.7	2.2	2.5
Total	7.6	5.8	1.9	3.7	9.5

Source: AASA data (1998).

The concession company quickly took onboard the problem of urban poverty in order to confront the challenge of providing water services to poor neighbourhoods in the concession area. However, the group that won the concession contract (Suez-Lyonnaise des Eaux, later Suez-Environnement) initially favoured sustainable development programmes, but later became less enthusiastic as it became more aware of the investment dynamics in developing nations such as Argentina.

In 1999 the *Community Development Unit* (CDU) was created within AASA in response to the demand of the regulator.[25] The original goal was to define and implement a **social back-up methodology** for network expansion in the poor neighbourhoods[26] of the concession. Little by little, the scope of these activities and responsibilities was enlarged until it encompassed, among other lines of work, the **regularization**[27] of services in the poor neighbourhoods[28] and the **professional training** of company staff (on issues linked to company activities: sustainable development, direct communication, management of community meetings, management of conflicts, participatory management of projects and so on). Since the beginning of 2002, the goal of the CDU has been to define the concessionaire's policy for low-income neighbourhoods by having the communities understand the value of the public/private participation model. With this in mind, it defined a series of almost forty projects entitled **Participatory Management Models** (MPG), the purpose of which is to achieve the full-scale expansion or regularization of services in a particular neighbourhood. The MPGs are built on a three-party agreement, institutionalized by a contract between the company, the community of the neighbourhood[29] in question

---

[25] Henceforth called the Sustainable Development Unit.

[26] The vast majority of poor neighbourhoods that the operator was concerned with in terms of expansion were precarious neighbourhoods since slums and peripheral cities were not included in the contractual goals of the operator.

[27] In the water sector, the term regularization concerns the establishment of a standard technical and commercial relationship with certain neighbourhoods. It can take very different forms. For example, regularization can mean organizing the expansion of the service to a neighbourhood that is not connected but is located inside an area that is served by the concession, or it may mean taking a closer look at the unpaid bills of certain customers in order to organize workshops focusing on commercial issues.

[28] The scope of the CDU was enlarged to include all of the poor neighbourhoods in the operational projects and not just the precarious neighbourhoods, as was initially the case.

[29] The neighbourhood community, according to the term used by AASA, comprises all the inhabitants of the neighbourhood. This community appoints its representatives, elected or not, to sign the contract.

Box 8.2 Participatory management models (CDU-AASA)

The criteria to be fulfilled in order to carry out an MPG are valid for all participants:

The neighbourhood community must request the service (following the concept of an 'informed request' established by the company). The project will be undertaken only if 80% or more of the neighbourhood agrees. The community must be able to organize itself and select its representatives, and must also provide the human resources for the work.[30]

The municipality must commit itself contractually to fulfil its responsibilities as regards the works (opening trenches in the streets and so on), distributing the necessary tools (gloves and shovels), and organizing distribution of the funds. These are subsidies of 150 pesos per month[31] allocated by the government to the *planes jefes y jefas de hogar* (heads of households) participating in community labour programmes.[32]

The company is in charge of the technical feasibility of the project. It must provide the necessary materials (pipes, wrenches, and so on) and technical training (workshops for people to familiarize themselves with techniques and safety issues), as well as being responsible for communication with the community (workshops introducing commercial aspects and providing answers to questions or doubts from inhabitants).

---

and the municipality. It is then agreed upon by the regulatory agency, whose role it is to supervise the process and who authorizes the consolidation of the partnership between all the stakeholders.

What were the main results of the programme? The installation of participative management models (MPGs) in 2002 marked the beginning of the company's operational stage. This came about together with a continuous growth in number and dimensions of operational projects. Twelve MPG projects were carried out in 2003, enabling 8,000 people to be connected. In 2004, there were twenty-one projects involving the connection of 30,000 people, whereas the projections for 2005 envisage the connection of more than 400,000 people, thanks in particular to the implementation of the 'Agua + Trabajo'[33] programme. Moreover, since January 2004, a specific resolution has been passed by the regulatory agency regarding a reduced water service tariff for poor neighbourhoods: it stipulates a bi-monthly (reduced) invoice of between 4 and 6.5 pesos (i.e. between US$1.5 and US$2.4) per service.

The profitability evaluation of the service extension projects to the poor neighbourhoods is still incomplete. However, it is already possible to see some degree of improvement in the bill collection rate following commercial efforts such as specific workshops. Likewise, the payment ratios are very good (much better than those in traditional neighbourhoods) when the community participates directly in management tasks such as handing out invoices and getting together with neighbours to pay. It is worth noting that water supply costs before connection to the urban network were

---

[30] Which is largely criticized by some social researchers who consider that the company is taking advantage of the social needs of the population to exploit free manpower (see Aspiazu et al, 2004).

[31] In 2002, 150 pesos was the equivalent of US$50.

[32] In addition to these funds, inhabitants who participate in the works benefit from a reduction in their water bill for a number of years.

[33] Translator's note: Water + Work.

Box 8.3 The paradoxical impact of the Argentine crisis on the programme

Paradoxically, the December 2001 crisis did not slow down the development of poor neighbourhood projects. On the contrary, the year 2001 became an actual springboard for the operational stage of participative management models. Moreover, these projects were the only opportunity for the AASA to proceed with the extension of networks, since all the remaining projects negotiated for the five-year plan had been temporarily stopped. This strange situation was the result of a number of combined effects: the *maturity* effect (the crisis arrived right when the company was ready to establish projects for the poor neighbourhoods); the *cost* impact (expansion in poor neighbourhoods is generally less expensive than in traditional ones);[34] and finally, the *image* effect (during the contract renegotiation period, the poor neighbourhoods projects represented the *cara humana* (human face) of *Aguas Argentinas'* activities).

---

much higher for poor neighbourhoods. The benefits obtained by the Argentine society from the programmes, in both sanitary and social terms, have been encouraging with much better health indicators in connected neighbourhoods due to access to drinking water. The programme has also meant an improvement in the dialogue between the neighbourhood communities, the municipality, the regulatory agency and the operator, and improved community organization.

8.3.2 The social tariff: a response to the recent inability of Argentina's middle class to cope with the water bill

In the aftermath of the 2001 crisis, the majority of the middle class fell into poverty. Many people, that had been connected to the water and sewerage networks, suddenly faced a situation in which they were no longer able to pay their water bill. In a highly ideologized context, many users organized direct demonstrations (*escraches*) in front of the private companies' offices to denounce the possible raise of tariffs (following the peso devaluation).

The social tariff programme was not aimed at the poorest population in the concession. Because it proposed a reduction in the water bill (according to economic and social criteria), it was necessarily addressed to *already connected* people, mainly the impoverished middle class. This measure appears to be emblematic of the 'new poverty' phenomenon that appeared in the 1990s after the structural adjustment policy instituted by the Menem government. In order to face this new social tension due to the degradation of purchasing power, the company and the public regulator decided to institute a social tariff programme for water and sewerage services.

8.4 THE EVOLUTION OF THE 'ENVIRONMENTAL QUESTION' IN THE CONTEXT OF WATER SECTOR PRIVATIZATION AND THE CONCESSION PROCESS

The first agreement for the concession of the drinking water service, signed in 1993, did not include a clause for environmental protection. This was by virtue of its

---

[34] This is because labour costs are low or non-existent, workers are able to recover certain materials, and because of the transfer of certain costs to the municipalities (tools, heavy works etc.).

Box 8.4 The social tariff programme (STP)

The social tariff programme was initially a demand which emerged from the company. It was formalized by the public regulator (ETOSS) in collaboration with the users' commission it encompasses, some entities of the water company and NGOs working on the field. The technical team of the regulator has driven the programme since its initial application. The STP was introduced during the tariff revision (authorized by ETOSS resolution 2/02, 9 January 2002). It includes a system of economic assistance to low-income users which is supposed to be 'efficient, transparent, explicit and focalized'. It is not cross-subsidy in nature. The economic assistance is organized by reduction 'modules' on the bill (4 pesos by service and by bill). Households can benefit from one or two reduction modules, according to their situation.

The criteria to be elected for the programme are as follows: to be living below the poverty line, to be registered as a regular client, and to have bills inferior to 30 pesos (since 2002). The user should initiate the demand process (to the municipality). The municipality selects STP beneficiaries from among all the petitioners and then sends a list to the company, which applies the reduction modules directly to the bills. The company has to reserve an annual budget of 4 millions peso for the STP.

---

preponderantly economic approach and the absence of unified legislation concerning water resource management.

With the 1994 constitutional reform,[35] new opportunities opened up in Argentina with regard to the unification of environmental protection criteria. This situation, in which provincial legislation appeared to be quite underdeveloped and out of date – amidst a process of structural reforms aiming at privatization and market liberalization – impacted at the federal level, as jurisdictions have to incorporate strong conflicts of interest between sectors demanding resource management without adequate protection or accountability.

In the Buenos Aires Metropolitan Area, one of the main conflicts of interest emerged because of a rise in the water table level. The conflict occured at the end of the 1990s, although a considerable part of the definitions of service management and supply had already been defined in the concession contract.

Two decisions created the conditions for emergence of the conflict: first, the closing of the wells (that extracted water from the underground aquifer) and the switch in the water supply to surface water (extracted from the Río de la Plata); and second, the creation of an investment plan for network expansion in which significant weight was assigned to the drinking water network, at the expense of the sanitation network.

8.4.1 Water table rise, flooding and environmental conflict

Rising groundwater is a natural response that the hydrological system adopts in order to preserve the equilibrium between its different variables. In the case of the

---

[35] The 1994 Constitution produced a series of innovations regarding environmental protection. One of the most important consisted of obliging the water operator to produce environmental impact studies for any investment project.

Buenos Aires Metropolitan Area, many factors contributed to this process during the 1990s.[36]

According to the 1991 census, the supply deficit of the domiciliary water network was 55% and the sewerage deficit had reached 80%. Over the course of the last decade, the domiciliary water network extended rapidly in different municipalities of the *conurbano bonaerense* (suburban zone). The sewerage network did not grow in equal proportion. This means that, in net terms, a greater volume of water enters the region, and – because of the low coverage of the sanitation systems – is dumped again in the same territory.

The balance between the different phenomena of natural and anthropic origin varies in each of the sub-regions taken into consideration inside the Metropolitan Area. Researchers (Santa Cruz and Buzzo, 1999; INA, 2002a, 2002b) agree that the weight of the anthropic factors – amongst them the diminution of potable water extraction from aquifers – is more important than that of the natural ones.

Together, these factors provoked the elevation of groundwater in seventeen districts of the *conurbano bonaerense*: Almirante Brown, Morón, Lanús, Tres de Febrero, General San Martín, Quilmes, Avellaneda, San Fernando, Vicente López, Tigre, La Matanza, Ezeiza, Hurlingham, Esteban Echeverría, Lomas de Zamora, San Isidro and Ituzaingó. In the following section we concentrate on the Lomas de Zamora case, one of the districts with a higher level of demand from neighbours and social organizations.

8.4.2 Urban water conflict and environmental conflict: the Lomas de Zamora water forum

Actors' interests and the knowledge they possess with regard to the water service externalities and their consequences on health, environment or the local economy, play a central role in environmental conflicts in the territory (Sabatini, 1997). The case under analysis refers to a neighbourhood organization that changed its claim agenda throughout the years. It initially emerged as an organization of people who were victims of flooding. Between the mid-1980s and the early 1990s, it obtained the construction of different sanitation works in the watercourse (whose overflows produced the greatest impact during floods). Towards the mid-1990s, under the denomination of 'Water for Everyone', it joined in the demand for the construction of drinking water supply works and improvement of service quality. In 1992, the Federal Government established the construction of the 'third reinforcement' – an underground river for the Lomas de Zamora potable water supply. The works started under the management of the public company (*Obras Sanitarias de la Nación*) and developed and ended under the management of the privatized company (*Aguas Argentinas*) in 1997.

---

[36] (1) The increase in precipitation level; (2) the tubing of watercourses (this solution, which was adopted to avoid overflows produced by the sudestadas – rainy southeast winds – prevents them from acting as natural drainage sources during water excesses as a consequence of precipitation increase); (3) increase in imperviousness caused by acceleration of the urbanization process; (4) closing down of groundwater extraction wells for industrial use (as a consequence of industrial closures during the recession which lasted from 1994 to 2001); and (5) the change in the source of supply for human consumption (from underground to surface water).

The first cases of home flooding caused by the elevation of the groundwater table were recorded at the beginning of 2000. In the same year, another rain-caused flooding took place, affecting the lower zone of the district, in which water drainage is much lower compared to previous episodes. In this context, a new organization emerged, the 'Hydric Forum', which began to place claims before the provincial and federal authorities and *Aguas Argentinas*, arguing that the responsibility for the elevation of the water table lay with them, on account of the closing down of the underground water extraction wells and their replacement by a superficial water supply.

The conflict became acute between 2001 and 2003, as the organization earned public visibility, through organizing protests, writing petitions to the authorities, taking the company to court and demonstrating in the Plaza de Mayo. During the same period the organization articulated demands from other groups of victims living in other parts of the *conurbano*.

At present, the conflict is undergoing mediation, with the intervention of the Federal Government, and a new funds contribution for the cleaning and sanitation of the watercourses (which continues to overflow due to lack of maintenance). *Aguas Argentinas*, through an agreement with the Provincia de Buenos Aires Government, has purchased drainage pumps (to extract water from the water table and thus lower its piezometric level) for allocation to the municipalities. However, not all municipalities are equipped to undertake maintenance, so in many localities the pumps work for only a few months.[37]

The neighbours continue to claim for sanitation works, since the systems of domiciliary off-loading in certain neighbourhoods (settled in lower land) are overflowing, creating serious health risks for the population.

8.4.3 The 'environmental problem' and the need for a responding institution

In this case, the creation of the environmental problem derived from the aggregation of demands and changes in the groups' profile. If the original demand was for drinking water, the paradox was that the water supply generated unexpected consequences that further affected the environmental quality of the habitat.

Scientific and technical knowledge plays a significant role in the definition of environmental problems. Complaints to the *Aguas Argentinas* company tended to provoke answers that limited the Company's responsibility, placing causality down to other problems not linked to the 'third reinforcing' project. It therefore became important for the organization to have proof that, beyond their legitimate claims, grounded the attribution of causality and responsibility. The publishing of a report by the Instituto Nacional del Agua (INA) allowed the organization's claims to acquire a greater level of legitimacy.

The other significant stage in the acknowledgement of the environmental problem was the transition to a stage of greater mobilization (following the change in Federal

---

[37] This agreement was signed between the Buenos Aires province and the company on 28 January 2003 for the installation of 1,500 pumps in seventeen municipalities of greater Buenos Aires. The agreement establishes that 'the concessionaire takes charge of the machinery and of two-thirds of the installation cost and the Government of the remaining amount'.

Government at the end of 2003). Demands began to be redirected towards the Federal Government with the diagnosis that the new political winds would facilitate the obtaining of answers at the federal level, thus generating pressure on the provincial government (which has legal authority over water management). A further important element was the impact that massive demonstrations had on public opinion and media coverage.

Lastly, relations between the organizations, the Provincial Government, the Federal Government, the Regulatory Agency (ETOSS) and the concessionaire company (*Aguas Argentinas*) have passed through different phases. In the beginning, there was a process of 'relative acceptance' of the company's role because of its role in the expansion of the drinking water service. As the domestic flooding problem grew in importance, the organizations tactics became more confrontational, directed first towards the Regulatory Agency (ETOSS), then the company through legal actions, and lastly, towards the Federal Government. The most important element of this social apprenticeship that the organization underwent is likely its understanding of the water problem as a multi-causal phenomenon, assigning shared responsibility to the company and the Government for resource management.

8.5 CONCLUSION

The urban water conflicts in Buenos Aires, which concern economic, social and environmental sustainability issues, reveal the extreme incompleteness of the concession contract and the complexity of the interactions between all the actors concerned with regards to urban water management: connected users, people to be connected, public authorities, the regulator, private companies, and so on.

If the cases presented in this chapter (referred to as 'open conflicts' because of their public nature, including demonstrations, mobilizations, denunciations, resort to court, media declaration, etc.) provide an interesting panorama, they still do not constitute the totality of the existing conflicts or even of the conflicts to come. The 'solutions' to those conflicts are not, all of them, identifiable or, if it is the case, sustainable.

The Buenos Aires cases demonstrated the existence of user conflict 'cycles'. Examples include demands by non-connected people for a lower connection fee; when a solution was found with the creation of the SUMA fee, this led afterwards to a new kind of conflict, led by previously connected people. Such cycles highlight the revealing nature of an analysis based on *concrete* urban water management. Studying urban water conflicts is a very pertinent way to analyse large water systems and their social implications. It can help to understand, on the one hand, the social and political dynamics at stake in their complexity and, on the other hand, the concession governability issues due to lack of social solidarity within the specific territory of the Buenos Aires Metropolitan Region. From this perspective, the purpose of this chapter is to identify *possible* fields of analysis for urban water management.

REFERENCES

Alcazar, L., Abdala, M. and Shirley, M.M. 2002. The Buenos Aires water concession, M.M. Shirley (ed.) *Thirsting for Efficiency: the economics and politics of urban water system reform.* Washington: World Bank/Pergamon, pp. 65–102.

Arza, C. 2002. *El impacto social de las privatizaciones: el caso de los servicios públicos domiciliario*s. Segunda serie de documentos de informes de investigación, March. Buenos Aires: FLACSO.

Aspiazu, D., Catenazzi, A. and Forcinito, K. 2004. *Recursos públicos, negocios privados. Agua potable y saneamineto ambiental en el AMBA*. Collección Investigación, informe de Investigación, no. 19, March. Buenos Aires: UNGS.

Aspiazu, D. and Schorr, M. 2003. Privatizaciones: la renegociación de los contratos entre la administración Duhalde ¿Replanteo integral de la relación estado-empresas privatizadas o nuevo sometimiento a los intereses de estas últimas? *Realidad económica*, Jan.–Feb., no. 193.

Cravino, M.C. 2001. La propiedad de la tierra como un proceso. Estudio comparativo de casos en ocupaciones de tierras en el Area Metropolitana de Buenos Aires. *Land Tenure issues in Latin America*, SLAS 2001 Conference, Birmingham, 6–8 April 2001.

De Gouvello, B. 2001. *Les services d'eau et d'assainissement en Argentine à l'heure néolibérale. La mondialisation des 'modèles' à l'épreuve du territoire*. Collection 'Villes et Entreprises'. Paris: L'harmattan.

Hirschman, A. 1970. *Exit, Voice and Loyalty. Responses to Decline in Firms, Organizations and States*. Cambridge, USA: Harvard University Press.

Instituto Nacional del Agua. 2002a. Estudio cualicuantitativo del ascenso de la napa en Lomas de Zamora. Informe Final Buenos Aires: Convenio Ente Nacional del Agua y Ente Tripartito de Obras y Servicios Sanitarios. August.

Instituto Nacional del Agua. 2002b. Estudio para el diagnóstico del escenso de las aguas subterráneas en el conurbano bonaerense y la Ciudad Autónoma de Buenos Aires. Etapa I. Informe Final Buenos Aires: Convenio Ente Nacional del Agua y Ente Tripartito de Obras y Servicios Sanitarios. August.

Pirez, P. 1994. *Buenos Aires Metropolitana. Política y Gestión de la Ciudad*. Buenos Aires: Centro Editor de América Latina.

Prévôt Schapira, M-F. 2002. Buenos Aires, entre fragmentation sociale et fragmentation spatiale. F. Navez-Bouchanine (ed.) *La fragmentation en question: des villes entre fragmentation spatiale et fragmentation sociale?*Collection 'Villes et Entreprises'. Paris: L'harmattan.

Sabatini, F. 1997. Espiral Histórica de conflictos ambientales. F. Sabatini and C. Sepulveda (eds) *Conflictos Ambientales. Entre la globalización y la sociedad civil*. Santiago: CIPMA.

Santa Cruz, J.N. and Silva Busso, A. 1999. Escenario hidrogeológico General de los Principales Acuíferos de la Llanura Pampeana y Mesopotamia Septentrional Argentina. II Congreso Argentino de Hidrogeología y IV Seminario Hispano Argentino sobre Temas Actuales en Hidrología Subterránea, Santa Fe, Argentina. Actas, vol. I, pp. 461–71.

Schneier-Madanes, G. 2005. Conflicts and the rise of users' participation in the Buenos Aires water supply concession, 1993–2003. O. Coutard, R.E. Hanley and R. Zimmerman (eds) *Sustaining urban networks. The social diffusion of large technical systems*. London/New York: Routledge.

Thwaites Rey, M. and Lopez, A. 2003. *Fuera de control. La regulación residual de los servicios privatizados*. Buenos Aires: Temas.

Chapter 9

In search of meaningful interdisciplinarity: Understanding urban water conflicts in Mexico

José Esteban Castro

School of Geography, Politics and Sociology, Newcastle University, Newcastle, UK

9.1 INTRODUCTION

This chapter focuses on contemporary water conflicts in Mexico's urban areas. It is based on recent research on the relatively high level of such conflicts recorded in the country, as suggested by observations made since the 1980s.[1] These take a wide range of forms, from peaceful demands to the authorities directed at gaining access to safe drinking water, to violent actions involving the destruction of water infrastructure and armed confrontations between the police and members of the population. The research shows that the widespread and regular occurrence of these conflicts and the political underpinnings of the overall process of water unrest in the country have been scrutinized by the Mexican water authorities since at least the late 1970s. However, research on the topic has been slow and patchy, and we are still far from fully understanding the multidimensional character of the process. This chapter builds on previous and ongoing research efforts to achieve a genuinely interdisciplinary framework for understanding and explaining water conflicts, a crucial challenge for scientific work in the twenty-first century. In this regard, the study of water conflicts provides a unique opportunity to develop an interdisciplinary approach for the study of water governance and management, bringing together bio-physical, ecological, technical, socio-economic, political and cultural aspects. This chapter presents a brief historical background of water conflict in the Mexican context, followed by a more detailed description and analysis of the particular forms of urban conflicts over water recorded in Mexico during the 1980s and 1990s. The final section introduces a discussion on the obstacles and opportunities for developing a holistic and interdisciplinary framework for the understanding and explanation of water conflicts.

9.2 URBAN WATER CONFLICTS IN MEXICO FROM A HISTORICAL PERSPECTIVE

Urban conflicts over water sources and water services have been a feature of Mexican history for centuries. Undoubtedly, the hydrological and climatic characteristics of the

---

[1] Torregrosa (1988–97). We have examined these issues in more detail in Castro (1992, 1995, 2006). A recent PhD thesis confirmed most of the trends and processes discussed in this chapter (Kloster Favini, 2008).

country have played a significant role, given that Mexico is located within 19 and 31 degrees latitude, a region that concentrates the largest deserts and arid zones of the planet. Over half of the territory is arid or semi-arid, with two-thirds of the annual rainfall concentrated during the rainy season (June–September), while the overall rainfall pattern is characterized by very high interannual variability. Moreover, the population has tended to locate in water-scarce regions (CNA, 2001: 26). However, Mexican history suggests that aridity, uneven precipitation patterns and population growth are not sufficient to explain water conflicts in urban areas. Furthermore, this history also provides important clues for developing an interdisciplinary understanding of the conflicts, through the intertwining of physical-natural factors with socio-economic, political and cultural processes.

The historical evidence suggests that in pre-Columbian times, activities directed at establishing control over available water sources – a crucial factor in the accumulation of social and political power – fuelled intensive and recurring conflicts, especially in the Basin of Mexico (León-Portilla, 1984; Musset, 1991; Palerm, 1990). Since colonial times, water conflicts have been exacerbated by the introduction of water-consuming activities, which triggered rising demand and competition for scarce water sources. This was particularly important in relation to productive water uses such as irrigation agriculture, the development of water-powered industries, mining, fishing, and also the provision of urban settlements (Bakewell, 1984; Brundage, 1972; Florescano, 1984; Gibson, 1964; Musset, 1991). The latter required freshwater for essential services, systems for wastewater collection and disposal, and – notoriously in the case of Mexico City – flood prevention waterworks (Boyer, 1975; Ezcurra et al, 1999; Musset, 1991; Sahab Haddad, 1991). In the Basin of Mexico, this included the construction of El Desagüe (The Drain), a complex network of channels and tunnels to drain floodwater and wastewater from the basin, started in 1607 and only completed in 1900 (Connolly, 1991; DDF-SOS, 1975; Gurría Lacroix, 1978; Hoberman, 1980; Lemoine Villicaña, 1978; Musset, 1991). The basin's ecology was completely transformed as a result, with the progressive desiccation of the lake system and the demise of the water-centred social organization that had characterized pre-Columbian settlements (Gibson, 1964; Fox, 1965). This process was punctuated by social confrontations, not just between Spaniards and Indians, but also between the Spanish Crown and the colonial authorities, and between different factions within the colonial elite (Hoberman, 1980; Musset, 1991). Moreover, the expansion of Spanish control over natural resources, including water, in practice implied the expropriation of existing property rights (e.g. existing indigenous land and water rights, fisheries, etc.) that fuelled recurrent struggles (Chevalier, 1963; Gibson, 1964; Horn, 1997; Meyer, 1984).

Pre-Columbian and colonial water conflicts would foreshadow future developments. Thus, after Independence from Spain in 1821 water conflict in the nineteenth and twentieth centuries continued to flare up around the control of water for productive activities and the increasing consumption of rapidly developing urban areas. Major landmarks in this process were the massive concentrations of land and water that took place during the second part of the nineteenth century, and particularly under the dictatorship of General Porfirio Díaz (1884–1911); and the Mexican Revolution (1910–17), which formally reversed these trends by asserting the public ownership of land, water and other natural resources in the remarkable Article 27 of the 1917 Constitution (Tutino, 1986, 1988; Katz, 1988, 1994; Knight, 1990; Bazant,

Table 9.1 Evolution of water supply and sanitation coverage in Mexico's urban areas (1990–2005), in percentages and figures

Year	Urban population	Urban population with water supply (% and figures)	Urban population with sanitation (% and figures)
1990	57 300 000	89.4 (51 200 000)	79.0 (45 300 000)
1995	66 700 000	92.9 (62 000 000)	87.8 (58 500 000)
2000	71 100 000	94.6 (67 300 000)	89.6 (71 900 000)
2005	76 300 000	95.2 (72 600 000)	94.6 (72 200 000)

Source: Author's elaboration from CNA (2006: 25–31).

1994; Kroeber, 1994; Aboites Aguilar, 1998). However, despite the significant progress made after the Revolution, the twentieth century became characterized by 'internal dynamics of inequality' and social conflict (González Casanova, 1965: 87). In fact, the post-revolutionary period was notoriously marked by social struggles arising from the implementation of the principles enshrined in Article 27, in particular the redistribution of land and water (Bartra, 1978; Hewitt de Alcántara, 1978; Bartra, 1985; García de León, 1985; Tutino, 1986; Oswald et al, 1986; Gordillo, 1988; Knight, 1990) – a problem that remains at the root of much social injustice and unrest in twenty-first century Mexico. The urban water sector is no exception, and despite the substantial progress made in universalization of essential water and sanitation services since the 1970s, access to these services continues to be a major area of social conflict in the country (Perló Cohen and González Reynoso, 2005; Kloster Favini, 2008).

Yet, Mexico's urban areas have experienced a significant improvement in the level of coverage for water and sanitation services since the 1970s. Official data summarized in Table 9.1 below, suggest that the pace of improvement has been maintained, and between 1990 and 2005 coverage reached around 95% of the urban population.

However, these figures do not reflect the fact that the sanitary quality of the water distributed is highly uneven. For instance, a recent analysis based on data from the Health Ministry suggests that only 25% of the water distributed for human consumption is subject to purification processes other than simple chlorination, which is deemed insufficient given the high levels of pollution affecting the water sources (Jiménez and Torregrosa, 2007). Moreover, the aggregate data presented in Table 9.1 also obscure the variability of the quality of the services provided: for example, in the Federal District alone around 1.1 million people, or 14% of the population, have to buy unsafe and expensive water from unregulated water vendors (idem). Inequalities in access are also reflected in the official figures, which show that by 2005 some federal states had achieved almost universal coverage for water supply, with Colima at the top with 98.3% of the population served; conversely, others like Chiapas, Guerrero, Oaxaca, Tabasco and Veracruz continue to lag well behind with between one-quarter and one-third of the population unserved. The gap is even larger with regard to sanitation, with some states such as Colima having achieved 99.6% of coverage while others like Guerrero (64.6%), Oaxaca (60.8%) and Yucatán (68.3%) continue to suffer from lack of investment in expansion of these essential services (CNA, 2006: 33).

Unequal access to essential water and sanitation has been a major source of urban water conflicts in recent decades. However, other areas of water management have also been at the forefront of social and political confrontations, including competition for scarce and often shrinking water sources or the impact of water-related disasters. Such conflicts have been an object of concern for the Mexican government since at least the 1970s, probably as a result of the continued increase in the number and seriousness of the events. It is highly likely that this is related to the rapid process of urbanization and population growth experienced by the country since the 1950s. For instance, the Mexico City Metropolitan Area grew from around 3.1 million people in 1950 to 8.8 million in 1970 and then to 15.1 million in 1990 (INEGI, 1991). It is therefore understandable that the main explanations offered for the prevalence of urban water conflicts in Mexico tended to rely upon issues such as rapid population growth, chaotic urbanization, low capacity to expand water services infrastructure, and hydrological and climatic factors. In turn, socio-economic and political factors have not been properly accounted for, although these issues are often mentioned in official documents. We return to the causes of these conflicts later in the chapter, but first let us explore the *nature* of contemporary urban water conflicts in Mexico.

9.3 URBAN WATER CONFLICT EVENTS IN MEXICO

Our reference to 'urban water conflicts' in contemporary Mexico is grounded on empirical studies of 'water conflict events' recorded since the mid-1980s in Mexico (Torregrosa, 1988–1997). These events cover a wide range from complaints voiced by water users through the media and other channels to violent actions involving attacks on water infrastructure and representatives of water authorities and other power holders in relation to the provision of water services. Table 9.2 shows the distribution of events as recorded in the press for the Mexico City Metropolitan Area (MCMA) during the period 1985–92.

In the case of the MCMA, an important pattern in the intra-annual temporal distribution of these events is that they tend to be concentrated during the dry season (November–March), which may suggest that there is a cause-effect relationship between the rhythms of the hydrological cycle and the occurrence of water conflict. However, there are a number of factors which demonstrate that we cannot explain water conflict only by reference to hydrological and climatic drivers.

For example, within the MCMA, two-thirds of the cases occurred in the conurbated municipalities neighbouring the Federal District, which in 1990 housed only around 42% of the metropolitan population. Moreover, the distribution of the events across

Table 9.2 **Water conflict events: Mexico City Metropolitan Area, 1985–92**

Total	1985[a]	1986	1987	1988	1989	1990[b]	1991	1992[c]
Federal District 656	29	104	91	86	105	43	161	37
Con. Municipalities[d] 1303	39	120	322	284	221	62	178	77
Total MCMA 1959	68	224	413	370	326	105	339	114

a. September–December only
b. August–December only
c. January–June only
d. Conurbated Municipalities of the State of Mexico
Source: Author's elaboration from Torregrosa (1988–97).

different municipal units shows a wide variation, suggesting that other factors have to be incorporated in the analysis. For instance, among the conurbated municipalities we have a concentration of over two-thirds of the events in seven out of a total of sixteen municipalities, with Ecatepec at the top with 16.1% of the cases, and at the other extreme, Huixquilucan with only 1% of the events. Similarly, as shown in Table 9.3, the spatial distribution of events in the Federal District is heavily concentrated in three municipal sections: Gustavo A. Madero, Iztapalapa and Tlalpan. These characteristics provide strong evidence against reductionist explanations based purely on techno-scientific factors and support the argument for developing a more complex and inter-disciplinary understanding of urban water conflicts (see below). Let us concentrate now on the analysis of the events.

In the study of these events we have identified a number of key characteristics, particularly regarding the protagonists, their objectives and the types of actions taken. As shown in Table 9.4, the protagonists may be dwellers acting on their own to solve a particular problem, probably without institutional connections, as suggested by the fact that in 30.9% of the cases the actors did not have an explicit link with any organ-ization. For example, according to one press report in 1987 'women with children, elderly people and other dwellers [from the municipality of Cuautitlán Izacalli] held a virulent protest in front of the Municipal Palace demanding potable water' (*El Sol Satélite*, 19 August 1987).

However, while in the Federal District over 42% of the events considered were initi-ated by actors without signs of organization, this type of protagonist only accounted for 24.2% of the cases in the conurbated municipalities. In turn, when observed in more detail, the organizations participating in water conflicts events represent a wide range of interests including peasant organizations, neighbourhood committees, workers' unions, ecologist groups, small business associations and political parties (see Table 9.5).

Table 9.3 **Water conflict events: Federal District 1985–92 (by municipal section)**

Section	1985	1986	1987	1988	1989	1990	1991	1992	Total	%
G.A. Madero	6	17	21	25	31	2	21	10	133	20.27
Iztapalapa	6	15	14	11	10	7	21	7	91	13.87
Tlalpan	5	14	16	9	9	5	18	5	81	12.35
Xochimilco	2	13	5	5	5	7	10	4	51	7.77
A. Obregón	1	6	5	5	9	1	8	2	37	5.64
Coyoacán	0	4	5	5	8	2	23	2	49	7.47
Tláhuac	0	4	6	8	3	4	6	0	31	4.73
V. Carranza	4	2	6	3	4	0	2	2	23	3.51
Cuajimalpa	1	3	2	3	7	0	11	0	27	4.12
Iztacalco	2	5	5	0	3	5	8	2	30	4.57
M. Contreras	0	7	3	1	2	1	5	1	20	3.05
Cuauhtémoc	2	3	0	0	6	2	4	0	17	2.59
Azcapotzalco	0	5	1	1	3	1	8	0	19	2.90
Milpa Alta	0	3	2	6	1	3	10	0	25	3.81
M. Hidalgo	0	1	0	4	4	0	3	1	13	1.98
B. Juárez	0	2	0	0	0	3	3	1	9	1.37
Subtotal	29	104	91	86	105	43	161	37	656	100

Source: Based on Torregrosa (1988–97).

Table 9.4 Level and type of organization of the actors (number of events and comparative percentages): Mexico City Metropolitan Area, 1985–92

	Federal District	Conurb. Municipalities	Total
Representatives and associations of dwellers	26.4	30.4	28.9
	(165)	(326)	(491)
Popular organizations	8.5	16.7	13.7
	(53)	(179)	(232)
Local Government Representatives	2.4	7.4	5.5
	(15)	(79)	(94)
Political parties/unions	5.9	6.2	6.1
	(37)	(66)	(103)
Other	1.3	5.2	3.8
	(8)	(56)	(64)
Without organization	42.3	24.2	30.9
	(264)	(260)	(524)
Without information	13.1	10.1	11.2
	(82)	(108)	(190)
Total	100	100	100
	(624)	(1074)	(1698)

Source: Based on Torregrosa (1988–97).

Table 9.5 Sample of organizations participating in water conflicts, MCMA 1985–92 (in chronological order)

Organization	Date of event
Municipal Collaboration Councils	26-09-85
Independent Peasant Central	09-85
League of Agrarian Communities, Federal District	03-10-85
People's Revolutionary Movement (MRP)	22-10-85
National Action Party (PAN)	06-01-86
Confederation of Popular Colonies, State of Mexico	12-02-86
Popular Movement of Southern Peoples and Colonies, Federal District	01-03-86
Mexico's General Unión of Workers and Peasants	16-03-86
Mexican Workers' Party (POM)	16-03-86
Women's Regional (Tlahuac)	06-11-86
Union of Settlers, Tenants and Shelter Seekers (Tláhuac)	06-11-86
Mexican Workers' Party (PMT)	13-11-86
Federation of Proletarian Colonies, State of Mexico	11-86
Federation of Settlers, State of Mexico	23-02-87
National Confederation of Popular Organizations	23-02-87
Union of Settlers and Traders, Chalco-Ixtapaluca Valley	29-03-87
Civil Association Union of Towns and Colonies (Ecatepec)	15-06-87
Tacaba Pact Organization	19-05-89
Neighbourhoods Assembly	19-05-89
Mexico City's Neighbours Alliance	19-05-89
Independent Proletarian Movement (Tultitlán)	12-07-89
Federation of Popular Colonies (Iztapalapa)	12-07-89
Authentic Party of the Mexican Revolution (PARM)	08-08-89
Union of Established Traders (Coacalco)	30-09-92
League of Agrarian Communities and Peasant Unions (Colorines)	14-10-90
Ecologist Mexican Movement	08-11-90
Party of the Democratic Revolution (PRD)-ARDF	22-11-90
Francisco Villa Popular Front	14-12-90

Source: Based on Torregrosa Armentia (1988–97).

Table 9.6 Sample of the actors targeted in water conflicts, MCMA 1985–92 (in chronological order)

Target	Date of event
The Governor of the State of Mexico	23-09-85
The municipal authorities of Naucalpan	10-02-86
The municipal and private water vendors (Naucalpan)	10-02-86
The trans-national industries located in Azcapotzalco	02-03-86
The water tankers of the provincial water utility CEAS (Netzahualcóyotl)	16-03-86
Hotel owners and industrialists (Valley Cuautitlán Texcoco)	28-03-86
The Mayor's deputy delegate in Iztapalapa	24-04-86
The municipal office (Tláhuac)	06-11-86
Agitators, members of the National Action Party (Chimalhuacán)	14-03-87
Private water vendors (Cuautitlán Texcoco Valley)	18-03-87
The government of the State of Mexico (Chimalhuacán)	20-03-87
Water speculators (Netzahualcóyotl)	22-04-87
The Mayor (Ecatepec)	12-06-87
The Mayor (Atizapán)	07-08-87
The provincial water utility CEAS (Tultitlán)	26-10-87
The Mayor (Chalco)	01-12-88
Municipal and private water vendors (Iztapalapa)	30-05-89
The Mayor (Tultitlán)	12-07-89
The municipal authorities and private water vendors (Chimalhuacán)	21-07-89
Soft-drink industries	08-11-90
Purified-water industries	12-09-91
Bottled-water and ice industries	30-09-91
The Federal District Department	17-11-91
Public officers of the Benito Juárez municipal section	15-09-92

Source: Based on Torregrosa (1988–97).

For example, in December 1986 the General Union of Mexican Workers and Peasants publicly alleged 'that in the municipalities of Ecatepec, Tlalnepantla and Netzahualcóyotl there are 2 million people without access to potable water [. . .] living in the most severe insalubrity, which has caused that in the marginalised areas parasitic diseases are the main cause of death'. The union leaders also added that 'water vendors were taking advantage of people's needs by selling water at exorbitant prices' (*Universal*, 12 December 1986). In another event, which took place in March 1987, 'ejido leaders from Chalco and Ecatepec [stated] that the authorities and private interests were profiting from the drilling of clandestine wells to sell water in the marginal areas of the Cuautitlán-Texcoco Valley' (*Excelsior*, 12 March 1987). These examples also illustrate the kind of targets chosen by the participants, which include a diversity of powerholders and other actors deemed to be responsible for the situations prompting the action. Table 9.6 provides a sample of the type of antagonists targeted in the events.

In most of the events recorded the actions were directed against municipal authorities, water authorities, or the local representations of federal ministries in charge of some aspect of water management. In addition, the protagonists also targeted a number of other actors such as local leaders, water vendors and industries and businesses blamed for intensive water consumption in water-scarce areas. One notable aspect is that in many cases there is no clear target and the protagonists seem unable to identify the cause or culprit behind the situation which provoked them into action. For example,

Table 9.7 **Actors targeted in water conflicts, Ciudad Juárez and Tuxtla Gutiérrez (1986–91), percentages and figures**

Target	Ciudad Juárez	Tuxtla Gutiérrez
The authorities	53.9	35.7
Non-government power holders	9.8	15.8
Other users (horizontal)	7.1	5.3
Without clear target	29.1	43.3
Total	100	100
	(254)	(171)

Source: Castro (1992).

Table 9.8 **Stated reason for the actions, Ciudad Juárez and Tuxtla Gutiérrez (1986–91), percentages and figures**

Stated reasons	Ciudad Juárez	Tuxtla Gutiérrez
Gaining access to the service	32.1	52.3
Complaints about service standards	34.7	21.9
Pollution of water sources	33.3	25.7
Total	100	100
	(421)	(237)

Source: Castro (1992).

in the cities of Tuxtla Gutiérrez and Ciudad Juárez, the share of water conflict events with diffuse targets represented 43.3% and 29.1% respectively, as shown in Table 9.7.

Another key characteristic of events concerns the reasons given by the protagonists for carrying out the actions. There are several broad areas of contention that account for a large number of events:

- actions directed at gaining access to water and sanitation services
- actions concerning the quality of services, ranging from inadequate provision of water and sanitation to problems arising from price increases or management inefficiencies, and
- actions triggered by the impact of water pollution. In some areas, other issues are also important, such as disputes over the control of water sources and infrastructure in the Mexico City Metropolitan Area.

Table 9.8 shows the main reasons indentified in water conflicts events recorded in Ciudad Juárez and Tuxtla Gutiérrez.

Actions directed at gaining access to services were prominent in the recorded events. For example, in October 1987 the Union of Settlers of the Valley of Mexico claimed that

the problem of water supply is worse than ever before in the conurbated municipalities of the Federal District, where 364 settlements lack distribution networks and 136 settlements and quarters have water only twice a week. [...] Lack of

Table 9.9 Instruments used in the actions, Ciudad Juárez and Tuxtla
Gutiérrez (1986–91), percentages and figures

Instruments	Ciudad Juárez	Tuxtla Gutiérrez
Petitions	22.7	27.9
Denunciations	58.9	58.9
Mass mobilizations/rallies	4.5	0.8
Threats	4.9	9.3
Direct action	5.8	0.8
Other	3.3	2.3
Total	100 (421)	100 (237)

Source: Castro (1992).

water leads people to take desperate measures such as blockading highways and other communication routes as well as organizing parades and rallies in protest against the municipal authorities (*El Universal*, 15 October 1987).

As for conflict events over service standards, including administrative aspects, actions are normally triggered by irregularity or poor quality of delivery of services, price increases or perceived corruption and abuse from the authorities, water vendors and other actors. For example, in March 1987, neighbours from Naucalpan claimed that 'private water vendors have organized a black market for potable water in the popular districts taking advantage of the service interruption affecting a large part of the city's southern area'. They added that the service interruptions were planned in collusion with municipal officers in order to create a market for private vendors who 'lack hygiene and sell unsafe water, thus exposing men, women and especially children to gastrointestinal disease' (*El Día*, 23 March 1987).

We identified five types of instruments employed in the events: petitions, denunciations, mass mobilizations and parades, threats and direct actions. In practice, in most events the actors resort to a combination of two or more instruments, with petitions and denunciations being the most common instruments deployed by the protagonists. Table 9.9 shows the relative weighting of the different instruments in events recorded in Ciudad Juárez and Tuxtla Gutiérrez.

Petitions are formal requests to the authorities, as in the specific case of 'a commission of neighbours representing 30,000 people from Naucalpan', which requested local water authorities to normalize the water supply after suffering forty-five days service interruption (*Metrópoli*, 31 October 1985). In this case, protagonists also highlighted what they perceived as an anomaly, in that 'just steps away from our houses private water vendors have access to a water source where they refill their water tankers to resell the water to industries and shops' (idem). In another case, which took place in October 1987, 'representatives of the Municipal Collaboration Councils claimed that owing to the lack of [networked] potable water supply the population of the municipalities located in the Cuautitlán Valley suffer multiple abuses and harassment from the private water vendors who are in collusion with the authorities' (*Excelsior*, 9 October 1987).

Following petitions and denunciations, the next instrument utilized is the organization of mass parades, rallies and other forms of mobilization, often concentrated

in a relevant public space such as the central square or in front of water authorities'
or local governments' headquarters. For instance, in June 1987 'around 9,000
people from the town of Santa Clara (Ecatepec) organized a mass concentration
in the civic square to express their dissatisfaction with the Mayor, who wishes to
place their water supply system, that has been managed by the community for
over sixty years, under municipal control' (*Excelsior*, 12 June 1987). Earlier in the
same year,

> around 300 neighbours from Fuentes del Valle [Tultitlán] organized a concentra-
> tion and agreed to demand the sacking of the Mayor and the regularization of
> the water supply in twenty-three neighbourhoods where the service is currently
> suspended; they also decided to start a campaign of non payment of municipal
> taxes and to request the immediate intervention of the state governor (*Excelsior*,
> 23 February 1987).

Mass parades and mobilizations are frequently associated with two other instruments
identified in water conflicts: threats of direct action, and their actual implementation
in extreme cases. Thus, in February 1987

> neighbours from the popular areas of Naucalpan threatened to 'kidnap' the pri-
> vate water tankers if they continue to sell water arbitrarily. [. . .] Inhabitants
> from Benito Juárez, Chamapa, Las Huertas, Olímpica Radio and Casas Viejas,
> who belong to the People's Revolutionary Movement, stated that they plan to
> kidnap any water tanker entering their neighbourhoods if they sell water above
> 500 pesos, because the families cannot afford to pay more than that for this
> public service (*El Sol Satélite*, 27 February 1987).

In other cases, people actually did fulfill the threats: in the Cuatitlán Texcoco Valley in
March 1987, 'the inhabitants assaulted the water tankers who were selling the liquid
between 500 and 600 pesos, which provoked numberless conflicts and confrontations
between the neighbours because as a result of the assaults the water tankers had
stopped the service altogether' (*Excelsior*, 10 March 1987). In a similar situation,
which took place two years later in Chalco, a woman justified the assaults on the
private water tankers in an interview with the press:

> it was around 3pm, and we had dust in the nostrils because of the lack of rain,
> while the sun was burning the skin. Then several people who lived in Los Tejones
> saw the water tanker coming and went for it. Imagine, these bandits [the water
> sellers] had a conflict among themselves and did not want to sell water to us.
> And that afternoon we were desperate, do you know what is to live without any
> water? I just ask. [. . .] and what do you want us to do? We had to steal that
> water because any human being has thirst and need, don't you? Anyone would
> have done the same thing' (*Unomasuno*, 23 October 1989).

The events studied contain many other examples of direct confrontations, some
including the destruction of property and the loss of human lives, others consisting of
instances of civil disobedience such as non-payment of water bills and taxes, road

blocking, occupation of buildings, kidnapping of water officers and vehicles, and so on. However, we must now look at ways in which these conflicts may be explained and understood.

9.4 EXPLAINING URBAN WATER CONFLICTS

The events described and analysed in the previous section were recorded throughout the country during the 1980s and 1990s. Their occurrence had been anticipated by government experts, who developed a 'map' to predict water conflicts in the country's main urban centres between 1980 and 2000, as shown in Figure 9.1.

However, the discussion about what explains the occurrence of urban water conflicts in Mexico has been dominated by the notion that conflicts over water are the result of interplay between hydrological and climatic constraints, such as 'natural water availability' and economic and technical processes such as 'water abstraction', 'water demand' and 'water consumption'. Thus, for instance, in Figure 9.1 cities like Mexico (59) or Ciudad Juárez (6) were classified as experiencing 'current and future water conflict' on the basis of expected rising water demand against constant or deteriorating available water resources. In contrast, no conflicts over water were predicted for cities such as Oaxaca (80) and Tuxtla Gutiérrez (84) until the year 2000. Unfortunately, while the predictions were accurate for cities such as Mexico and Ciudad Juárez, which effectively were – and still are – subject to recurrent conflicts over water, the case of Tuxtla Gutiérrez and other cities located in water-rich areas demonstrates the need for research into the causes of urban water conflicts to progress beyond hydrological, climatic and other techno-scientific factors.

Indeed, urban water conflicts are part and parcel of wider social and political confrontations between alternative, often antagonistic, societal projects. Explanations of the causes of water conflicts thus require interdisciplinary coordination to capture the intertwining of physical-natural and social processes – given that water conflicts cannot merely be explained away by reference to factors such as low water availability, aridity or high population pressures. It is well-established that human beings have been able to develop cooperative arrangements grounded on solidarity and rational principles for the fair allocation of limited water sources, whether in conditions of aridity like medieval Valencia (Glick, 1970), or in more favourable hydrological situations such as Bali (Geertz, 1980), Ceylon (Leach, 1959) or the Philippines (Ostrom, 1990). Conversely, evidence shows that abundance of water can also be the context, if not the reason, for protracted social and political conflicts, as shown in Swyngedouw's study of Guayaquil (Swyngedouw, 2004) or by the aforementioned case of Chiapas (Castro, 2006, 1992).

One important and persistent obstacle is the fact that production of scientific knowledge about water in general, and water conflicts in particular, is fragmented by artificial epistemic divisions such as 'hard' vs. 'soft' or 'natural' vs. 'social' disciplinary entrenchments. I avoid the use of 'sciences' here to denote the fact that these entrenchments and artificial separations and oppositions are an 'epistemological obstacle', borrowing from Bachelard's classical critique, rather than a vehicle for scientific knowledge (Bachelard, 1938). This can be illustrated using the findings emerging from current research on water conflicts, which allows us to identify a number of 'epistemic subjects' that produce knowledge about water from diverse, often unconnected or

Figure 9.1 **Map of conflicts over water supply services in the main Mexican urban centres (1980–2000)**

Source: Adapted from SARH (1988: 50).

even irreconcilable perspectives.[2] Table 9.10 provides a schematic characterization of the divergent perspectives that coexist in the study and explanation of 'water conflicts'.

For example, in the map shown in Figure 9.1, 'urban water conflict' is understood as being the result of a mismatch between quantitative observables, such as the

─────────────

[2] We borrow the concepts of 'epistemic subject' and 'observable' from Jean Piaget (1971, 1977, 1978).

Table 9.10 Water conflicts and epistemic subjects

Epistemic subject	*Rationale*	*Observables*	
Water expert (hydrologists, water engineers, etc.)	Techno-scientific	– Quantitative indicators – Physical-natural and technical conditions and drivers – Water as a resource	
Business-financial expert	Market	– Quantitative indicators – Economic efficiency – Market criteria	
Institutional-administrative expert	Policy-administrative	– Bureaucratic norms – Organizational principles	'Water conflict'
Politics expert	Political	– Water governance systems – Electoral and party-political considerations	
Ecologist	Ecological	– Indicators of sustainability-unsustainability – Ecosystems	
Critical social scientist	Socio-political	– Power configurations – Structural inequalities – Social identities – Languages of valuation	

Source: Author.

predicted ratio between natural water availability and water demand between 1980 and 2000. The conceptualization of urban water conflicts underlying the map was limited to a techno-scientific perspective, while experts in water politics would have placed the emphasis on different observables, such as the potential impact of water conflicts on electoral results as a proxy for decisions on investments in water services infrastructure.

Thus, to give an example, the mass mobilizations around water supply problems or the kidnapping of drivers of private water tankers by desperate civilians in the outskirts of Mexico City often have entirely different meanings for water experts, administrators, politicians or social scientists. Likewise, water administrators and financial experts concerned with the implementation of policies directed at improving the economic efficiency of basin management may view the fierce resistance of the population to handing over a community-managed water system to the municipal authorities in legal-bureaucratic terms, as a matter to be resolved by the police. In contrast, experts in electoral politics would probably interpret the event in terms of its potential impact on voters' preferences at the local and regional level, while others may read the confrontations as indicators of the correlation of political forces in the struggle for territorial control over the basin. In turn, a political ecologist would look at how confrontation over control of a local water system contributes to unearthing long-term structural conditions that have determined the exclusion of a large sector of the population from access to safe and affordable water services. Table 9.10 illustrates how different epistemic subjects may construct 'water conflict' as an object of knowledge (the table is not exhaustive and the subjects are not listed in hierarchical order).

In summary, Table 9.10 offers a schematic illustration of some of the key obstacles facing the production of scientific knowledge about water and particularly water conflicts, which may be grounded in genuine interdisciplinary coordination. It is important to emphasize that use of the term 'epistemic subjects' refers to bodies of knowledge and traditions of thought, not individuals or collective actors, as the latter may actually embody one or more epistemic subjects. The fundamental point here is that we need to give centrality to the identification of conceptual frameworks, rationalities and observables operating in the field of water research in order to achieve meaningful interdisciplinarity. On the positive side, the practical urgency for achieving a more holistic and scientific understanding of water conflicts offers unparalleled opportunities for developing truly interdisciplinary efforts, which bring together still largely unconnected and even divergent disciplinary fields involved in research on water conflicts and water issues in general.

9.5 CONCLUDING REMARKS

Undoubtedly, the main motivation behind the actions of the majority of protagonists involved in those water conflict events considered here is securing continued access to the essential services of safe water supply and sanitation. However, we have argued that such events cannot be explained away through reference to their techno-bureaucratic or administrative dimensions, or to the impact of physical-natural or socio-demographic determinations on the management of water and water services. These events form part of structural social confrontations to overcome qualitative and quantitative inequalities preventing millions from attaining full access to the conditions of life in a civilized society. Comprehension and understanding of the multidimensional character of this process has been hindered by the overriding techno-scientific and bureaucratic rationality characterizing water governance and management, which has historically contributed to rendering unobservable the social character of water. Unfortunately, the incursions of social science that have been most influential in the design of water policy since the 1980s have also tended to reinforce this unbalanced understanding of water problems. Thus, the far-reaching water sector reforms informed by these policies have strengthened technocratic tendencies in water governance and management and continue to shift the focus away from fundamental socio-political considerations that are at the root of water conflicts. In consequence, problems ranging from widespread water inequality and poverty to depletion of aquatic ecosystems are reduced to their techno-scientific and bureaucratic aspects. This diagnosis leads to 'technical' fixes such as the conversion of essential water services into private goods or re-centering water governance around 'non-political' free-market principles. This chapter attempts to contribute towards developing a conceptual framework that may help overcome this prevailing reductionism and develop meaningful trans-disciplinary coordination across the natural and social sciences in order to better grasp the multidimensionality of the processes involved.[3]

---

[3] See Castro and Heller (2009) for a recent contribution towards the critical examination of prevailing water management and policy.

This conclusion draws on a long-standing tradition in the social sciences concerned with developing appropriate cognitive structures for making observable such structural regularities as structural social conflicts – whether in relation to water or not. However, the task of elaborating adequate explanations of the causes and consequences of water uncertainty and inequality requires the development of further interdisciplinary coordination between the intellectual domains of, for instance, water engineers, hydrologists and social scientists, which to date has been a slow and relatively fruitless endeavour. The gap between the intellectual domains developed by techno-scientists and social scientists concerned with social inequality and struggle remains a major obstacle to achieving this goal. The persistence of this obstacle continues to hamper our full understanding of 'water conflicts', and consequently diminishes the chances of avoiding their negative consequences, which almost systematically affect the most vulnerable sectors of the population.

A truly inter-disciplinary approach to understanding and explaining water conflicts must strive to make observable those processes which create and reproduce the structural socio-economic and political inequalities that continue to preclude a large sector of the world's population, not only from participating in the governance of water, but also from accessing essential volumes of safe water for daily survival. This kind of approach requires addressing 'water conflicts' as an object of knowledge in its own right, and constitutes a crucial step towards transforming the unacceptable conditions characterizing the global 'water crisis'.

REFERENCES

Aboites Aguilar, L. 1998. *El Agua de la Nación. Una Historia Política de México (1888–1946)*. Mexico City: Secretaría de Educación Pública (SEP) and Centro de Investigaciones y Estudios Superiores en Antropología Social (CIESAS).

Bachelard, G. 1938. *La Formation de l'Esprit Scientifique*. Paris: J. Vrin.

Bakewell, P. 1984. Mining in colonial Spanish America. L. Bethell (ed.) *The Cambridge History of Latin America*, Vol. 2. Cambridge: Cambridge University Press, pp. 105–51.

Bartra, A. 1985. *Los Herederos de Zapata*. Mexico City: Ediciones Era.

Bartra, R. 1978. *Estructura Agraria y Clases Sociales en México*. Mexico City: Ediciones Era.

Bazant, J. [1985] 1994. From Independence to the Liberal Republic, 1821–1867. L. Bethell (ed.) *Mexico since Independence*. Cambridge: Cambridge University Press, pp. 1–48.

Boyer, R.E. 1975. *La Gran Inundación. Vida y Sociedad en México (1629–1638)*. Mexico City: Secretaría de Educación Pública (SEP).

Brundage, B.C. 1972. *A Rain of Darts. The Mexica Aztecs*. Austin/London: University of Texas Press.

Castro, J.E. 1992. El Conflicto por el Agua en México. Los Casos de Tuxtla Gutiérrez, Chiapas y Ciudad Juárez, Chihuahua, 1986–1991 (unpublished thesis, Master in Social Sciences). Mexico City: Latin American Faculty of Social Sciences (FLACSO).

Castro, J.E. 1995. Decentralization and modernization in Mexico: the case of water management. *Natural Resources Journal*, 35(3): 461–87.

Castro, J.E. 2006. *Water, Power, and Citizenship. Social Struggle in the Basin of Mexico*. Houndmills, Basingstoke/New York: Palgrave-Macmillan.

Castro, J.E. and Heller, L. (eds) 2009. *Water and Sanitation Services: Public Policy and Management*. London and Sterling, VA: Earthscan.

Chevalier, F. [1952] 1963. *Land and Society in Colonial Mexico: The Great Haciendas*. Berkeley/Los Angeles: University of California Press.

Comisión Nacional del Agua (CNA). 2001. *Programa Nacional Hidráulico 2001–2006.* Mexico City: CNA.

Comisión Nacional del Agua (CNA). 2006. Situación del Subsector Agua Potable, Alcantarillado y Saneamiento. Mexico City: CNA (available at: http://www.cna.gob.mx/eCNA/Espaniol/Publicaciones/Subsector2006/DSAPAS%202006.pdf).

Connolly, P. 1991. *El contratista de Don Porfirio. La construcción del Gran Canal de Desagüe,* 3 vols. Mexico City: Universidad Autónoma Metropolitana Azcapotzalco, División de Ciencias Sociales y Humanidades.

Departamento del Distrito Federal, Secretaría de Obras y Servicios (DDF-SOS). 1975. *Memoria de las Obras del Sistema de Drenaje Profundo del Distrito Federal,* 4 vols. Mexico City: DDF.

Ezcurra, E., Mazari-Hiriart, M., Pisanty, I. and Aguilar, A.G. 1999. *The Basin of Mexico. Critical Environmental Issues and Sustainability.* Tokyo/New York/Paris: University of the United Nations.

Florescano, E. 1984. The formation and economic structure of the hacienda in New Spain. L. Bethell (ed.) *The Cambridge History of Latin America,* Vol. 2, Cambridge: Cambridge University Press, pp. 153–88.

Fox, D. 1965. Man-water relationships in metropolitan Mexico. *Geographical Review,* 55(4): 523–45.

García de León, A. 1985. *Resistencia y Utopía. Memorial de agravios y crónica de revueltas y profecías acaecidas en la provincia de Chiapas durante los últimos quinientos años de su historia,* 2 vols. Mexico City: Ediciones Era.

Geertz, C. 1980. *Negara. The Theatre State in Nineteenth-Century Bali.* Princeton, NJ: Princeton University Press.

Gibson, C. 1964. *The Aztecs under Spanish Rule. A History of the Indians of the Valley of Mexico, 1519–1810.* Stanford: Stanford University Press.

Glick, T.F. 1970. *Irrigation and Society in Medieval Valencia.* Cambridge, US.: Harvard University Press.

González Casanova, P. 1965. *La Democracia en México.* Mexico City: Era.

Gordillo, G. 1988. *Campesinos al Asalto del Cielo. De la Expropiación Estatal a la Apropiación Campesina.* Mexico City: Siglo XXI.

Gurría Lacroix, J. 1978. *El Desagüe del Valle de México durante la Época Novohispana.* Mexico City: Universidad Nacional Autónoma de México, Instituto de Investigaciones Históricas (IIH).

Hewitt de Alcántara. C. 1978. *Modernización de la Agricultura Mexicana, 1940–1970.* Mexico City: Siglo XXI.

Hoberman, L.S. 1980. Technological change in a traditional society: the case of the Desagüe in colonial Mexico. *Technology and Culture,* 21(3): 386–407.

Horn, R. 1997. *Postconquest Coyoacán. Nahua-Spanish Relations in Central Mexico, 1519–1650.* Stanford: Stanford University Press.

Instituto Nacional de Estadística, Geografía e Informática (INEGI). 1991. *Censo General de Población y Vivienda 1990.* Mexico City: INEGI.

Jiménez, B. and Torregrosa, M.L. 2007. Water services in Mexico: are they a public priority? J.E. Castro, L. Heller and M. Drakeford (eds) special issue on 'Comparative Experiences in the Provision of Water and Sanitation Services: Challenges and Opportunities for Achieving Universal Access', *Journal of Comparative Social Welfare,* 23(2): 155–65.

Katz, F. (ed.) 1988. *Riot, Rebellion, and Revolution. Rural Social Conflict in Mexico,* Princeton, NJ: Princeton University Press.

Katz, F. (ed.) [1986] 1994. The Liberal Republic and the Porfiriato, 1867–1910. L. Bethell (ed.) *Mexico since Independence.* Cambridge: Cambridge University Press, pp. 49–124.

Kloster Favini, K. 2008. La Determinación de lucha por el agua en México – un análisis de los procesos nacionales y locales (The determination to struggle for water in Mexico – an analysis of national and local processes). PhD in Political and Social Sciences (Sociology). Mexico City: National Autonomous University of Mexico.

Knight, A. [1986] 1990. *The Mexican Revolution*, 2 vols. Lincoln/London: University of Nebraska Press/Cambridge University Press.

Kroeber, C.B. 1994. *El Hombre, la Tierra y el Agua. Las Políticas en Torno a la Irrigación en la Agricultura de México, 1885–1911*. Jiutepec, Morelos: Instituto Mexicano de Tecnología del Agua and CIESAS.

Leach, E.R. 1959. Hydraulic society in Ceylon. *Past and Present*, 15: 2–26.

Lemoine Villicaña, E. 1978. *El Desagüe del Valle de México durante la Epoca Independiente*. Mexico City: UNAM-IIH.

León-Portilla, M. 1984. The early civilizations of Mesoamerica. The Mexicas (Aztecs). L. Bethell (ed.) *The Cambridge History of Lain America*, Vol. I, Cambridge: Cambridge University Press, pp. 3–36.

Meyer, M.C. 1984. *Water in the Hispanic Southwest. A Social and Legal History, 1550–1850*. Tucson, US: University of Arizona Press.

Musset, A. 1991. *De l'Eau Vive à l'Eau Morte. Enjeux Techniques et Culturels dans la Vallée de Mexico (XVIe-XIXe Siècles)*, Paris: Éditions Recherche sur les Civilisations (ERC).

Ostrom, E. 1990. *Governing the Commons: The Evolution of Institutions for Collective Action*. Cambridge: Cambridge University Press.

Oswald, U., Rodríguez, R. and Flores, A. 1986. *Campesinos Protagonistas de su Historia: (la Coalición de los Ejidos Colectivos de los Valles del Yaqui y Mayo: una Salida a la Cultura de la Pobreza)*. Mexico City: Universidad Autónoma Metropolitana (Xochimilco) (UAM-X).

Palerm, A. 1990. *México Prehispánico. Ensayos sobre Evolución y Ecología*. Mexico City: CONACULTA.

Perló Cohen, M. and González Reynoso, A.E. 2005. *¿Guerra por el Agua en el Valle de México? Estudio sobre las Relaciones Hidráulicas entre el Distrito Federal y el Estado de México (Water over Water in the Valley of Mexico? A Study of the Hydraulic Relations between the Federal District and the State of Mexico)*. Mexico City: National Autonomous University of Mexico and Friedrich Ebert Foundation.

Piaget, J. 1971. *Structuralism*. London: Routledge and Kegan Paul.

Piaget, J. 1977. *The Grasp of Conciousness*. London: Routledge and Kegan Paul.

Piaget, J. 1978. *The Development of Thought. Equilibration of Cognitive Structures*. Oxford: Blackwell.

Sahab Haddad, E. 1991. La lucha por el agua y contra el agua en el Valle de México. CEHOPU, *Antiguas Obras Hidráulicas en América. Actas del Seminario (México, 1988)*. Madrid: CEHOPU, pp. 153–64.

Secretaría de Agricultura y Recursos Hidráulicos (SARH), Comisión del Plan Nacional Hidráulico 1981. Plan Nacional Hidráulico. Mexico City: SARH.

Secretaría de Agricultura y Recursos Hidráulicos (SARH) 1988. *Agua y Sociedad: una Historia de las Obras Hidráulicas en México*. Mexico City: SARH.

Swyngedouw, E. 2004. *Social Power and the Urbanization of Water. Flows of Power*. Oxford: Oxford University Press.

Torregrosa, M.L. (coord.) 1988–97. Research Reports, *Programa de Investigación Agua y Sociedad*. Mexico City and Jiutepec, Morelos: Facultad Latinoamericana de Ciencias Sociales (FLACSO) and Instituto Mexicano del Agua (IMTA).

Tutino, J. 1986. *From Insurrection to Revolution in Mexico. Social Bases of Agrarian Violence 1750–1940*. Princeton, NJ: Princeton University Press.

Tutino, J. 1988. Agrarian social change and peasant rebellion in nineteenth-century Mexico: the example of Chalco. F. Katz (ed.) *Riot, Rebellion, and Revolution. Rural Social Conflict in Mexico*. Princeton, NJ: Princeton University Press, pp. 95–140.

Chapter 10

Conflict *versus* cooperation between the state and civil society: A water-demand management comparison between Cape Town and Johannesburg, South Africa

Laïla Smith

Water and Sanitation, Africa, AusAID, Pretoria, South Africa

Ensuring universal access to water and sanitation services is one of the most pressing issues facing countries across the developing world. Since South Africa's first demo-cratic elections in 1994, the country has extended water services to over 18 million people (Eales, 2009). The new climate of delivery, however, is occurring within the context of cost-recovery and, as such, a vast proportion of households have been disconnected from access to water during the first decade of democracy (McDonald and Pape, 2002: 9). Given the cost-recovery principles that mediate who and how low-income households can access water, the relationship between the state and civil society is a critical variable in influencing household ability to maintain services. This chapter demonstrates that strong civil society organizations play an important role in mediating the tensions between the state's cost-recovery imperatives and low-income households' right to not only access but to *maintain* access to services.

Municipalities in South Africa have developed water demand management (WDM) approaches that vary to a great degree depending on whether they view water as a valued commodity or as a scarce resource. There are clear trade-offs in choosing economic over environmental objectives or both environmental and economic objec-tives over social objectives, such as ensuring basic access to poor residents. Cities that prioritize economic objectives by selling water without demand management, risk depleting the long-term supply that feeds their urban populations. Cities that prioritize conservation through public education at the expense of developing cost-reflective pricing may fall short in securing the revenue necessary for investing in the infrastructure required to meet long-term demands.

In the South African context, regardless of the trade-offs, all local governments are constitutionally bound to ensure equity in access to water, and as such must find a way to carefully balance each of these competing objectives. At the same time, domestic water use is the second largest consumption sector (27%) in the country and is growing faster than any other category of use (Yako, 2008). Minimizing water wastage in domestic consumption is therefore a key priority for local governments.

Water demand management has been one of the tools used by cities to balance these trade-offs. A central theme in water demand management is reducing unaccounted for water (UAW), which deals with both commercial losses due to non-payment and unbilled water (metered but not billed), as well as water losses stemming from infrastructure and household leakages. Concerning commercial losses, South African municipalities have much in common regarding the structural problems of non-payment for services. This is due to a number of reasons. The country has a 29% unemployment rate that makes it very difficult for millions of households to generate the income necessary to be able to pay for services. Hundreds of thousands of households (predominantly from previously disadvantaged communities) have been shackled with historical arrears. The bulk of these arrears stem from the 1980s linked to a period of service delivery boycotts, a form of contestation used by township communities against the poor quality of services delivered by the apartheid state (Tomlinson, 1994). A large proportion of this debt was initially based on estimated readings and was exacerbated by annual interest. Given the high sums that accumulated on low-income household bills, many poor households cannot possibly repay these debts and as such, have given up paying for their current accounts. The state has been unable to bridge the communication gap with service users in order to effectively address the underlying reasons why 'non-payment' prevails to such a significant degree in previously disadvantaged areas (Turok, 2001). The commercial challenges facing municipalities due to non-payment are compounded by the issue of water wastage due to a combination of both leaks in eroding water infrastructure networks and wasteful domestic water use, be it through community standpipes being left running or household water leaks being left unattended.

In addition, municipal neglect of infrastructure maintenance has exacerbated physical losses. Approximately one-third of the country's bulk water purchased is lost before it reaches the tap (Seago and McKenzie, 2007). Considering South Africa is the world's thirtieth driest nation due to low annual rainfall, the country can hardly afford to waste this water.

These two universal structural problems in water service delivery in cities in the south have mobilized some South African cities to experiment with various water demand management approaches in order to encourage (if voluntary) or coerce (if mandatory) households to use water wisely. This chapter will compare how two of South Africa's largest metropolitan areas have balanced these economic, environmental and social trade-offs through discrete WDM approaches. It examines the efforts of both municipalities to establish a set of values amongst predominantly low-income black communities about how to manage water wisely. The manner and conditions associated with the introduction of water reducing devices into previously disadvantaged areas – areas that have never had the benefits of citizenship under apartheid but are suddenly expected to be responsible consumers in a democratic South Africa – shapes relations between the state and civil society.

The City of Johannesburg has historically viewed water as a commodity, which has influenced its institutional choice: it set up a public water utility with a five-year management contract with Suez. This underlying approach to water as a way of bringing in municipal revenue has shaped the city's water demand management approach. The city's flagship WDM project, Gcin' Amanzi, typified in the main by the introduction of prepaid meters, is noted for achieving significant savings for the utility, but at the cost of conflict between the municipality and organized civil society. In contrast, the City of Cape Town has historically viewed water as a scarce resource and, as such, has focused on water conservation in its WDM approach. Cape Town's prioritization of a conservation approach to WDM, at the time of writing, was reflected through the intentions and policy outcomes of one of its pilot projects in a low-income black township called Mfuleni. The focus on conservation and the underlying intention of reducing household debt initially created a cooperative and non-conflictual interaction with low-income consumers, but did not deepen its relationship with organized civil society. The city failed to address the social issues, such as stakeholder engagement, in relation to the introduction of a technological device because it was viewed by the city as the means of addressing consumer education on the importance of WDM. This approach inevitably led to deep-seated tensions with civil society, and the latter was not successful, in spite of efforts to engage Parliament, in changing the city's approach.

10.1 BACKGROUND TO THE SOUTH AFRICAN WATER CONTEXT

The exclusion of the poor from access to water in rapidly growing cities in developing countries is a significant aspect of urban inequality. South African municipalities today bear the legacy of apartheid caused by extreme inequality in access to public services. A primary goal of the government led by the African National Congress during the post-apartheid period has been to redress the impacts of apartheid through a more equitable distribution of public services. After the first democratic elections in 1994, the Reconstruction and Development Programme symbolized the country's efforts towards a more egalitarian society. This policy espoused a bottom-up approach to development in terms of 'putting people first' in the remaking of the state.

In 1996, a currency crisis prompted a dramatic reversal of national policy through the 'Growth Economic and Redistribution Programme (GEAR), which espoused a fiscal discipline view of public expenditure and was 'top down' in its approach to development (Davids, 2005). The consequence was the decentralization of significant functions to local authorities, but with a simultaneous reduction in intergovernmental transfers from national to local government. Local government, during this period was mandated by the national sphere to achieve efficiencies starting with greater cost-recovery in the delivery of services (Bond, 2001; McDonald and Pape, 2002).

The impact of GEAR's drive for fiscal discipline (1996–2000) on the water sector in particular, led municipalities to experiment with various forms of public private partnerships: concessions, Build Operate and Transfer (BOT), and lease and management contracts (Hemson and Bakker, 2000; Smith et al, 2004; Ruiters, 2006). These PPP contracts have undergone significant difficulties for a variety of reasons. First, the contracts were designed with poor quality information thus making the projected targets for improvements inaccurate. Second, municipalities underwent enormous territorial and administrative transformation during this period. This created a very fluid and ever-changing municipal environment which conflicted with the underlying assumption of a static municipal context within the contracts. Third, the contracts preceded a host of municipal legislation meant to enforce the authority function of local governments and, as such, many of the contracts operated in an un-regulated environment to the detriment of both the municipalities in question and low-income residents – the intended beneficiaries of these partnerships (Smith and Morris, 2008).

By the twenty-first century, the significant taxation reforms in the South African Revenue Services (SARS) contributed to a windfall of revenue for the National Treasury (Smith, 2004). This served as a catalyst, moving the National Treasury's mandate of fiscal discipline towards expanding its role, through significant increases on public expenditures. One of the outcomes of this national shift in fiscal policy was the introduction of free basic services to assist municipalities in addressing the fiscal challenges to ensure access for the poor, regardless of (in)ability to pay. In the water sector, this translated into a Free Basic Water Policy (DWAF, 2001), which guaranteed free access to 6 m^3 a month, estimated at 200l/household/day (25l/capita/day (lcd) for a household of eight. This contextual policy background is important for understanding the efficiency and equity imperatives facing the two South African cities examined in this chapter.

10.2 CASE STUDY 1: JOHANNESBURG

Johannesburg is the financial epicentre of South Africa, the provider of 13% of the country's gross domestic product (GDP). Nevertheless, the legacy of inequality in access to services has left the city deeply polarized. Having received the bulk of Johannesburg's resources during the apartheid years, affluent white people living in the 'leafy' northern suburbs today enjoy a standard of municipal infrastructure and services on par with the world's wealthiest city districts. Conversely, the urban poor, predominantly African, live mainly in township areas to the south of the city and on the periphery to the north (City of Johannesburg, 2001).

Much of the service delivery challenge of the 1990s, post-democratic elections, was to extend infrastructure to historically unserved and underserviced neighbourhoods (predominantly African), while also improving maintenance on sunk infrastructure in historically advantaged areas of the city (predominantly white). By the 1990s, infrastructure decay was also posing an increasing threat to the return on tradable services. For example, substandard water infrastructure in many low-income areas of the city caused by historical underinvestment was contributing a high UAW rate, estimated at 43% in 2001, leading to a loss of approximately US$231 million[1] of potential annual sales at the time (Savage et al, 2003).

10.2.1 Institutional profile of water service provision in Johannesburg

The financial crisis in Johannesburg and the burden of gross inefficiencies in the operation of core services prompted a major institutional reform in the late 1990s, entitled 'Igoli 2002'. Core and non-core services were decentralized to municipal entities, where the city retained ownership.

As part of this transformation, Johannesburg Water (Pty) Ltd (JW) was set up as a utility company mandated to provide water and sanitation services to Johannesburg residents. A management contract was established at the outset of the utility's creation with the aim of integrating these different management structures, improving on operational efficiency and branding the newly corporatized water company as part of a strategy to become more customer focused. The five-year contract was awarded to a joint venture formed by Suez Group and was managed through an entity called JOWAM (Johannesburg Water Management).

Johannesburg is situated amid two catchment management areas but gets all of its bulk water from Rand Water, a Gauteng Water Board. This region is chiefly supplied

---

[1] An exchange rate of 7 Rands to the dollar is used throughout the chapter.

by the Vaal River, and additionally by water piped from the Lesotho Highlands. Both resources are stored in the Vaal Dam. There is an enormous electricity cost to pumping water 80km from the Vaal Dam to the City of Johannesburg. As such, the cost of purchasing bulk water represents approximately 80% of JW's total unit cost.

As of 2006, approximately 35% of the water purchased by the city was still lost or unaccounted for. Despite a corresponding annual loss of $29.6 million, the City of Johannesburg had not developed a formal WDM policy to encourage JW's households to consume less in order to prolong the use of existing infrastructure and defer the higher costs of building new infrastructure to pump in raw water (City of Johannesburg, 2005: 90). This was a surprising omission given the relatively high costs incurred in the purchase of bulk water.

The city's hesitation to impose WDM in more affluent and developed areas is due to the social and economic trade-offs. The tariff structure charges high rates to large water consumers to subsidize lower tariffs and the free allocation of water to two-thirds of the city's consumers. As such, Johannesburg Water's approach to water has been to sell as much as possible in order to obtain the revenues necessary to allow the current levels of cross-subsidies within the tariff system. This equity principle is critical given the high concentration of poor in the city. Nevertheless, given the considerable costs incurred in transporting water across such vast distances in order to service its residential and industrial users, neglecting the promotion on water conservation amongst affluent users is short-sighted in terms of the eventual cost-implications of having to invest in new infrastructure to augment the city's water supply.

The focus of the utility's WDM has been on low-income African areas historically neglected by the city's infrastructure investment programme. This was also where the bulk of the city's physical water losses were located. This is an important consideration given that the biggest challenge facing JW at the outset of the management contract in 2001 was a UAW rate of 43%. This problem was linked to both commercial and physical losses. With regard to physical losses in metered areas, investigations into the network indicated a loss of $10\,m^3$/km pipe/day on average. In deemed consumption areas, predominantly located in low-income black townships, the loss was double. For this reason JW focused on replacing decaying infrastructure in Soweto in particular. With over 1 million people, this area comprises approximately one-third of the city's population.

10.2.2 The Gcin' Amanzi Project

The core of the turnaround strategy for JW to become commercially viable was the eradication of its deemed consumption areas through metering in order to reduce water losses in those areas. Operation Gcin' Amanzi (OGA), which means 'conserve water' in isiZulu, was targeted as a massive infrastructure repair and upgrade of Soweto in order to address both water loss problems and non-payment. According to JW, 90% of the losses in deemed areas were concentrated in Soweto, totalling an annual loss of approximately 7 million m^3 a month. Soweto was also targeted with a prepaid meter solution as a means of addressing the low payment levels of 13% (City of Johannesburg, WSDP, 2005).[2]

---

[2] Residents living in the deemed areas of Soweto had previously been charged a flat rate of $21 a month for an unlimited amount of water.

The Gcin'Amanzi project came about as a result of the city's failure to invest historically in disadvantaged areas. The utility realized that one-third of the city's water consumers cross-subsidizing the other two-thirds was not sustainable and, thus, focused on an area of the city where improvement of water losses and payment issues would help even out the level of cross-subsidization across the city.

The project had a three-pronged strategy:

1. reduce leaks due to on-property losses through a once-off repair and replacement of domestic plumbing[3]
2. address physical losses by replacing old leaking infrastructure, installing valve pressure reduction equipment, and replacing/resizing more than 500 large meters, and
3. address the commercial losses through the introduction of a prepaid metering system across Soweto.

OGA has been bold in its objective, which is to improve the infrastructure affecting water and sanitation services to 170,000 households, translating to approximately 1 million people.[4] The operation's efficiency objectives were to reduce the UAW rate in order to decrease the purchase of bulk water from Rand Water, and to proportionally reduce inflows into sewers (hydraulic loading), lowering sewage purification costs to JW (Smith, 2005: 29).

The company's decision to combine a once-off repair of indoor plumbing fixtures to reduce leaks with the installation of prepaid meters was a novel approach in addressing both commercial losses through non-payment as well as physical losses. JW claimed that unless these two issues were addressed simultaneously, the company would continue to face high financial risks. The company believed that the introduction of prepaid meters, as opposed to conventional meters, would ensure that households took the issue of WDM seriously since they would have to bear the costs of neglecting to repair leaks on their property and wasteful use. Despite the fact that the cost to repair plumbing on private property is usually borne by the homeowner, JW was willing to cover these costs of repair as a once-off effort to complete the infrastructure upgrading in the area. The company determined that the installation of conventional meters might prevent households from assuming the responsibility to maintain plumbing fixtures over time, thus diminishing future efficiency gains.

Given the social risks around introducing this technology, JW developed pro-poor components for the introduction of prepaid meters. First, households that agreed to have prepaid meters would remain at LOS 3[5] with in-house connections and waterborne

---

[3] It should be noted that water providers in South Africa hold the responsibility for repairing all water leaks that are beyond the erf. It is the responsibility of the household to repair leaks within the house and on the erf. Both water leaks projects discussed in this chapter have been innovative in having the water provider repair on-site leaks in low-income areas.

[4] Personal communication with Kirsten Harrison, former Human Development Specialist, Corporate Planning Unit, City of Johannesburg, 10 June 2004.

[5] The City of Johannesburg established three levels of service (LOS) delivery as part of its overall strategy of addressing its water and sanitation backlogs in a manner financially sustainable for the city as well as affordable for households. Level 1 is basic with access to a communal standpipe within 200 metres and a Ventilated Pit Latrine for sanitation. Level 2 provides a yard tap with a pour-flush toilet. Level 3 is full waterborne sanitation and an in-house water connection.

sanitation, or if at a lower level of infrastructure would be upgraded to Level 3. Second, households that received prepaid meters were also offered a subsidized tariff of 20% during the project duration of four years. Third, JW argued that the free water policy could be more easily implemented through prepaid meters by automatically configuring the meters to disburse $6\,m^3$ of water a month. The company was the first in the country to implement free sanitation by offering an additional $6\,m^3$ to households that agreed to have a prepaid meter. Fourth, the policy sought to address the issue of high levels of household debt by introducing an incentive to comply with the rules associated with prepaid meters, that is, not tampering with the technology and/or not reconnecting them illegally, in return for a gradual debt write off, structured over a three-year period. The policy was endorsed by the city council in February 2004 when it agreed to write off US$1.5 million in arrears owed by the residents of Soweto due to years of non-payment for water and sanitation.

The importance of consumer 'buy-in' for the success of the OGA initiative cannot be underestimated. Thus, a significant effort was developed by the utility around consumer education in order to inform and persuade residents of the benefits of using prepaid meters. A programme focusing on the institutional and social development (ISD) of the initiative was developed in 2002 to deal with stakeholder consultations. The ISD component of the programme consisted of community liaison, awareness campaigns and marketing campaigns targeting both political and social stakeholders. A consulting company, Nemai, was contracted to manage public participation for the utility's flagship initiative. The focus of Nemai's work was to obtain political approval for the project through widespread consultation with ward committee councillors. With regard to social stakeholders, several workshops were carried out targeting the Soweto Development Forum, Community Policing Forum, Business Forum, Women's Forum, Pensioners, Youth groups and so forth. A total of 138 people attended these workshops, which remains under 5% of the residents living in Phiri, the pilot area for the programme (Nemai Consulting, 2003).

'Consultation' was also carried out by the utility's Community Liaison Officers within Soweto prior to the project being rolled out. Through these programmes, CLOs conducted door-to-door campaigns that offered training on how to use prepaid meters and handed out pamphlets explaining how prepaid meters work in order to encourage residents to sign a prepaid meter agreement form. These marketing techniques should not, however, be confused with genuine public participation processes. The utility's approach to public consultation was more akin to employing persuasion tactics, particularly when it began providing CLOs with a US$8 commission for every signed agreement completed.[6] This amounts to more money than most people in the pilot area, Phiri, receive in a single day.

Despite the questionable form of community consultation, the economic outcomes from OGA to date have been dramatic in terms of reduced water losses and subsequent financial savings to the city from a bulk water purchase perspective. The

---

[6] Knowledge of this practice was revealed to the author during a question and answer period following a Johannesburg Water presentation on Gcin'Amanzi at a public forum held by Rand Water on 17 August, 2005. It was strange that this practice was not announced in Johannesburg Water's monthly Gcin'Amanzi reports to council, of which the author would have taken note had it appeared, given that she was the municipal regulator for the utility during this period.

pre-intervention consumption for the areas piloted was 66.7 m³/property/month. Over a two-year period, this was brought down to 11.28 m³/property/month, resulting in an 83% water consumption reduction of 49 m³/property/month. The consequences of this in terms of savings for the city was a total accumulated reduction of water purchases of 58,039,120 m³ (achieved by Operation Gcin'Amanzi as it approaches 50% of its completion target in October 2007) (Johannesburg Water, 2007). This equates to a savings through reduction in water purchases of US$11.5 million. As of this date, a total of 82,845 prepaid meters were been installed out of the 170,000 required to complete the project (Johannesburg Water, 2006) contributing to an increase in water sales in the area of US$1,495,000 million over the project period. The increased sales, combined with the reduction in water losses, led to an overall saving of US$12.5 million. The project's return on investment as of 2007 was 38% for every US$1 spent to date.

10.2.3 Community response

The predominant political challenge facing JW in choosing this technology is that the city has some of the most vocal and well-organized civil society organizations (CSOs) in the country, who are adamantly opposed to the privatization of essential services. These groups see cost recovery measures such as prepaid meters as inherent to privatization. Organizations such as the Anti-Privatization Forum (APF) and the Soweto Electricity Crisis Committee have actively campaigned in Soweto against the introduction of prepaid meters and have, to a certain degree, contributed to the anxiety among some residents about the impact of prepaid meters. Such social mobilization against the introduction of prepaid meters has been an ongoing threat to JW's plans for OGA due to increasing security costs to prevent vandalism of the newly installed meters and to protect workers from harassment. JW's fears that the APF would create confusion within Phiri as to the benefits of the prepaid metering programme contributed to a position of intolerance regarding the mere presence of APF members near or on project construction sites in the pilot area. As a consequence, a number of APF members were arrested for wearing 'anti-privatization' t-shirts at construction sites and holding placards, as they were deemed by JW as 'interfering' in JW project construction plans in Phiri, during the pilot phase of the project.

This two-year sustained community mobilization against the introduction of prepaid meters was met with open hostility by the city and utility and was seen as interfering in a 'flagship' project. In essence, the Gcin'Amanzi project aimed to address the manner in which the tariff structure was being cross-subsidized, but the bigger questions of development got subverted through narrow cost-recovery imperatives. The authoritarian manner of project delivery sent a paternalistic signal to civil society that the administration would rather coerce than guide low-income households on how to manage their water supply if they could not afford to be responsible, paying consumers.

The inability of organized civil society and the city and/or utility to establish a dialogue culminated into a class action suit in July 2006 brought by five residents of Phiri, Soweto, all of whom are unemployed and living in poverty.[7] The application

---

[7] The lead applicant lived in a household of twenty people.

brought against the City of Johannesburg, JW and Department of Water Affairs (DWA) was launched under the auspices of the Coalition Against Water Privatization, a collection of community organizations and non-governmental organizations struggling against the negative effects of current water policies on the poor, and was supported by the Centre for Applied Legal Studies (CALS) at Witwatersrand University and the Freedom of Expression Institute.

The grounds for the case against DWA were that the policy of providing 6m^3 free per month (or 25 lcd in a household of eight) was insufficient to cover household needs. It was claimed that this amount is incommensurate with international norms that stipulate that at least 50 lcd is necessary to ensure basic needs (Gleick, 1996).[8]

The grounds of the court case against Johannesburg Water was that the procedural requirements of the service provider to communicate, ensure fair process, warning and prior restrictions, before cutting off water services was not followed with the introduction of prepaid meters (Mazibuko et al, 2006: 20). The complainants indicated that after the first 6 m^3, provided free every month with the prepaid meter, had been used, the prepaid meter automatically shut off without the procedural requirements of notification as established in the National Strategic Framework (DWAF, 2003) and the City of Johannesburg's bylaws.

The grounds of the court case against the City of Johannesburg were that its indigent (pro-poor) policy, which promises a service subsidy for refuse removal, sewerage and leniency on credit control provisions, including the removal of interest charged on arrears, was amended so that these benefits would be subjected to the compulsory installation of prepayment water meters. (Mazibuko et al, 2006: 23).

In July 2006 the application to the Johannesburg High Court requested that the court declare that the decisions of Johannesburg Water to limit free basic water supply to 6 m^3 per household per month and to unilaterally install prepayment meters were unconstitutional and unlawful. The court was asked to order Johannesburg Water (Pty) Ltd. to provide a free basic water supply of 50 lcd, and the option of a credit-metered supply installed at the cost of the City of Johannesburg, to the residents of Phiri, Soweto.

It is important to note that civil society's use of legal means to force an interaction with the ANC-led city council was employed after failing to establish a dialogue over two years. Numerous legal opinions were sought by non-profit organizations within the CSO coalition, but these could not be presented as evidence to influence a policy review in council matters without first being presented in a court case. Between March 2004 and July 2006, letters of demand were sent to JW, and settlement meetings were pursued without any response (Dugard, Mckinley and Setshedi, 2006). Demonstrations and protest marches were organized throughout this period with the City Council responding by: initially refusing the right to march; threatening to arrest residents should they march with a prepaid meter in hand; photographing all marchers so as to create a state file on them; and as mentioned above, arresting protesters should they appear on construction sites. Given the acrimonious relations between organized civil society and the City Council, at times represented through Johannesburg Water, it is not surprising that legal means were pursued in order to force the council into a reconsideration of the social implications of its policies relating to WDM.

---

[8] The United Nations recognized this standard in General Comment 15, November 2002, footnote 14.

The High Court judgment was given on 30 April, 2008 and argued in favour of the claimants, suggesting that the prepaid meters had been unlawful in terms of the city's bylaws. It argued that the principle of equity in the delivery of services was not applied in the city in that prepaid meters were being used as a credit control mechanism in Phiri, a socio-economically deprived area of the city, and not in more affluent parts of the city, such as Sandton. It suggested that given its affluence, the city should provide 42 lcd in order to bring greater accuracy to the notion of 'sufficient' in city's free basic water policy. The ruling put a moratorium on any further implementation of prepaid meters until the bylaws were amended.

An equally interesting dimension to the High Court ruling was the dismissal of the city's evidence that it had adequately 'consulted' the public in relation to the installation of prepaid meters, thus claiming that the procedures for installing prepaid meters were faulty in terms of administrative justice. The ruling suggested that a more meaningful form of consultation was necessary between the city and its stakeholders, thus emphasizing the concern of poor public participation mechanisms in relation to water services in the city.

Johannesburg Water and the National Department of Water Affairs appealed and the matter was taken to the Constitutional Court. The Appeal Judgment by this Court was issued in October 2009, over four years after the case was launched. The court ruled that providing $6\,m^3$ per month per household was constitutionally acceptable. The Judge made it clear that the role of the court was to test the reasonableness of state policies and the implementation of prepaid meters, and not to quantify what the 'minimum core' should be in relation to the realization of socio-economic rights.

10.3 CASE STUDY 2: CAPE TOWN

Cape Town is located in the Western Cape Province and is prone to water scarcity due to a Mediterranean climate: winter rainfalls alone supply the city's year-long water demands. Cape Town obtains approximately 75% of its raw water from DWA-owned dams and the remainder from its own sources, 15% of which come from sources within the Cape Metropolitan Area (DWA, 2006: 4). DWA identified the Western Cape Region as the first major urban region where demand for water will exceed total potential yield by 2030, depending on how fast WDM interventions are introduced (City of Cape Town, 2001: 4). Statistics from the City Engineer's Annual Reports indicate that between 1960 and 1998, demand for water escalated at double the rate of population growth. With the current implementation by DWA of the Berg River Scheme, the city's water service development plan estimates that existing water resources supplying water to Cape Town will be sufficient until 2013. As such, since the first democratic local government elections, the water department, has introduced conservation as the fundamental principle driving day-to-day operations. This approach has been facilitated in part by the governance proximity between bulk water provision and distribution – both are part of the City of Cape Town and, as such, the distribution division has a very clear sense of the short-term and long-term costs associated with increased demand.

After a decade of consumer education programmes and imposition of water restrictions, the department has successfully contributed to changes in consumer behaviour, particularly in affluent areas. Examples include fixing leaks when they arise, installing water fixtures in homes and complying with water restrictions when they arise. In 2006, Cape Town had an UAW rate of approximately 19%, which was within the range of the international norm for metropolitan areas. In line with this conservation approach, the city has tried to structure tariffs to ensure affordability, and has had the lowest tariff rates in the country for over a decade. This has in part been facilitated by cheaper bulk water, representing approximately 45% of the total unit cost of production of water,[9] that is, just less than half of what it costs in Johannesburg.

The city's continued effort to keep tariffs affordable has in part been a response to the fact that more than 39% of the city's households live below the poverty line (City of Cape Town, 2008). Low tariffs, combined with the bulk of water losses occurring within low-income areas, have significantly drained the revenues coming from the sale of water. The city's social objectives are beginning to conflict with the ability of the department to finance the institutional transformations necessary to ensure its ability to deliver services in a sustainable manner.

10.3.1   The state of municipal debt: the driver behind Cape Town's WDM approach

The City of Johannesburg created Johannesburg Water in 2001, just before the passing of the Municipal Systems Act, which created stringent steps for any municipality to establish an external entity as a water service provider. The City of Cape Town's water officials tried to emulate this institutional model but was unable to get the decision passed by council. Ultimately, council took the decision not to go the utility route but to keep this core service in-house. Unfortunately, the political instability in the city – there have been five different administrations in the last twelve years – has contributed to a lack of leadership in driving public sector reform. As a result, the department has been suffering from high levels of non-payment rates, while a growing water debt has seriously eroded revenues. Cape Town's water debt in 2007/08 was US$130 million with a provision for bad debt at US$25 million for water stemming from indigent (low-income) areas alone. The city's payment rate for 2007/08 was 76%.[10] As the water department has no autonomy over the management of its finances, the interest on this debt was automatically transferred by corporate finance on a yearly basis. This left less and less for covering basic capital expenditures to deal with the city's rapid growth, let alone the much-needed repairs and maintenance of old and collapsing infrastructure.

This financial background is necessary to emphasize the pressure on the municipality to:

1.   write off bad-debts in order to clear its balance sheets and free up revenues for capital and operational expenditures
2.   minimize water wastage given its financial implications, and
3.   increase payment levels. In low-income areas, it is implausible to expect households shackled with historical debt to pay current accounts, thus the need to write off their debts.

---

[9] Pers. comm. with Arne Singels, Water Department, City of Cape Town, 20 December, 2006.
[10] Pers. comm. with Jaco De Bruyn, Head of Water Services Planning for the Water Department, City of Cape Town, 15 November, 2006.

10.3.2 Mfuleni pilot project

In 2005, the City of Cape Town's Water Department developed an Integrated Water Leaks Repair Strategy aimed at indigent households affected by debt stemming from water leaks. The goal was to inform the council of potential synergies in the various projects underway since the last decade, and how to best bring together different departments such as billing, housing, the call centre and finance in order to ensure a holistic approach to WDM. Out of this strategy, Mfuleni was identified as a pilot project area to test and improve the strategy before rolling out the initiative across the city.

The Mfuleni pilot began in 2006 and was designed to incorporate the three objectives stated above in a small and isolated manner. The reason for starting small was to learn from the community dynamics, as well as the technical challenges of a water leak reduction initiative, in order to incorporate these lessons into the project methodology before moving into bigger and more politically difficult township areas of the city.

Mfuleni is a small isolated area on the outskirts of the city and consists of approximately 8,000 households (32,000 to 40,000 people), half of whom had leaks. Approximately 90% were home owners, 55% being in the state of site and service and therefore unbilled (Frame and Jacobs, 2006: 1). Most of the households living in the area fell within the indigent category as their household property was valued at less than US$65. As such, these households receive a rebate of US$13 which equates to 4.2 m^3 of water; with the additional free 6 m^3, supply was deemed by the department as commensurate with basic sanitation needs.

The goal of the project was to 'strive for a leak free, affordable, equitable and sustainable water supply in Mfuleni' (ibid, p. 1). During the short duration of the project (six months) approximately 42% of the properties were repaired, 75.2% of which had cistern leaks. The repair work undertaken, combined with the community education programme, resulted in a reduction of 11.4 to 18.9 m^3/household per month, representing a saving of 7.5 m^3 per month. The consultants' report on the project claimed that the key financial benefit for this initiative was the postponement of new water resource and infrastructure development, which would result in a cost saving for the city reaching up to US$3,107,000 million over a twenty-year period (ibid, p. 15), and therefore a lower tariff in the long term (ibid, p. 8).

A community empowerment programme was developed to train local plumbers to play an ongoing role in plumbing maintenance. The pilot developed a strong community development approach aiming to provide the tools necessary to enable people to start their own businesses and thus earn a livelihood by playing an ongoing role in the maintenance of water and sanitation infrastructure in the area. According to the project's consultant, in order to meet these development objectives, approximately 25% of the pilot's resources went to repairing water leaks, while 75% was allocated to changing people's habits. Emphasis has been placed on engagement with the beneficiaries of the city's upgrading effort, as this component disappeared when the pilot turned into a programme.

As with the Gcin'Amanzi project, the Mfuleni pilot had the following characteristics:

1. council would repair all infrastructure leaks in the area
2. council would repair all leaks within the household

3. if a household consumed over the free allocation of water but paid their water accounts on a regular basis for a period of six months (in Johannesburg it was three years), thus demonstrating good payment habits, the council would write off water accounts resulting from leaks on the property.

The Cape Town project team assessing the results of the pilot learnt that without imposing consequences for water wastage, people's habits would not necessarily change. While the outcomes of the water leaks reduced household consumption to $11.6\,m^3$ a month, there were no guarantees that households would continue to take responsibility in repairing leaks on a voluntary basis. In comparison, by virtue of introducing prepaid meters, JW ensured that once the city invested in repairing leaks within the household, the latter would then have an incentive to maintain the household infrastructure for water and sanitation as it would have to bear the immediate cost for failing to do so. The rationale is that a prepaid meter rapidly reveals to the consumer the household cost of not repairing a leak when it arises. With conventional meters, the costs associated with household leaks, for which the water bill will arrive some time later, make the matter somewhat less urgent for the household to address. This has implications for low-income households who cannot afford to incur debt.

10.3.3 Water saving device options

The lesson learnt from the Mfuleni project was that water wastage in a city facing scarcity of water supply must be addressed. This was translated into a water demand management policy where the city would repair water household leaks in low-income areas, but in return for this investment households had to bear consequences for not attending to water leaks. The city explored a number of different options before adopting widescale implementation.

First, disconnections were no longer permissible as they did not permit implementation of the free allocation of $6\,m^3$. Second, as part of the council's credit control policy, flow restrictors were installed in households as a consequence of non-payment in order to ensure the free allocation of water. The restrictor provides 25 litres a day but on a drip system. This technology has been criticized by numerous civil society organizations in terms of contravening national norms regarding water pressure (10 litres per minute).[11] Third, prepaid meters had been explored but for a number of reasons had not been applied to low-income areas. In 2001, prepaid meters were implemented in a low-income coloured township, Klipheuwel, as a pilot and encountered a number of operational problems. The use of prepaid meters was explored under the Democratic Alliance for roll-out in several other low-income, high non-payment areas at the beginning of 2006 (Bond, 2006: 11). An anti-prepaid meter campaign was created through a coalition of non-governmental organizations, such as the South African

---

[11] This restrictor has been applied in municipalities across the country as a means of enforcing credit control for non-payment. This practice has displaced water disconnections in most municipalities. The latter, under current legislation, would be illegal as it prevents a household, if unable to pay, from receiving free basic water.

Municipal Workers Union (SAMWU), the Anti-Eviction Campaign, the Coalition of South African Trade Unions (COSATU)[12] and the Environmental Monitoring Group (EMG). The campaign strategically targeted the then ANC mayor, Nomaindia Mfeketo. Towards the later part of 2005, she publicly announced that 'the Council, [has] taken a policy decision not to implement the prepayment water meter system in the City [. . .] Access to water is a critical right for our people and no measures which will have a negative social impact on our communities will be implemented by this administration as long as I am mayor'.[13]

This announcement constituted a significant victory for the coalition of non-governmental organizations and their ability to influence council policy through strategic political targeting, having previously had no success in trying to build a sustained dialogue with the water department.

Predictably with Cape Town politics, the ANC lost two months later and with the opposition party ruling council, the water department was tasked in mid-2006 with introducing prepaid meters in order to improve payment rates for the city. The administration was devious in its interpretation of this mandate and given its sensitivity to the negative implications of prepaid meters on the poor, implemented prepaid meters in middle to upper income new townhouse complexes only.

Flow limiters, rather than prepaid meters, were selected as the optimal water saving device by the water department, imposed as a punitive measure on indigent households that did not pay for water.[14] Flow restrictors work with conventional billing systems, but the meter itself is integrated into the household meter and obfuscates the ability of the household members to read their own meters. The flow limiter was set at 350l/day in line with the city's free basic water and sanitation policy. If the household consumed over the 350kl amount or failed to pay for consumption over and above this, a flow trickler to reduce water pressure kicked in.[15] The water department viewed the flow restrictor as an ideal device as it alerted both the department and the household to a water leak in terms of average monthly consumption rates suddenly increasing. This would then allow the city to intervene if the household failed to address the situation by restricting the water flow.

This device was not inexpensive. At US$32 per unit, the department invested an initial US$65 million in capital expenditures to install this device to all indigent households across the city in order to 'assist' them in keeping up with their ability to pay for current accounts by not incurring high water bills associated with water leaks. By 2008, 30,000 devices had been installed.

The department's view was that flow limiters would help low-income households to remain debt free by providing the capability to detect the existence of a water leak and

---

[12] COSATU forms part of the tri-partite alliance governing the country along with the South African Communist Party and the ANC.

[13] Cape Argus, 21 August, 2005, p. 2.

[14] These just happened to be concentrated in township households with high water debts, largely resulting from a legacy of inferior quality infrastructure.

[15] The trickler is commonly used across the country as a credit control mechanisms for households that do not pay for water services. It is illegal for municipalities to disconnect households for non-payment as, according to national policy, all households, and in some municipalities, low-income households only, are eligible to a free 6 m^3 per month.

swiftly repair it in order to avoid high water bills. The real motivation was pressure from the finance department to reduce the city's bad debt (largely due to poor revenue collection). Nevertheless, the water department estimated that once these devices were installed across indigent households, the city would see a further economic benefit of approximately US$130,000 per month by virtue of eliminating monthly meter reading and invoicing of accounts. The symbol that so often characterizes the relationship between the city and a citizen, a service delivery bill, would simply be eliminated. This was the administration's crowning technological solution to a situation that was essentially about people too poor to pay for water services who required a much deeper interface with the council in order to resolve the problem of accumulated debt associated with inability to pay.

10.4 CSO RESPONSE

Civil society organizations protested against the water demand management device on several grounds. Opposition was led by the Western Cape Water Caucus and included the SAMWU, the EMG, Illitha Lomso, the Environmental Justice Network Forum and WESSA. First, they claimed that the meter was faulty as it did not allow households to actually read the meter, and thus play an active role in managing their water demand. Second, there was no city-wide engagement with civil society active in the water sector as to the reasons why this device was selected over other WDM approaches. Third, they claimed gross negligence in terms of administrative justice regarding the lack of consultation with the majority of households prior to the installation of these meters. The city had outsourced the implementation of the device to small private companies and had neglected the oversight required to ensure these companies followed just procedure by consulting with households as to whether they wanted the devices installed or not.

The mobilization of civil society escalated the matter to Parliament where the national regulator and the city were drawn into meetings with the portfolio committee responsible for water. Both the national regulator and Parliament expressed concern about the lack of public consultation around the implementation of the device. They imposed a moratorium on implementation of the device in late 2008 pending further research. When the research report was conducted, the water officials of the city viewed it as biased, ignored the moratorium, and proceeded to roll-out the devices, thus creating tension between the national regulator and the city.

10.5 THE POLITICAL TERRAIN OF THE TWO CITIES

In order to understand why these different approaches to WDM were selected in both Cape Town and Johannesburg, it is important to briefly review the political environment of both city councils. This can help shed light on why the commercial drive in Johannesburg was politically endorsed despite growing community conflict, while the uncertainty of the Cape Town political environment gave the administration greater authority in devising technical solutions to poverty-related problems. The political uncertainty in Cape Town also forced officials to tread more cautiously, and often deviously, as to how they introduced water conservation policies and plans for the installation for devices into the political system and historically disadvantaged areas.[16]

In a sense, comparing the WDM approaches of two municipalities through the Gcin'Amanzi and Mfuleni initiatives is like comparing apples and oranges. At the time of writing, the former entailed an annual US$42 million investment from the City of Johannesburg as part of a three-year strategy to turn around the dismal financial situation of the city's water services. Yet, by ignoring CSO opposition, even though the city won the constitutional court case, the continued roll-out of pre-paid meters was indeed stalled because of the Mayor's understanding of the social consequences of continuing to install pre-paid meters in the lead-up to local government elections in 2011.

In contrast, the Mfuleni pilot in Cape Town was a US$780,000 project over a six-month period – a tiny pilot in comparison to Johannesburg's. Yet, the lessons from this small pilot laid the foundations for a city-wide installation of flow restrictors as a water demand management device that has continued as a council policy and bylaw, despite civil society's discomfort with this approach. Clearly the scale and magnitude of Johannesburg's WDM initiative makes a technical comparative analysis of achievements rather inappropriate. There is, however, value in comparing the policies underlying these two initiatives and how these policies have shaped state/civil society relations.

The ANC is the dominant political party in South Africa and has had a majority in Johannesburg since the first democratic local government elections. The City of Johannesburg is blessed with mature politicians by virtue of the decade-long dominance of ANC rule. This has enabled the city administration to be bold in formulating a long-term vision for the transformation of the city, and to have council politicians work in concert to translate this vision to the grassroots level through the ward committee system, a formally prescribed method for public engagement. This political maturity and unity is also cursed with an arrogance that translates into little space for dissension from the majority view. Ward committee structures in Johannesburg tow the party line and do not allow for debate within the ward committees that extend beyond the party's policies. For example, ward committees are not spaces for residents of Soweto to debate whether they believe prepaid meters are appropriate for their community or not. The ward committee would perceive any perspective critical of prepaid meters as anti-ANC, since the party endorsed the use of this technology in the city.

It is out of this context of arrogance that the Anti-Privatization Forum was born. Trevor Ngwane, a former ANC councillor from Pimville, Soweto, did not believe in the city's project to commercialize basic services and was forced out of the party because of his dissenting views. He went on to form the Anti-Privatization Forum, which has become one of the most significant thorns to the ANC in Johannesburg and

---

[16] It must be noted that this piece was written just as the main opposition party, the Democratic Alliance, came to power through a coalition in 2006. In 2009, the DA also took over the Western Cape Province, thus for the first time in the country, an opposition party is ruling one of the country's most wealthy provinces and its main metropolitan area. Since this article was written, a greater political certainty and confidence in taking longer-term decisions has begun to emerge.

operates nationally as a coalition of civil society organizations against the commercialization of services (Ballard et al, 2006).

Given the magnitude and levels of investment required by the city, JW successfully enlisted support not only from the Mayoral Committee, but from councillors across Soweto. This solid political support may well have contributed to the utility seeing involvement or engagement with organized civil society as less important. Because the utility had political support for Gcin'Amanzi, regardless of community unease relating to the introduction of prepaid meters, it may well have chosen to ignore negative responses from organized civil society. While ongoing civil society conflict had serious costs to the project in terms of delays in implementation and increased security costs, council support meant that JW responded with no tolerance to these skirmishes, police intervention, and an unwillingness to listen to the issues being raised.

This unwillingness of the ANC-led city council to meet with CSOs was in part due to the latter being represented and manipulated by the political agenda of the Anti-Privatization Forum. The APF's dogmatism in conflating council's cost-recovery efforts with privatization and silencing a plurality of perspectives within the coalition of Community-based organizations (CBOs) and NGOs that look at service delivery issues beyond the privatization lens, has helped to delegitimate itself as a true 'representative' of civil society. While numerous NGOs and CBOs were associated with the coalition that mounted the court case, the APF was the single driver behind the social mobilization against Gcin'Amanzi. Its unconstructive method of engagement, such as stoning the Mayor's house in opposition to the city's commercialization drive, resulted in the city and utility losing interest in meeting with civil society organizations. The city perceived these requests for engagement to be largely politically driven by the APF, which was overtly anti-ANC.

The political support given by the ANC to the Gcin'Amanzi project has enabled the utility to weather the storm of CSO conflict by simply escalating the costs to the council: JW asked the council to finance the increased annual costs for running the Gcin'Amanzi project. This strategy worked until the point where CSOs took the conflict into the legal realm, challenging the manner in which current legislation was being implemented. The civil society conflict that had emerged from the utility's commercial approach to the delivery of a basic need threatened to force the utility and the city to pay heed to the legislative requirements that they are bound to, regardless of to the importance of turning around the financial situation of the water services. The CSOs in Soweto were reacting not only to the dire implications of prepaid meters being employed in a predominantly poor area, but also to the city's authoritarianism towards dissent against its 'flagship' project; they claimed it contravened legislation regarding how the city is required to interface with its residents. In essence, the CSO outrage is that first generational rights – political rights, were being limited by the state.

While the court case was not won by the claimants, it forced the City of Johannesburg to do its homework in terms of researching additional provisions to handle contingencies when the provision of $6\,m^3$ of water, disbursed through prepaid meters, was insufficient for poor households to cope. This research forced the city to revise its policies in terms of financing a social package that subsidized JW to cater for households facing unfortunate circumstances, such as HIV-affected households, large family size and households that have to cater for funerals, enabling them to access higher amounts of free water when using prepaid meters. To quote a Johannesburg official responsible for managing the implications of the court case for the council, 'the court case has forced us to do the research homework that we should have done before the start of the Gcin'Amanzi project'. Unfortunately, political pressure to turn around Johannesburg Water's financial situation did not permit such social considerations. Unfortunately, Cape Town also did not learn this important lesson.

The outcome of the Constitutional Court's ruling in relation to water services in Johannesburg was that the courts should not dictate the content of socio-economic rights. The case highlighted the point that rulings in relation to complexities of budget decisions determining 'access within reasonable means', policy research and review processes are beyond the court's institutional capacity (McKaiser, 2009). Yet at the same time, the court is meant to ensure government accountability by allowing citizens to come before it to test the state policies against standards of reasonableness. In sum, the manner in which the case was mounted drew the courts too deeply into commenting on state policy and thus it retreated to the form in which socio-economic rights should be tested, rather than deepening its function in facilitating the enactment of these rights.

In the case of the City of Cape Town, the political situation was anything but stable in the post-apartheid period. The ANC won a majority in the first democratic elections in 1997 and lost control of the City in 2000[17] to the Democratic Alliance (DA). Since 2000, the balance of power has shifted back and forth from the DA to ANC in 2003, and back in 2006 to the DA which leads the city since then. This is perhaps normal in most democratic countries, but it has been an anomaly in South Africa, where the ANC has dominated all metropolitan areas since the country became democratic in 1994.

This political instability has been disastrous for the city in terms of developing a long-term vision and thinking strategically about the state of its infrastructure services. To quote the *Mail and Guardian* 'Cape Town has functioned in spite of its government, not because of it. It is a governance tragedy' (Edwards, 2006: 10). Against this climate of uncertainty, no political party has been willing to take bold steps to address structural or latent crises in infrastructure services. This environment, in the face of looming crisis, has led to an increasingly demoralized cadre of officials that have begun to work by stealth in trying to bring innovative ways to address growing service delivery problems, rather than openly engaging with politicians to seek their support for major structural reform.

It is perhaps because of this situation that the small location of Mfuleni was chosen to start a rather sophisticated and far-reaching initiative from what was an innocuous pilot project. At the time, political instability prevented the water department from putting forth a far reaching strategy to address these structural problems, as this would require significant investment from the council fiscus plan. Until the DA took control of the city in 2006 and the province in 2009, no one party in Cape Town would commit itself to sacrificing long-term public investments to resolve service

---

[17] This was the result of the former New National Party (the National Party had governed the country for sixty years under apartheid) forming a coalition with the Democratic Party, a young liberal party that largely represented a middle-class white constituency, to create the Democratic Alliance. The DA has strong support in both the coloured (majority population in CT) and white areas of the city. The Democratic Alliance has had a much more overt orientation towards the commercialization of council services than the ANC.

delivery problems, such as infrastructure investment (where there are no ribbon-cutting photo opportunities), if it did not know whether it was going to remain in power beyond the next year or two. This situation offers the lesson that perhaps significant public sector reform requires a degree of political stability in order to move forward. Now that the DA party is firmly ensconced in the Western Cape, it may bring about a measure of political stability that the province has not known since the advent of democracy in 1994. However, it is yet to be seen whether this will have redistributional impacts.

10.6 CONCLUSION

This chapter has clearly not addressed the access to services' issue: both metropolitan areas have good coverage rates by international standards. In 2006, Johannesburg had reached 95.7% coverage for water and 87% for sanitation,[18] while Cape Town had 100% for water and 95%[19] for sanitation.[20] While access is a critical issue, this chapter has focused rather on looking at conditions under which low-income households are able to *maintain* access to services. Both JW and Cape Town's water departments see the use of water reducing devices as a way of assisting 'uneducated' low-income residents to better manage their water use. In Johannesburg and Cape Town residents and CSOs responded conflictually to this paternalism.

Both Johannesburg and Cape Town have put in place a set of policies that aim to assist the poor in terms of accessing free basic water and managing household consumption through the introduction of water-saving devices. Both cities have structured their tariffs so that the cost of water for low-end consumers is significantly subsidized. While JW has higher tariffs than those of Cape Town, it subsidizes the cost of water for the first $30 \, \text{m}^3$ of consumption. In Cape Town, the cost of free basic water is only recouped within the tariff after the first $20 \, \text{m}^3$ of water, thus keeping water relatively affordable for poor households that manage to keep their demand within this level of consumption.

The efforts of these two cities to develop water policies that cater to some of the needs of low-income residents has not translated into a willingness to engage with civil society organizations, which are at times the best equipped to voice the needs of the poor. There has been no genuine engagement with CSOs in either city with regard to

---

[18] These figures were provided by Johannesburg Water, October, 2007. Coverage rates for sanitation in Johannesburg are defined according to the number of households with access to a ventilated pit latrine (Level of Service 1, according to national standards set by the National Water Regulator). This figure excludes households serviced by tankers or chemical toilets.

[19] Coverage rates for water are defined in Cape Town as twenty-five households accessing a single tap within 200 metres. For sanitation it is five households sharing one pour-flush toilet. According to this definition, the City of Cape Town does not count backyard shacks as part of their backlogs because they can access water through the main account holder. The city admits that backyard shacks remain excluded from accessing the free basic water policy but do not see this as a priority. Pers. comm. with Jaco De Bruyn, official for the Water Department, City of Cape Town, 15 November, 2006, and updated for statistics, 5 October, 2007.

[20] Pers. comm. with Jaco De Bruyn, Head of Water Services Planning for the Water Department, City of Cape Town, 15 November, 2006.

water services. In the case of Johannesburg Water, organized civil society did have a voice and requested a dialogue with the service provider and the city, but was denied this because the APF was not recognized by the city as a legitimate representative of civil society. This dialogue may well have not taken place because the manner in which the APF chose to engage was dogmatic in conflating the utility's cost-recovery mechanisms as synonymous with privatization. Nor was the APF terribly strategic in devising proactive mechanisms to enlist the council and utility to engage. The council left the resolution of this problem to its water utility, which has to date not developed any sort of water user forum to engage with civil society, even though this is part of its contractual obligation through its service delivery agreement with the city (Smith and Morris, 2008: 15). The utility was also limited by local ANC politicians from efforts to engage more widely with civil society, as these politicians saw themselves as the legitimate voice in representing the public in Soweto. Local politicians, during this period, certainly played a gate-keeping role in preventing the utility from engaging more widely.

Johannesburg council, after repeatedly being told that their approaches to cost-recovery are synonymous with privatization and are exacerbating inequality in access to services, chose not to engage. The result of this breach in communication between the public, utility and council was a court case that forced a dialogue on the policies of basic provision of water services to the poor. It is questionable whether the court case could have emerged without the strong support of a capable coalition of organized civil society organizations, namely a rights-based law centre, Centre for Applied Legal Studies, associated with an academic institution, Wittwatersrand University. Nevertheless, in a country where socio-economic rights are in their infancy within the judiciary, there is perhaps a need for social justice litigation to be prepared more strategically by civil society organizations.

In the case of Cape Town, the water department, showed initial cautiousness in experimenting with different approaches to WDM and the introduction of water-saving devices. It learnt from these experiments that there need to be consequences for not using water wisely and proceeded with a technological approach without any broad-based engagement with civil society organizations, nor meaningful consultation with ordinary citizens as to why or where these meters were being installed. The Water department was unwilling to meet with organized civil society within the sector to discuss broad-based strategic issues relating to WDM, let alone the social limitations of the water-saving devices it had introduced. Despite its efforts to develop policies that are sensitive to low-income households, the refusal to engage with CSOs in a regular and sustained form, illustrated a split between the state and civil society manifested in the absence of a relationship between CSOs and the water department.

Johannesburg and Cape Town city councils are mandated by the constitution and national legislation, such as the Municipal Systems Act, to meaningfully engage with communities with regard to how services are delivered. The most unfortunate outcome of the Constitutional Court decision in relation to Phiri is that it ignored the Johannesburg High Court questions as to whether consultation with the public where 'recipients are informed of the new policy' is sufficient to be called meaningful participation. The national regulator's research on the Cape Town water demand devices came to a similar conclusion: that the process of implementation of the WDM policy was flawed, primarily because it failed to enable adequate public consultation or household choice. The prescribed and paternalistic manner in which community consultation has occured in both cities, precludes CSOs from effectively engaging in strategic decision-making on service delivery issues. Both cities are using technology to mediate the state's hard hand in forcing low-income households to conserve water with a cost-recovery spin, be it through prepaid meters or flow restrictors. The inability or unwillingness of both cities to engage with civil society to find a middle-ground in forcing households to take greater responsibility for water wastage has left the public in both cities with a bad taste regarding rising levels of authoritarianism. This consequence has led to increasing frustration among low-income residents at the poor levels of communication by local government. It is no wonder then, that by 2008, 2009 and 2010, this frustration has spread across the country and has fuelled a significant increase in service delivery protests.

REFERENCES

Ballard, R., Habib, A. and Valodia, I. 2006. *Voices of Protest: Social Movements in Post-Apartheid South Africa*. Durban: University of Kwa Zuli-Natal Press.

Bond, P. 2001. *Elite Transitions*. Johannesburg: Zed Press.

Bond, P. 2006. Affidavit, High Court of South Africa, Case no. 06/13865, July, pp. 1–15. Available at: http://www.law.wits.ac.za/cals/phiri/BondaffidavitFinal.pdf

City of Cape Town. 2001. Water Services Development Plan. Comprehensive WSDP, November, pp. 1–82.

City of Cape Town. 2008. Characteristics of Households Living in Poverty.

Document available at: http://www.capetown.gov.za/en/stats/CityReports/Documents/Characteristics_of_Households_Living_in_Poverty_18_09_2008.pdf.

City of Johannesburg. 2001. *Johannesburg, an African City in Change*. Cape Town: Zebra Press.

City of Johannesburg. 2005. *Water Services Development Plan*. Johannesburg.

City of Johannesburg. 2006. Portfollio Committee Report for Municipal Entities. September.

Davids, I. 2005. Voices from Below – Reflecting on ten years of public participation: the case of local government in the Western Cape Province. Foundation for Contemporary Research, pp. 1–128. Available at: http://www.fcr.org.za/publications/list-of-publications/voices-from-below.pdf/view

DWAF. 2001. *Free Basic Water Policy*. Johannesburg: DWAF.

DWAF. 2003. *Strategic Framework for Water Services*. Johannesburg: DWAF.

DWAF. 2006. *Western Cape Water Summit*. Johannesburg: DWAF.

Dugard, J., McKinley, D. and Setshedi, V. 2006. *Media Summary*, Johannesburg, 12 July, pp. 1–2. Accessed at: http://www.law.wits.ac.za/cals/newsitems/phiri.pdf.

Eales, K. 2009. *Strengthening the provision of sustainable water services: a concept paper, commissioned by Integrated Water Sector Support*, Department of Water Affairs, pp. 1–36.

Edwards, M. 2006. The mismanagement of the mother city, *Mail and Guardian*, 3–9 November, pp. 9–10.

Gleick, P.H. 1996. Basic water requirements for human activities: meeting basic needs. *Water International*, 21(2): 83–92.

Hemson, D. and Bakker, K. 2000. Privatising water: BoTT and hydropolitics in the new South Africa. *South African Geographical Journal*, 82(1): 3–12.

Frame, J. and Jacobs, R. 2006. *Mfuleni Integrated Water Leaks Repair Pilot Project*. City of Cape Town, Water Services, WDM.

Johannesburg Water. 2007. *Progress report: Operation Gcin'amanzi*. Entities Portfolio Committee. Johannesburg, 15 March.

Johannesburg Water. 2006. *Water Loss Reduction Report for August.* Entities Porfolio Committee, Johannesburg, October.

Mazibuko, L., Munyai, G., Makoatsane, J., Sophia, M. and Vusimuzi, P. 2006. Founding Affidavit, Johannesburg High Court, case no. 06/13865, pp. 1–48. Available at: http://www.law.wits.ac.za/cals/phiri/MAZIBUKO_Founding_affidavit_Final.pdf.

McDonald, D.A. and Pape, J. 2002. *Cost Recovery and the Crisis of Service Delivery in South Africa.* Cape Town: Human Sciences Research Council Publishers.

McKaiser, E. 2009. *Court strikes right balance on water for poor people.* Business Day, 13 October. Accessed at http://www/businessday.co.za/articles/Content.aspx?id=83847.

Nemai Consulting. 2003. Public Participation report, Operation Gcin' Amanzi, Johannesburg.

Ruiters, G. 2006. The political economy of public-private contracts: urban water in two eastern cape towns. D. McDonald and G. Ruiters (eds) *The Age of Commodity: Water Privatization in Southern Africa.* London: Earthscan, pp. 148–65.

Savage, D., Gotz, G., Kihato, C. and Parnell, S. 2003. *Strategic Review of iGoli 2002.* External Review Team report, Office of the City Manager, Johannesburg.

Seago, C.J. and McKenzi, R.S. 2007 An assessment of non-revenue water in South Africa. WRC TT 300/07.

Smith, L. 2004. Cooperative governance: relations between the South African Revenue Service and the Ministry of Finance in addressing the redistributive agenda. *Tax Notes International*, 12 April issue.

Smith, L. 2005. Neither public nor private: unpacking the Johannesburg Water Corporatization model: Programme on Social Policy and Development, Paper No. 22. Geneva: UNRISD.

Smith, L., Gillett, A., Mottiar, S. and White, F. 2004. Public money, private failure: the limits of market-based solutions in water delivery. D. McDonald and G. Ruiters (eds) *The Age of Commodity: Water Privatization in Southern Africa.* London: Earthscan, pp. 130–47.

Smith, L. and Morris, N. 2008. Are municipalities entities a panacea to service delivery constraints. M. Van Donk, M. Swilling, E. Pieterse and S. Parnell (eds) *Consolidating Developmental Local Government: Lessons from the South African Experience.* Cape Town: UCT Press.

Tomlinson, R. 1994. *Development Planning.* Oxford: Oxford University Press.

Turok, I. 2001. Persistent polarization post-apartheid? Progress towards urban integration in Cape Town. Discussion Paper. Cape Town: Urban Change and Policy Research Group.

Yako, P. 2008. Water Sector Leadership Group Meeting, presentation made by the Director General of the National Department of Water, 14 November, 2008.

Chapter 11

Conflicts of influence and competing models: The boom in community-based privatization of water services in sub-Saharan Africa

Sylvy Jaglin[1] and Anne Bousquet[2]

[1]University of Paris-Est Marne-la-Vallée, Latts, France
[2]Global Water Operators Partnerships Alliance, UN-HABITAT, Nairobi, Kenya

In the cities of sub-Saharan Africa, conventional water supply networks serve only a minority of the population. Although aggregate data suggest that approximately 60% of city dwellers have access to potable water, these figures have only limited significance and cover a whole range of different methods of supply (private connections, standpipes and enhanced water supply points at a 'reasonable' distance of 200 m from the home).[1] At the middle and lower levels of the urban hierarchy and even in certain capital cities (see Table 11.1), for example, conventional water supplies reach less than 30% of the population. Three main configurations can currently be identified.

In large cities (of more than 500,000 inhabitants),[2] public services operating with a monopoly remain unable to supply the entire population, despite being the subject of delegation contracts for the last twenty years. In the suburbs, unofficial water supply markets (resale and door-to-door vendors), are dynamic and thriving, but expensive (prices are sometimes prohibitive, and always higher than those of conventional networks, even though water quality varies tremendously). These markets target those excluded from the network, as well as those who are connected but suffer from poor service (cuts or extended drops in pressure).

In medium-sized cities (50,000 to 500,000 inhabitants), conventional networks serve an even smaller percentage of the population and suffer from serious profitability problems. In countries with a centralized service, national operators, whether public or private, usually struggle to fulfil their responsibilities. Geographical distance and functional shortcomings, as well as the weight of geopolitical factors, produce systems in which certain people are neglected. In decentralized countries, the development of

---

[1] These figures also include theoretical coverage rates (number of water supply points multiplied by the number of people supposed to use them divided by the total population), even though the list of infrastructure rarely takes into account the actual condition of the installation and the level of service provided (Valfrey, 2005: 7).

[2] The demographic boundaries are far from stable and are given here as a rough indication. They correspond partly to those used, for example, by PsEau in its presentation at Africities 2006 (Nairobi) www.pseau.org/event/africites_4/index.htm.

Table 11.1 Network coverage rates and water access methods in some African cities (% of households, 1999)

	Abidjan	Nairobi	Dakar	Kampala	Dar es Salaam	Conakry	Cotonou	Ouagadougou	Bamako
Population (millions)	2.8	2.0	2.2	1.1	2.8	1.1	1.1	1.0	1.0
Operator	SODECI	NCC	SDE	NWSC	DAWASA	SEEG	SBEE	ONEA	EDM
PC	76	71	71	36	31	29	27	23	17
SP	2	1	14	5	0	3	0	49	19
IS	22	27	15	59	69	68	73	28	64

PC = private connections; SP = standpipes; FI = independent suppliers and free traditional sources.
Source: Collignon, Vézina (2000).

supply networks is being held back by a lack of public authority funding and resources. Here also, local operators (NGOs, private entrepreneurs) have taken over, but are rarely linked to the official service.

Finally, as distance from the conventional network increases – in small urban centres (3,000 to 50,000 inhabitants) and on the edges of larger cities often outside the area served by the national operator, and certainly beyond the scope of short- and medium-term plans to extend the network – independent supply systems delivered as part of community development projects[3] are being presented as a sustainable method of collective access to drinking water. These systems are the subject of this chapter.[4]

Studies on water services in developing cities barely mention this issue, largely focusing on the 'privatization' of services that benefits large firms (Lorrain, 1999, 2001; Shirley 2002; Bakker, 2003, 2007; Breuil, 2004). Having failed to make conventional networks accessible to all, reforms have significantly altered the general context outside those networks by making it easier to adopt the principle of community-based management, not only in rural zones where it has a long history, but also, more recently, in urban areas (Page, 2003). Over the past thirty years, projects in this field have nevertheless shown a high level of dynamism and have given rise to numerous innovations, both technical and organizational. Although they have yet to radically transform the geography of access to potable water, they nonetheless point the way towards an alternative method of water supply – one which is decentralized, involves multiple operators and is independent of the slow progress of conventional urban services, while broadening accessibility to services in geographical terms, although not improving social equity.

In order to analyse these projects and the services they have produced, this chapter deliberately concentrates on projects that have attracted little media coverage compared to recent highly publicized reforms. We have examined the diversity of identifiable developments in water services, in small towns in particular (Etienne,

---

[3] The term 'community' is used here in a broad sense to refer to villages or urban districts, depending on the situation. It includes the notion of proximity (social and geographical), which is given high priority in the implementation of many projects, particularly by NGOs.

[4] As well as those listed in the references, this article was written with reference to the field work carried out by S. Jaglin (2005a), Anne Bousquet (2006) and Anne Belbéoc'h (Jaglin et al, 2011).

1998; Coing et al, 1998; Page, 2003; UNDP, 2006), focusing on the work of international solidarity organizations and small informal entrepreneurs rather than large international firms.

The first part of the chapter describes some of the main characteristics of drinking-water supply systems set up under 'community-based privatization'[5] schemes. Acknowledging the spread of these schemes, the second part examines the sustainability of water services in this context. The third part discusses the dissemination of the community-based privatization model[6] and its ideological foundations. It points out, for example, that this model is thriving in sub-Saharan Africa, funded by major sponsors and NGOs alike in a context of decentralization, in contradiction, if not in conflict, with the idea of strengthening local public authorities and municipal project ownership. Nevertheless, practical analysis on the ground highlights the limitations of community-based management of urban amenities, particularly the instability and unsustainability of services (Ayee and Crook, 2003; Bousquet, 2006; Coing et al, 1998; Jaglin, 2005a, 2005b; Manor, 2002; Meï, 2008). Although the community-based privatization model is dominant, it is certainly not always suitable for the development and appropriation of the skills necessary for sustainable service management. Why, then, is it becoming so common in the urban water sector, marginalizing municipal services in the process?

11.1 WATER SUPPLY SYSTEMS IN URBAN AREAS WITH NO NETWORK CONNECTION

The projects we have studied and their alternative water supply systems demonstrate, over a fairly long period (around thirty years), an attempt to consolidate a water supply model that differs from that of the large technical network, which has dominated for many years. Although the latter model has not been totally abandoned, recent reforms of conventional networks aimed at modernizing their management by involving the private sector, have failed to produce undeniable progress in terms of supplying all urban areas. Other collective water supply models have therefore been developed.

11.1.1 Dispelling a myth: the spread of public-private partnerships from large cities

In the main, the participation of the private sector in recent decades has meant large delegation contracts involving international private companies (public-private

---

[5] The expression is deliberately provocative and is meant to emphasize the fact that community-based projects, rather than providing an alternative water supply management system, are used, on the contrary, as a form of public-private partnership (in this case between a sponsor, a resident community and small private entrepreneurs) and a vehicle for the introduction of market principles (purchase of water based on volume consumed, constitution of renewal provisions, keeping of operating accounts, subcontracting and purchase of services from the competing private sector, etc.) in communities in which water generally used to be free and rationed.

[6] The term 'model' refers here to an abstract structure that enables us to distinguish some of the main approaches in the water sector, based on their own characteristics and action concepts, thereby highlighting the differences between them. We are proposing these models, which are both provisional and refutable, in order to organize our observations, but they are not necessarily recognized by the relevant stakeholders.

partnerships or PPPs). The success of these contracts is currently fairly limited in sub-Saharan Africa (Jaglin, 2006). Firstly, the disadvantages of delegation are often magnified in these countries (information imbalance, lack of transparency, insufficient investment, etc.) and contractual provisions have proved to be an inadequate tmethod for ensuring that risks are shared appropriately over the long term.[7] Secondly, specialist, centralized regulatory bodies set up to resolve some of these issues have performed fairly poorly, or even aggravated factors of instability.[8] Finally, results and standards are lower than those promised in terms of prices (which should have fallen as a result of competition and the efficiency of private companies), investment in expansion projects, and attention to the needs of the poor (Leborgne, 2006; Trémolet, 2006).

These difficulties have caused some contracts to be terminated (Mali, Tanzania) (see Table 10.2). While some others are performing quite well (Sodeci in Côte d'Ivoire, SEEG in Gabon, SDE in Senegal), others are causing great concern and overall results, even in the opinion of experts from the World Bank and OECD, are mixed:

> The privatization process in Africa is still far from complete and has led to mixed results. The successful cases of the *Compagnie Ivoirienne d'Electricité*, *Sonatel*, and *Société d'Energie et d'Eau du Gabon* cannot hide the dramatic failures. Most privatizations have been imposed from the Bretton Woods Institutions – sometimes without taking into consideration the country's specificities – and suffered therefore from a lack of government commitment, which, in turn, led to significant opposition by the population and difficult implementation [. . .] Privatization reform should involve a constant dialogue between the different actors engaged (citizens, State, private sector and trade unions) in the process, in order to increase the awareness of the population, broaden local participation, build support and foster the State's accountability (Wegner, 2005: 3).

These days, there is little talk of new large contracts for public services and the focus is mainly on preserving those that already exist. These do not follow a fixed pattern and much has been said and written about the economic and, more recently, social efficiency that could result from better governance of these public-private partnerships (Breuil, 2004). Recent developments tend to suggest that a second generation of reforms is on the way, with the emphasis on an invigorated public sector and small local private players (Trémolet, 2006). These reforms may go at least some way to solving the main problems that have been identified.

In countries with a strong public sector, the decline of large PPPs has paved the way for more endogenous reforms. This is the case in Cape Town (South Africa), where the debate on 'privatization' (Smith, 2002) has given way to a quieter modernization of the municipal water service, involving traditional forms of cooperation between the

---

[7] Particularly commercial risks, bearing in mind the unreliability of consumer data, the low revenue levels and, in some cases, the reluctance of public authorities to honour their pricing commitments (Trémolet, 2006).

[8] They have added extra demands and were considered as intruders in the already complex relationship between authorities and operators.

Table 11.2 Examples of delegation contracts cancelled or not renewed and negotiations suspended or in difficulty related to African water services

Country	Company	Main shareholder	Reason for cancellation or suspension
South Africa (Fort Beaufort) 1995–2001	WSSA (Water and Sanitation Services South Africa)	Northumbrian Water (subsidiary of ONDEO) – Group 5 (50% of capital)	Ten-year water concession contract signed in 1995 and cancelled for legal reasons in December 2001.
Gambia 1993–?	MSG (Management Service Gambia)	Subsidiary of Vivendi Water	Complex situation in which the government is said to have objected to the ten-year leasing contract signed in 1993.
Ghana 2000	Azurix	Enron	BOOT project cancelled after funding withdrawn by the World Bank in 2000 for lack of transparency in the awarding of the contract.
Guinea 1989–1999/ 1999–2000	SEEG	Saur/Vivendi Water	At the end of the ten-year leasing contract, negotiations to renew the contract failed —> SONEG
Kenya (Nairobi) 2001	WSD (Water and Sewerage Department)	Seureca Space (subsidiary of Vivendi Water)	Negotiations for a ten-year management contract suspended in 2001.
Mali 2000–2005	EDM	Saur International (then Finagestion)/IPS-WA	Renationalization in 2005 after private partners transferred their capital shares to the Malian State —> EDM
Mozambique 1999–2001	AdM (Aguas de Moçambique)	Saur/Aguas de Portugal	SAUR withdrew in 2001 for reasons that have remained confidential, leaving Aguas de Portugal with 73% of the shares in the consortium.
Uganda (Kampala) 2002–2004	NWSC	Gauff Engenieure then Ondeo	Two consecutive contracts. The second with Ondeo (2002–2004) was not renewed.
Tanzania (Dar es Salaam) 2003–2005	City Water Services	Biwater/Gauff Ingenieure	Termination in 2005 of the leasing contract signed in 2003 with an Anglo-German consortium (Biwater and Gauff) —> DAWASA
Zimbabwe	–	Biwater	Company pulled out of negotiations for commercial reasons, considering the project unviable.
	Gweru	Saur	Negotiations suspended

Source: Compiled by the author.

public sector and private businesses, particularly local firms (procurement or service contracts), as well as new social control methods that give users and citizens a more prominent role in the definition of negotiated strategies (see Laïla Smith's contribution in this book).

These are interesting developments, but we are still a long way short of fulfilling the aim of providing a universal service. In recent decades, the debate has been partly aimed at the wrong target: the main problem with these PPPs is not so much the involvement of the private sector as the potential limits to its efficiency.

In decentralized countries (a large proportion of English-speaking Africa), contracts rarely cover urban areas outside large cities (see Table 10.3). Companies that operate water supply networks do not service urbanized zones: under current conditions, large companies will remain reluctant to invest in unprofitable areas (poor suburbs, small and medium-sized towns), particularly in Africa, where the financial and institutional risks are high.

Of course, in some cases, these contracts have been adapted for medium-sized towns: Kisumu (Kenya), Copperbelt (Zambia), and Dolphin Coast, Fort Beaufort, Nelspruit, Queenstown and Stutterheim (South Africa). However, these examples pale into insignificance in view of the extent of the needs. In South Africa, for example, where the main challenge is to supply towns in former homelands, poor villages and small municipalities, the handful of contracts signed in medium-sized towns have turned out to be unsuitable for the task, if not totally disastrous, such as in KwaZulu-Natal, where privatization was blamed for a serious cholera epidemic (Bakker and Hemson, 2000; Pauw, 2006).

In centralized countries (mainly French-speaking Africa), the geographical coverage of water services could be improved through national contracts which extend the operator's area of responsibility to a highly variable number of towns determined by the public authorities[9] (see Table 10.3). However, published analysis of the extent and efficiency of cross-subsidies, designed to promote the spread of services among the lower levels of the urban hierarchy, remains inadequate. It is known that in Côte d'Ivoire centralization has had some interesting stimulating effects (Collignon et al, 2000), but elsewhere, evaluations have been quick and often superficial. Moreover, they do not cover small towns outside the area served by the official operator.

Without explicitly admitting that some characteristics of the PPP model make it suited – aside from a few notable exceptions – only to large cities in emerging countries, the poor inhabitants of small and medium-sized towns have, in a way, been forgotten on two fronts: firstly by sponsors and secondly by researchers mobilized by controversies that have resulted from the ostentatious growth of the dominant PPP model.

In Mali, for example, unrealistic expectations have proved costly. The privatization of EDM and the resolution of conflicts disproportionately mobilized the Malian authorities and the community of financial backers to support a company that provides water to less than 10% of the country's population (Trémolet, 2006: 68). This example demonstrates the imbalances that have been created over the past twenty years by the widespread use of large delegation contracts, even though many other current projects also need financial and technical support in order to broaden access to potable water.

---

[9] Nearly 600 in Côte d'Ivoire, around 60 in Senegal, 50 in Niger and 15 in Mali.

Table 11.3 Main public-private partnerships for urban water services in sub-Saharan Africa (2003) – in chronological order

Country	Date	Company	Service	Type of contract	Area concerned	Main investors
Côte d'Ivoire	1960 (renewed in 1987)	SODECI	Water	Renewable twenty-year concession contract	Abidjan + secondary cities (550 locations in 2000)	SAUR (46% of capital)
Guinea	1989–1999 + two years	SEEG	Water	Ten-year leasing contract (+two one-year extensions before renewal negotiations failed)	Conakry + sixteen secondary cities	SAUR-Vivendi Water (51% of capital)
Central African Republic	1991	SODECA	Water	Fifteen-year leasing contract	All cities	SAUR (100%)
South Africa	1992	WSSA (Water and Sanitation Services South Africa)	Water Sanitation	Twenty-five-year concession contract	Queenstown	Northumbrian Water (subsidiary of ONDEO) – Group 5 (50% of capital)
	1993	WSSA	Water Sanitation	Ten-year concession contract (cancelled in 2001)	Stutterheim	Northumbrian Water (subsidiary of ONDEO) – Group 5 (50% of capital)
Mali	1994	EdM (Électricité du Mali)	Water Electricity	Four-year management contract	Sixteen cities with more than 10,000 inhabitants	SAUR-EDF
	2000–2005	EdM	Water Electricity	Twenty-year concession contract with acquisition of 60% of assets (cancelled in 2005).		SAUR (39%) IPS (21%) govt (40%)
South Africa	1995–2001	WSSA	Water Sanitation	Ten-year concession contract	Fort Beaufort	Northumbrian Water (subsidiary of ONDEO) – Group 5 (50% of capital)
Guinea-Bissau	1995	EAGB	Water Electricity	Public service concession	Bissau	Lysa (Suez-Lyonnaise)-EDF
Senegal	1996	SdE (Sénégalaise des Eaux)	Water	Ten-year leasing contract	Dakar + Forty-six secondary cities	SAUR (51% of capital)

(continued)

Table 11.3 (Continued)

Country	Date	Company	Service	Type of contract	Area concerned	Main investors
Djibouti	1996	GdE	Water	–	Djibouti	Vivendi Water
Gabon	1997	SEEG (Société d'énergie et d'eau)	Water Electricity	Twenty-year concession contract	National	Vivendi Water (51% of capital)
South Africa	1999	SWC (Siza Water Company)	Water Sanitation	Thirty-year concession contract	Dolphin Coast	SAUR (58% of capital)
		GNUC (Greater Nelspruit Utility Company)	Water	Thirty-year concession contract	Nelspruit	Biwater-NUON (40% of capital)
Mozambique	1999	AdM (Aguas de Moçambique)	Water	Fifteen-year leasing contract (Maputo and Matola) and five-year management contract for five other cities	Maputo, Matola + five secondary cities (Beira, Dondo, Quelimane, Nampula, Pemba)	Aguas de Portugal (73% of capital) (SAUR withdrew in December 2001)
Cape Verde	1999	Electra	Water Electricity	Fifty-year management contract	Main cities	Aguas de Portugal (51% of capital)
Chad	2000	STEE (Société tchadienne d'électricité et de l'eau)	Water Electricity	Thirty-year management contract (with, after the initial phase, acquisition of majority shareholding in the STEE and conversion into leasing contract)	National	Vivendi Water
Cameroon	2000	SNEC (Société nationale des eaux du Cameroun)	Water	Provisional contract (offer to buy back 51% of assets + twenty-year concession contract)	National (only six main urban centres)	ONDEO Services (thought to have withdrawn offer, in favour of Vivendi Water)

Country	Year(s)	Company	Sector	Contract type	Location	Private operator
Kenya	2000–2004	Malindi Water Company	Water	Four-year management contract	Malindi	Gauff Ingenieure
	2004			Twenty-year management contract		
Burkina Faso	2001	ONEA (Office national de l'eau et de l'assainissement)	Water	Five-year assistance and service contract	Capital + thirty-five secondary cities	Vivendi Water
Niger	2001	SEEN (Société d'exploitation des Eaux du Niger)	Water	Renewable ten-year leasing contract	Capital + fifty secondary cities	Vivendi Water (51% of capital)
South Africa	2001–2006	JOWAM (Johannesburg Water Management Company)	Water Sanitation	Five-year management contract (connections and standpipes)	Johannesburg	ONDEO Services (20% of capital)
Zambia	2001	AHC-MMS	Water Sanitation	Three-year management contract, renewable for one year	Copperbelt (Kiwte, Luancha, Mutulera, Chingola, Chililabonbwe)	Saur-l
Congo	2002	SNDE (Société nationale des eaux du Congo)	Water	Leasing contract	National	Biwater
Uganda	2002	NWSC (National Water and Sewerage Corporation)	Water Sanitation	Two-year service contract	Kampala*	ONDEO Services
Tanzania	2003–2005	City Water Services	Water Sanitation	Ten-year leasing contract terminated by the government in May 2005.	Dar es Salaam + coastal districts of Kibaha and Bagamoyo	Biwater + HP Gauff (51%) STM of Tanzania (41%)

*The NWSC manages the water services of eleven cities, but the service contract signed with ONDEO only concerns its commercial operations in the capital, Kampala.
Sources: compiled by S. Jaglin; PSIRU database (www.psiru.org/companydetails); internet sites of countries and firms; Water Utility Partnership, 1999, (www.wupafrica.org/REFORMS/contents/html); miscellaneous.

11.1.2 Project-based rationale and systems involving decentralized players: behind the profusion, a model

Enthusiasm of the 1990s is behind us: everyone admits that approaches to 'privatization' need to be more carefully designed. However, this caution has not generated renewed interest in public procurement, nor curbed the distrust in its ability to change. Rather, it widened public-private partnerships to new forms involving NGOs, small private companies and organized user groups (Jaglin et al, 2011). These arrangements, in which the private sector and civil society are the main stakeholders, are being promoted – for different reasons and with different ideological considerations – both by international aid agencies and by NGOs.

They are translated into numerous water supply projects carried out in an environment marked by politico-administrative decentralization (Jaglin, 2005a). Despite being truncated and incomplete, the decentralization process has, on the one hand, loosened the straightjacket of national monopolies by encouraging a plurality of mechanisms and operators, and on the other, lifted the restrictions of national service regulations, thus creating opportunities to offer different types of service. Nevertheless, this decentralization has been accompanied by a chronic shortage of resources and skills, which means that local players remain structurally dependent on help from abroad (Otayek, 2005). This structural weakness results in numerous support measures being offered, creating competition between various initiatives that involve different service models and intervention mechanisms. Although coordination has been attempted, the persistent rivalries between sponsors hinder a real programme-based approach to projects and investments (Belbéoc'h, 2006; Valfrey, 2005).

The projects involve the creation or repair of a small-scale local infrastructure for water production and distribution (boreholes with a water tower and standpipes, mini-networks carrying pressurized water to standpipes and private connections, etc.).

Box 11.1 Decentralization and water service reforms: cooperation difficulties in Benin

In Benin, the decentralization policy adopted in principle back in 1990 was gradually implemented after the municipal elections at the end of 2002. In the face of the poor financial and human capacities of the new local authorities, the influence of foreign project sponsors has led to:

*   competing external models with priority given, on the one hand, to the construction of representative local governments (France, Germany) and, on the other, to the establishment of effective governance and management based on public-private partnerships and the active participation of civil society (World Bank; Denmark)
*   competing cooperation projects, some supporting municipal project ownership (Agence française de développement) and others assisting village community-based organizations and their representatives (decentralized cooperation agreement between the French Picardy regional council and the Beninese *département des Collines*), creating rivalry between the latter and elected representatives.

Source: Belbéoc'h (2006).

These are funded initially from abroad. Management responsibility during the project phase is delegated to an NGO or a northern and/or southern consultancy firm. The population and local businesses are involved at the implementation stage.

Once the construction or repair work has been carried out, the main concern is continuity of the service, with increasing levels of money and priority given for management and maintenance of public installations, which are the weak links in these projects. Inspired by two quite distinct reference models, the rural model (committee) and the urban model (management contract), the mechanisms converge, resulting in a gradual standardization process based on a number of principles of 'good management' imposed by the project organizers (Coing et al, 1998): commodification of water (priced according to volume), contractual delegation of certain functions to private companies (billing, maintenance), and a regulatory mechanism provided by a group of organized users (association, committee).

Although they bear certain similarities to public-private partnerships, these projects are very different to large PPPs. The private sector concerned is part of the local economic fabric – entrepreneurs with few qualifications and small, usually informal businesses. The project operators organize this sector, setting up a group of small businesses approved by the public authorities into which candidates are then recruited through a tendering process. They endeavour to make it as professional as possible in order to consolidate and harmonize practices, and try to devise contracts that are suited to the local socio-economic situation.

In terms of coordination and supervision, these projects officially promote service governance by a community-based institution, which operates an installation or a group of installations in partnership with small, local, private entrepreneurs. The social engineering that forms part of these projects therefore attempts to perpetuate a

---

Box 11.2 A community-based water supply system: Zamcargo (Dar es Salaam)

Zamcargo is a poor informal district around 4km from the centre of Dar es Salaam (Tanzania). Having been gradually urbanized since the 1970s, it now has a population of approximately 3,000. Following a drought in 1997 and the resulting cholera epidemics, the Tanzanian government called on the international community to finance a water-drilling project for Zamcargo. It was not long before the British NGO WaterAid offered to help the local residents manage the drilling programme, which had been abandoned by the city's water company. Although it owned the programme, the water company had preferred to delegate operational responsibility to the local residents due to a lack of funds and appropriate technical expertise. With users contributing 10% of the total cost (financial contributions and practical assistance), the NGO equipped the district with a micro-network of three 'water kiosks', where water is sold. Fed by pumps and a raised reservoir, the kiosks are managed by an elected committee of water users. At the request of the most well-off residents, private connections were also fitted and equipped with a meter. Income from the sale of the water ($€0.8/m^3$) should enable the committee to cover running costs (electricity for the pump, payments to standpipe operators) and maintenance charges (plumbing work, repairs or replacement of pump, etc.) without foreign funding. The committee is responsible for managing the micro-network in a transparent and democratic way: its members must present accounts at meetings where important decisions are put to a vote.

Source: Bousquet (2006).

form of service co-production (Ostrom, 1996), with the aim of creating, on an appropriate scale, institutions capable of managing the common property of a group of users.

The lack of capacity of local public authorities serves here as an argument (or pretext) for the development of local private and community-based solutions. This notion is generally based on two strong beliefs: firstly, that the local community makes it easier to reach compromises where joint management decisions are concerned, since it provides an environment where relations of trust can be developed – a type of middle ground for cooperation in which credible commitments can be entered into; and secondly, that community-based management better meets the needs of heterogeneous populations than centralized systems, since it is more inclusive and democratic.

11.2 THE LIMITATIONS OF COMMUNITY-BASED PRIVATIZATION

In the future, a growing proportion of African urban households are expected to depend on these alternative water supply systems. However, the term 'alternative' needs closer consideration. Although they provide a technical alternative to the main networks, these systems are, on the contrary, very much in line with the prevailing reforms in terms of the relationship they create between a service and its users, insofar as they endeavour to link the notions of individual and collective responsibility with that of payment.[10] In promoting awareness of the 'user pays system', they are different both from public water management, which is reputed to be inefficient, and from forms of free community-based water provision. Rather than contradict dominant market principles, community-based privatization therefore represents a compromise between recognition of the right to a minimum volume of potable water and the principle that the service should be paid for by its users.

11.2.1 Dysfunctions of the model . . .

The systems that have resulted from these projects are rarely satisfactory, and the actual governance of decentralized mechanisms differs from the official variants of the co-production model. In achievement terms, on the short run the latter produces relatively satisfactory results, since it is usually possible to find a consensus sufficient for the definition of a project and the management of investments. However, it is much less reliable in terms of long-term management and regulation of the service.

While the ownership of infrastructure often remains problematic, its management creates numerous misunderstandings. For example, in the small towns of Guinea studied by Jean-Pierre Olivier de Sardan (2000) and in those of neighbouring countries analysed by Janique Etienne (1998) and Anne Belbéoc'h (2006), the local community often appears too deeply divided and interests too disparate to form the basis of true 'community-based' management. Fluctuating between the commercialization of water supply and semi-privatization by 'hoarders' on the one hand, and the granting of

---

[10] Ben Page provides an excellent illustration of this using the example of the water service in Kumbo, Cameroon (Page, 2003).

virtually free access with sporadic financial mobilization to cover maintenance costs on the other, the community-based model struggles to find a sustainable compromise between a profitable water service that generates financial revenue (water as a market commodity) and a social service that is accessible to all (water as a public commodity) (Olivier de Sardan, 2000: 82).

Yet local regulatory bodies are inefficient: water supply point committees, which are often just factional or semi-private groups, are dysfunctional; the transparency of accounts and decision-making is rarely checked, while user participation sometimes boils down to an initial 'election' that is never repeated. With the commercial risk, these community-based systems de-localize the regulation of conflicts to collectives that have neither the resources nor the skills for the task (Manor, 2002). Promotion of the central role to be played by users' associations in the co-production system also places municipal authorities in an awkward position, since their local jurisdiction in this area is often enshrined in recent decentralization legislation, as is the case in Mauritania (Carlier, 2001) and Benin (Belbéoc'h, 2006).

In outlying urban areas, or even old peri-central working-class districts, which are by no means among the most unstable or recent ones, community-based management schemes no longer seem capable of managing collective services in a sustainable way. In her studies of poor districts of Nairobi, Dar es Salaam and Lusaka, Anne Bousquet questions the peculiar way in which these projects attempt to combine commodification of water with community-based governance and regulation at the level of local neighbourhood collectives, which are organized in an ad hoc way and on flimsy socio-demographic foundations. She shows that community-based privatization schemes rarely have the regulatory capacity required to ensure the long-term future of these services, while the transfer of responsibility for conflict resolution, solidarity or water conservation issues from the very local scale to urban or metropolitan levels is problematic (Bousquet, 2006). Similar observations have been made in peri-urban villages around Ouagadougou (Meï, 2008).

11.2.2 . . . or an unsuitable model?

The difficulties encountered could be considered as teething troubles (even though the subject has been talked about for many years) but, at a deeper level, the model is failing to support sustainable water services because it is, in our opinion, structurally unsuited to this objective in many African societies.

NGOs have three possible courses of action: a contract-based service provided on behalf of a public authority; independent provision of services to fill the cracks (or huge gaps) left by the public service; or provision of a service on behalf of a foreign sponsor (Hulme, 2001). Due to the structural weakness of local public authorities, the third of these options was often imposed from the 1980s onwards, creating vertical relationships between the sponsor and the private service provider, bypassing local public authorities. As far as the regulation of water services is concerned, this creates three main problems.

Firstly, local service providers' financial dependence on northern sponsors brings into question the sovereignty of poor countries, reducing their level of responsibility for service provision and weakening public responsibility in the recipient countries (Hulme, 2001). Entrusting the task of arbitration, which is political by nature, to

---

Box 11.3 Conflicts and regulation linked to the management of a community-based micro-network (Zamcargo, Dar es Salaam)

The crisis affecting the community-based water service in Zamcargo extends beyond the institution that was set up to manage it in 2003, even though residents complained that the committee lacked transparency and even honesty. As a result, the whole project has been called into question. The coproduction element of the service is not limited to its technical and operational aspects, since it also involves mechanisms for the local definition of rules and regulation, the shortcomings of which become apparent in times of crisis. First of all, the water kiosks are not used in accordance with the guidelines laid down in the project charter drawn up by WaterAid for several of its community-based micro-networks: the water meters do not work, the standpipe operators do not keep the accounts up-to-date, and the virtual absence of public meetings fuels residents' distrust. The water committee, for example, is suspected of diverting some of the revenue from the sale of water. Repairs to the pump costing Tsh 5 million (€4,000) may have shown that funds are available, but the people responsible did not produce any documentation to show where the money had gone. The charter promotes the principle of 'participatory' functioning, through which residents are supposed to exploit and regulate the micro-network themselves. However, this mechanism has failed because meetings that were initially planned have not been held and no appeal mechanism has been designed. Nevertheless, in 2002, when the committee split into two opposing factions, with the chairperson and his supporters on one side and reformers new to the scene on the other, the residents requested external arbitration from the first tier of the traditional politico-administrative structure, which was supposed to resolve small-scale conflicts (the *Mtaa*). Unfortunately, the *Mtaa* did nothing (its leader seemed to have been bribed by the water committee chairperson). The residents then struggled to involve higher authorities before finally appealing to the supreme body of the municipal administration ('District Commissioner'), who decided to send in an investigation committee in 2003. The process ended with the committee resigning and new elections being held.

Source: Bousquet (2006).

---

parochial[11] organizations seriously dilutes the process of developing a public sense of responsibility. Great importance is attached to this in the debate on 'good governance', since the governance structure is fragmented with the higher hierarchical levels (sponsors and international cooperation agencies) based outside the country. This exterritorialization process[12] hinders the creation of a local socio-political regulatory system capable of democratically and sustainably establishing the values and rules that are necessary for services to function (Piveteau, 2005; Ribot, 2007).

Secondly, since there is no adequate regulation of competition linked to the resources generated by projects at the local level, the most powerful groups and

---

[11] In the sense that they only operate in the interests of clearly defined groups, in competition with other NGOs and associations.

[12] A process whereby service regulation, no longer the responsibility of local stakeholders, is confiscated, knowingly or otherwise, by outside bodies in the name of project efficiency and the profitability of investments. This insidious process of 'recolonization' operates under the guise of local democratic forums which function, *de facto*, as intermediaries of this exterritorialization process (Piveteau, 2005).

Box 11.4 Diversification of water supply areas in Bamako (Mali)

In 1997, a basic water system equipped with nine standpipes funded through Japanese development aid was opened in the district of Kalabancoro, on the western side of the Malian capital city. The management of the system, initially entrusted to the district's association of water users, ended in failure and the service was withdrawn. In 1999, it was revived by the CCAEP[13] (*Cellule de conseil aux AEP* – water service advisory unit), a body attached to the urban water division of the national water directorate.

The CCAEP appears to have tried to set up a dynamic local management system by extending the network and adding thirty or so private connections to the standpipes. However, user satisfaction levels seem very mixed and the independence of the service comes at a high cost, causing long-term damage:

- water selling prices are very high (FCFA 350/m^3): three times as expensive as the lowest price band (0–20 m^3/month) of the national supplier EDM (also for standpipes), FCFA 114/m^3
- the strong taste of the water is thought to be the reason for low consumption levels and the closure of some standpipes in districts where the wells do not run dry
- the water company (EDM), which was not consulted about the creation and management of the water supply system, knows nothing about the technical specifications of the installations: it will therefore be difficult for it to take over the equipment in the future
- the system is still running at a loss, in particular due to the misappropriation of takings at the standpipes and the exorbitant cost of operating and maintaining the generator.

Source: AFRITEC (2000: 30–31).

---

individuals are able to influence decisions disproportionately for their own profit. In the management of natural resources and local services, for example, mechanisms are often developed for the capture or even confiscation of resources for the benefit of dominant groups. These intra-community tensions are all the more acute when communities are poor and the quest for outside funding is literally a matter of life and death (Manor, 2002). Conflict resolution mechanisms are rarely taken into account in these projects, which wrongly assume that the local community is depoliticized.

Thirdly and lastly, these projects fail to take sufficient account of the issues of social equity and fair distribution. They follow economic principles designed to promote sustainability of the infrastructure and its use, based on local coverage of costs. These small-scale mini-networks aimed at poor populations cannot hope to achieve economies of scale similar to those of large networks. As a result, water prices, which are often high and usually much higher than those of the conventional service, do nothing to resolve the problem of social injustice that is often raised (the poor pay more for their water than the rich), or even confirm the inequalities amongst the poor themselves, since each project sets prices in accordance with highly variable local costs.

Furthermore, the question of redistribution is rarely addressed: prices are usually calculated so as to cover micro-local running costs without any cross-subsidies either within the community or between different mini-networks.

---

[13] Since renamed the GCS AEP: see Box 11.8.

Box 11.5 The water mafia in Kibera (Nairobi, Kenya)

Kibera is one of the largest shanty towns in sub-Saharan Africa: an estimated population of between 500,000 and 1 million people live there in extremely poor sanitary and urban conditions, despite its proximity to the city centre and industrial zone of Nairobi. With no public standpipes, the residents collect their water mostly from private kiosks or, since the 1990s, from kiosks managed by user collectives under NGO supervision. The number of water retailers is estimated at around 2,000, generating turnover of US$15,000 to $39,000 per day. Corruption and criminal practices, stimulated by fierce competition, are commonplace among all players in the water sector. The cascade of pressure exerted ranges from the most powerful, including the employees of the Nairobi water service (the conversion of which from a direct municipal service to a publicly owned company has not stopped the corruption), to the most vulnerable – the poor shanty town residents who pay the final cost in terms of the price they pay for water. For example, water vendors, both legal and clandestine, suffer from constant racketeering of water service staff (bribes to obtain a connection or avoid disconnection) and other government departments (sanitary services in charge of monitoring water quality, police and justice). In turn, whether working on their own or in syndicates (approximately 1,500 in the *Maji Bora Kibera* association), the vendors indulge in criminal practices in order to force the price of water up: artificial shortages (obtained by bribing the water service staff) and price-fixing. They also frown on community-based initiatives, which try to escape these practices and, by blackmailing them with violence (threats, physical attack, damaging equipment), force them to raise their prices. Finally, the Kamjeshi and Mungiki gangs (types of politico-ethnic mafia-like sects), primarily involved in protection rackets targeting minibus (*matatu*) drivers, are also interested in this lucrative business and extort money from water vendors, particularly those working independently.

Source: Bousquet (2006).

From the social equity point of view, these projects are doing little to reopen the debate on how to combine the commercialization of water services with new forms of solidarity. Communities are often left to solve the problems of their poorest members even though local solidarity and social equity do not necessarily go hand-in-hand. They are also powerless against violent forms of 'regulation' in cities like Nairobi, where real 'water mafias' are thriving.

11.3 THE COMMUNITY-BASED MODEL *VERSUS* THE PUBLIC MODEL: IDEOLOGICAL DOMINATION AT STAKE

It is intriguing that, while critical studies of water service reforms devote plenty of attention to 'market privatization', they barely refer to 'community-based privatization' and the basic principles of the underlying collective action models. However, these are no less ideological. It is particularly important to stress this, since the practical results of the projects are not always convincing.

Below we have listed three principles involved with the rapid spread of this model. While the first is rather utilitarian, the subsequent two are very clearly ideological in nature and more broadly convey a model for society.

11.3.1 The need for efficient sectoral approaches and depoliticization of management

In order to achieve their sectoral objectives and spend the large quantities of money at their disposal, sponsors need to act fast: if they cannot find the institutions they need to implement their projects locally, which is often the case, they favour ad hoc structures both further up (creation of parallel administrative and financial agencies) and lower down the chain (committees, users' associations). In doing so, they pose a threat to elected local public authorities, or even boycott them completely (Leclerc-Olive, 2003a, 2003b; Jütting et al, 2005).

Coupled with the promotion of the community-based model, this practice is far from neutral. Firstly, it deprives the local public authority of the chance to learn, and therefore of some of the cognitive and instrumental skills it needs to make coherent, 'open' choices related to the creation and management of local public services. Secondly, it maintains the myth of a depoliticized local governance system (Jobert, 2002), depriving local authorities of the public forum and debate[14] necessary for the maturing and legitimization of decisions. Finally, it results in the accumulation of knowledge by NGOs rather than by African local public authorities, which are consequently held captive by a dominant body of expertise created elsewhere.

While it hampers the development of municipal authorities and their project management opportunities, this approach creates a disconnection between service management, which is treated as a technical matter, and the implementation of local public policies. This depoliticization is problematic because it gives the impression that management decisions have nothing to do with the specificities of the local community. Historically, local management systems elsewhere have not been developed in this way: management decisions taken in response to difficulties encountered have always been compromises based on local legal, economic and political circumstances. These decisions have then been gradually adapted and altered in accordance with the limitations and shortcomings that become apparent. Nowadays, in small African towns, there appears to be no choice at all, but simply the need to find a 'good' model. Besides, who will take responsibility for its development and adaptation if the public authorities do not have the time or resources to learn how to do so?

11.3.2 The supposed advantages of community-based regulation

Supported by the conditions attached to the intervention of development aid agencies, which are very anxious not to be accused of interfering, the power of the myth of apolitical urban management can also be explained by their understanding of the role played by so-called civil society and collective action in development. The projects we are considering here demonstrate a renewal of interest in community-based institutions, informal standards and interpersonal networks, which are capable of fostering efficient, sustainable collective action in the local management of common resources and commodities (Evans, 1996), or even in the emergence and consolidation of standards

---

[14] Or deliberation, defined as a democratic process through which public debate transforms preferences and prepares, develops and criticizes public action (Leclerc-Olive, 2003b).

of cooperation between public service providers and residents' organizations (Ostrom, 1996; Tendler, 1997). The role played in this by institutional bodies, both formal and informal, based on the intensive use of information, trust and social capital,[15] is particularly important because it appears likely to form the foundation of collective action with a strong cognitive dimension, mobilizing a variety of endogenous resources as well as local knowledge. By promoting shared standards and rules and using flexible sanction mechanisms, community-based management would also make it difficult for any individual or group to ignore the common rules of reciprocity and would limit *free-riding*. Although probably underestimating the sources of local conflicts and, conversely, overestimating the crucial role played by external stakeholders in resolving them and moving on, this approach is very common nowadays in the construction of frameworks for the governance of water supply mini-systems in sub-Saharan Africa.

Experiments of this type go hand-in-hand with the rehabilitation of 'community-based' local authorities, in conformity with certain neo-institutionalist ideas about the inherited colonial administration (Hibou, 1998: 15). However, numerous African examples have proved local communities to be powerless to deal with the tensions caused by competition between and within different communities for access to limited resources, and have found it hard to become 'modern' regulatory bodies for market services (Jaglin, 2005b; Ribot, 2007).

The deficit in analysis is partly linked to the fact that collective action mechanisms do not have any models specific to urban communities, and do not take into account the characteristics of those communities. Where water is concerned, most analyses focus on rural communities, water resource management and irrigation rather than household water supply (Ostrom, 1992; Subramanian et al, 1997). Whether it can be adapted to urban environments is dependent on whether cities contain stable, well-informed communities with shared interests; whether these communities form a suitable basis for service management and are able to express their preferences; and whether a willingness to take on responsibility encourages politicians and administrators to listen to them, and so on. These are conditions that are rarely met in African towns and cities. Furthermore, by insisting on social capital as a resource for collective action, these analyses underestimate its negative external effects in unequal, hierarchical societies (social exclusion), or some of its perverse forms in violence-dominated urban areas, such as the shanty towns of Nairobi (see Box 11.5).

In this context, weaknesses within communities justify various forms of 'recentralization'. Sometimes, initiatives of this kind are carried out at the municipal scale (Lusaka, Ouagadougou). This is usually because local public authorities worry about the marginalization of the poor and want a minimum level of harmonization and integration of service management. However, the reforms have not prepared the public authorities and the residents for this. When the dysfunction of community-based services causes public authorities to take action *a posteriori*, these attempts at municipal

---

[15] Social capital is defined as a resource created by interaction between agents, by the endogenous development of behavioural standards, and by the institutionalization of these standards as rules. Based on social networks, standards of reciprocity (or trust) and common values and beliefs, it is likely to generate sustainable external effects which influence the economic situation of individuals and groups, particularly by promoting collective action.

Box 11.6 The LWSC and the recentralization of the water service in Lusaka

Having seen numerous community-based water projects fail in the outskirts of Lusaka, the LWSC, a water company owned entirely by the municipal authority, established a recentralization strategy in 1997 for the management of micro-networks. The strategy was inspired by a model introduced by the Japanese development aid body at the request of the Zambian Government, following serious cholera epidemics in 1997 in George Compound. This vast informal semi-official district has autonomous boreholes that feed an intricate network of communal stand-pipes, managed as part of the traditional municipal network. Although a technical division and a monthly card payment point (entitling the holder to consume 200 litres/day/household) were specifically created *in situ* with standpipe operators recruited locally, the operating deficit is subsidized thanks to geographical cross-subsidies from more well-off districts within the same management division. The standpipe operators and cashiers are employed by the LWSC, while the community is involved in resolving water-related conflicts through local courts.

The LWSC would like to extend this model to include failing community projects, sometimes at the request of the users themselves, who have been deprived of water by the serious shortcomings of the community-based management system. It also targets new projects funded by NGOs and sponsors. To this end, it proposes to set up a PPP, in which it would retain a dominant influence, through its new community water supply department. However, the NGOs in the water sector, which have sustained the urban management in a sort of archipelago through their micro-projects, are resisting this recentralization process. One of them, Care International, reacted by launching a counter-model based on 'water trusts', which aim to professionalize the water committees that previously managed water supply points on a (supposedly) independent and unpaid basis.

Source: Bousquet (2006).

Box 11.7 NWASCO: Zambian water regulator

The Zambian water regulator NWASCO, operational since 2000, has two main tasks: firstly, to introduce competition in the water sector ('yardstick competition' and 'sunshine regulation' between the newly created commercial utilities), and secondly, to serve the general interest, particularly the provision of access to water for poor populations. To this end, it promotes alternative technical systems and the harmonization of investment funding mechanisms. It has its own special fund and structure, the Devolution Trust Fund (three people: a manager, a socio-economist and a water and sanitation engineer), which grants incentive loans to suppliers who comply with the regulator's guidelines. (For example, €360,000, provided particularly by the GTZ and Danish and Irish development aid bodies, has been spent on providing access to water for 90,000 people since 2003.) NWASCO is using this tool to promote a programme-based policy at the national level, the main aspects of which (inspired by the George Compound pilot project) are contained in the 2002 'Guideline on water supply to peri-urban areas'.

Source: Bousquet (2006).

regulation are often seen as a form of interference or even dispossession by the communities and the NGOs, which try to resist this 'recentralization' process.

Processes of recentralization are also organized at a national scale, thus leading to greater exterritoriality. In Zambia, this has resulted in the creation of an independent national regulatory body, while in Mali, two private national structures (2 AEP and GCS AEP) monitor and advise providers of water supply systems.[16]

Both exterritorialization and recentralization illustrate, in their own ways, the limits of community-based systems. However, neither seriously address their limits or their contradictions with regards to the strengthening of local public authorities and their role in service regulation, claimed as an essential part of decentralization reforms.

11.3.3   A deep distrust of local public authorities

The role devolved to local public authorities is also influenced by more general ideological factors. Firstly, that of a neoliberal understanding of democracy and the State that focuses on 'civil society', which organizes decentralization for the benefit of community-based collectives and private entities, including NGOs, which are considered a 'substitute private sector' (Leclerc-Olive, 2003a: 172). Secondly, that of development theories, founded on a community-based, participative rhetoric that is particularly strong in the epistemic communities of developers in Africa (Otayek, 2005: 10). Since the short-term development goals of the United Nations also seem easier to achieve through this approach, it creates a deep distrust of local public authorities in numerous project operators and sponsors and diverts them away from action tools that could, on the contrary, strengthen a model for public municipal service management.

Box 11.8 GCS AEP: a private advice and monitoring body in Mali

The GCS AEP (*Groupement de conseil et de suivi des adductions d'eau potable* – water service advisory and monitoring group) is a private body, which receives a tax of FCFA 20/m^3 sold (€0.03). Created to audit water supply systems managed by users' associations in small towns and villages, it monitors technical and financial aspects of these projects. Every six months, GCS AEP checks the accounts and the condition of the equipment: after each field visit. It presents the results to the community (the mayor, users and the Regional Director of Hydraulics and Energy attend the meeting), and provides advice to the users' association to improve management. It also supports operators through daily communication by radio, and provides training as well as other optional services (feasibility studies, purchase of equipment, production of management tools, etc.).

Comprising six people at the end of 2001, it monitored fifty-nine water services (of a total of 200) spread across the eight regions of Mali. Constructed in villages with between 1,200 and 17,000 inhabitants, these systems served around 400,000 people, producing almost 1.7 million m^3 and generated an overall turnover of FCFA 35 million (€53,340).

By producing and disseminating standardized information, the GCS AEP is helping to harmonize different practices and, more broadly, to define and delimit public water services in these urban environments.[17]

Sources: www.pseau.org/outils/ouvrages/dnh_suivi_conseil_2003.pdf; AFRITEC (2000).

---

[16] In 2009, 153 rural and semi-urban water services (approximately 25%) were regulated through this system, called STEFI (Suivi technique et financier). Some of them have been monitored since 1994.

[17] According to some sources, Mali's GCS AEP focuses its expertise on 'large' systems (59), from which it earns most of its income, and is disinterested in numerous small projects which are often in a catastrophic state (pers. comm., Antéa, 30 August 2006).

Such a model requires a solid local project management structure and regional institutions capable of making management, investment and pricing decisions in a democratic way. Yet, this is clearly not always the case and distrust of politicians in general and of local ones in particular, as well as of public institutions, is based on objective reasons: there are numerous stories of mayors' incompetence, of their lack of involvement in development projects abandoned in favour of spending on unnecessary luxuries, and of their indifference towards their official duties (Belbéoc'h: interviews, 2006). Although these failings may be interpreted in different ways,[18] there is little doubt that local public authorities face many different obstacles as they carry out their duties. But that is not the issue here. Rather, the question is how to respond to this situation. Should municipal authorities be bypassed, in favour of the private sector and civil associations, or do they need to be given the opportunity and the time they need to learn how to manage public affairs and all the associated pitfalls?

The contrast between strategies illustrating each of the two main models identified was clearly evident at the Africities Summit in Nairobi in September 2006. While the mayor of Dschang (secondary city in Cameroon with 70,000 inhabitants) presented the concept of a concerted municipal strategy, developed as part of a programme run jointly by Ps-Eau and the PDM (Sonkin, 2006), the chief executive officer of the Kisumu water company described, in a different speech, the process of privatizing the service in a Kenyan city of 450,000 inhabitants (Ombogo, 2006).[19]

The first speaker highlighted some of the key aspects of the strategy: the presence of a motivated, competent municipal leadership, active communication with residents, diagnostic work (particularly the identification of conflicts) and, in particular, the development of a shared, recognized definition of the water service. He also explained that one year of consultation had been required to draw up an action plan. This example served as a reminder that assisting municipal project management is inevitably a slow and uncertain process, which is difficult to manage and reconcile with the quantitative objectives of short-term sectoral policies.

The second speaker mentioned neither the local authority, except to disparage public management, nor the population, except to stress that serving the poor was a major challenge. The city's public authorities seem to be excluded from the institutional project, which is based on three key players: a national regulator, an organizing body which owns the regional infrastructures and a private local operator. Although the strategic role of the local authorities was emphasized in the final declarations of the summit,[20] this model does little to strengthen public institutions or help them learn about urban management.

---

[18] For example, Jean-Pierre Olivier de Sardan believes that the notion of local public ownership is attested to in many West African village communities, but 'there is no longer a connection between public ownership (for the benefit of everyone) and public management (generally unknown in the sense of 'projects')' (Olivier de Sardan, 2000: 80). Conversely, Michèle Leclerc-Olive notes 'a lack of "public spirit" among elected representatives' and, more generally, stresses that 'the notion of public ownership is absent, replaced by the simple notion of common ownership – or shared ownership – which is clearly not to be confused with that of public ownership, which is inalienable' (Leclerc-Olive, 2003a: 183).

[19] See the communications of session 3 of Africites 4 (2006). Available at: http://www.pseau. org/event/africites_4/session_ss3_fr.htm.

[20] See: www.africites.org/index.php.

Behind its operational aspects, the dominant 'apolitical' model of public services is not devoid of normative objectives. In medium-sized towns, it tends to promote private enterprise rather than municipal ownership. In villages and peri-urban districts, it fosters the construction of civil forums 'characterized by the defence of a common good (rather than responsibility for a public good) and experimentation with negotiation rather than with deliberation' (Leclerc-Olive, 2003b). This leads to the development of community-based rather than public water services. These services, however, struggle to attain a level of consolidation that would enable them to be considered sustainable. Moreover, it remains to be shown whether their long-term survival, if it could be maintained, is desirable.[21] Indeed, many examples in Senegal and elsewhere suggest that the existence of a water supply mini-network rapidly generates a demand for private connections. This has a profound impact on management of the service, unless an artificial quota system is introduced – as it has been in some parts of Benin. A large number of private connections increases consumption levels (from 8 l/inh/d on average to more than 35 l/inh/d), thus raising the potential profitability of the system, while individualized consumption and billing makes it easier to collect payment. The sustainability of the infrastructure and service ceases to be a structural problem and the system can be entrusted to a market operator (public or private). However, in view of the skills required and the costs, sanitary controls and service regulation (price, quality) cannot be organized at the community level, and the marginalization of the municipal authorities then becomes a serious obstacle to the necessary process of service adaptation and transformation.

11.4 CONCLUSION

Part of radical Anglo-Saxon geography denounces the ethnocentrism and economicism of numerous contemporary urban studies that focus on northern cities and problems (Robinson, 2002, 2005). Yet in this chapter, field studies based analyses show more complex and diverse situations than those generally referred to in academic debates. For example, current forms of privatization are, *de facto*, varied and involve many different forms of 'private', particularly informal ones. More attention is paid to the behaviour and views of ordinary city-dwellers who are not served by the network than is often claimed: in their projects, NGOs use their knowledge of these practices to break away from the dominant technical paradigms, and the network standard is neither exclusive nor even a short-term reference point in every case. The projects studied thus illustrate practical efforts carried out in recent years in order to invent new ways of linking supply and demand in collective services.

Yet, not all the results of these projects are positive. They do not all achieve their objectives. Although they take note of the elitism of the colonial model of the modern city and its technical networks, they are not devoid of power-related issues; they also impose standards and techniques that are not agreed by consensus and, through them, ways of organizing society. By increasing the involvement of local communities without taking sufficient account of the inequalities they harbour, and ignoring the social

---

[21] The following observations were inspired by discussions with staff from Antéa, a consultancy firm involved in the field covered by this article (30 August 2006). Any interpretations and possible errors are entirely the authors' responsibility.

conservatism that these communities sometimes reflect, they also foster the reproduction of dominant power relationships that, in some of the projects studied, disadvantage the most vulnerable.

Although they have managed to get community-based water services up and running, they have not demonstrated an ability to integrate these services into the decentralization process, on the one hand, and to monitor changes in water demand on the other. Under the combined effect of localization of forms of governance (at the community level) and recentralization of regulation[22] (at the national level), the provision of water services in these urbanized areas is partly shielded from local political decision-makers, contradicting administrative reforms and the requirements of local authorities, which have legal responsibility in these matters. In addition to the fact that this separation between governance and service regulation creates serious cooperation difficulties, in its current form, it also nurtures a dangerous vacuum at the municipal level, which is no longer able to analyse issues independently and is therefore deprived of some of its legitimacy in resolving conflicts – a task which is therefore left to those further up in the hierarchy. As in other fields, these findings once again bring into question the role of local public authorities and the need for them to 'return' to the overall regulation of water services, including those provided through technically decentralized systems.

REFERENCES

AFRITEC. 2000. *Partenariat Eau et Assainissement (WUP) Projet No. 5: Étude de cas du Mali – Rapport provisoire*, s.l., Abidjan, Partenariat Eau & Assainissement-Afrique de l'Ouest et du Centre/Water Utility Partnership-Africa. www.wupafrica.org.

Ayee, J. and Crook, R. 2003. 'Toilet wars': urban sanitation services and the politics of public-private partnerships in Ghana. Brighton (Sussex), Institute of Development Studies, December (IDS Working Paper, 213).

Bakker, K. 2003. Archipelagos and networks: urbanisation and water privatisation in the south. *Geographical Journal*, 169(4), pp. 328–41.

Bakker, K. 2007. 'Trickle Down'? Private sector participation and the pro-poor water supply debate in Jakarta. *Geoforum*, 38: 855–68.

Bakker, K. and Hemson, D. 2000. Privatising Water: BOTT and Hydropolitics in the New South Africa. *South African Geographical Journal*, 82(1): 3–12.

Belbéoc'h, A. 2006. *Le département des collines au Bénin, laboratoire pour la Maîtrise d'Ouvrage Communale*: rapport de mission, Cérève, September.

Bousquet, A. 2006. *L'accès à l'eau potable des citadins pauvres: entre régulations marchandes et régulations communautaires* (Kenya, Tanzanie, Zambie). Paris: Université de Paris I-Sorbonne, 2 vols (doctorate thesis).

Breuil, L. 2004. *Renouveler le partenariat public-privé pour les services d'eau dans les pays en développement: comment conjuguer les dimensions contractuelles, institutionnelles et participatives de la gouvernance?* Paris: ENGREF (doctorate thesis).

Carlier, R. 2001. *Concessionnaire de réseaux d'adduction d'eau potable, naissance d'un métier*. Paris: GRET (Traverses, 9).

Coing, H., Conan, H., Etienne, J., Jaglin, S., Morel, À., L'huissier, A. and Tamiatto, M. 1998. *Analyse comparative des performances de divers systèmes de gestion déléguée des points d'eau collectifs*. Paris: BURGÉAP, April, 2 vols.

---

[22] We draw a distinction here between governance (all cooperative aspects of the collective management of the service) and regulation (all processes aimed at the dynamic reproduction of a system).

Collignon, B., Taisne, R. and Sie Kouadio, J. 2000. *Analyse du service de l'eau potable et de l'assainissement pour les populations pauvres dans les villes de Côte d'Ivoire*, Paris: Hydroconseil, April.

Collignon, B. and Vézina, M. 2000. *Independant Water and Sanitation Providers in African Cities*. Full Report of a Ten-Country Study, Washington: World Bank.

Étienne, J. 1998. Formes de la demande et modes de gestion des services d'eau potable en Afrique subsaharienne: spécificité des 'milieux semi-urbains'. Paris, ENPC (doctorate thesis).

Evans, P. 1996. Government action, social capital and development: reviewing the evidence on synergy. *World Development*, 24(6): 1119–32.

Hibou, B. 1998. *Économie politique du discours de la Banque mondiale en Afrique subsaharienne: du catéchisme économique au fait (et méfait) missionnaire*. Paris: CERI. (Les Études du CERI, 39).

Hulme, D. 2001. Reinventing the third-world state: service delivery and the civic realm. W. McCourt and M. Minogue (eds) *The Internationalization of Public Management: Reinventing the Third World State*. Cheltenham (UK)/Northampton (USA): Edward Elgar, pp. 129–52.

Jaglin, S. 2005a. *Services d'eau en Afrique subsaharienne: la fragmentation urbaine en question*. Paris : CNRS Éditions (coll. Espaces et Milieux).

Jaglin, S. 2005b. La participation au service du néolibéralisme? Les usagers dans les services d'eau en Afrique subsaharienne. M-H Bacqué, H. Ray and Y. Syntomer (eds) *Gestion de proximité et démocratie participative: une perspective comparative*. Paris: La Découverte, pp. 271–91.

Jaglin, S. 2006. Introduction de la table ronde GRET-Latts 'les multinationales de l'eau et les marchés du sud: Pourquoi Suez a-t-elle quitté Buenos Aires et La Paz?'. Paris, 13 November.

Jaglin, S., Repussard, C. and Belbéoc'h, A. 2011. Decentralisation and governance of drinking water services in small West African towns and villages (Benin, Mali, Senegal): the arduous process of building local governments, *Canadian Journal of Development Studies*, 32(2), 119–138.

Jobert, B. 2002. 'Le mythe de la gouvernance antipolitique', report to the VII congress of the French Association of Political Science, Lille: 18–21 September.

Jutting, J., Corsi, E. and Stockmayer, A. 2005. Decentralisation and poverty reduction. *Policy Insights*, 5. www.oecd.org/dev/reperes.

Leborgne, F. 2006. La privatisation de l'eau au Mali. *Responsabilité & Environnement*, 42: 44–58, April.

Leclerc-Olive, M. 2003a. Mondialisations et décentralisations: complémentarités ou alternative? Éléments de réflexion pour une enquête au niveau local. A. Osmont and C. Goldblum (eds) *Villes et citadins dans la mondialisation*. Paris: Karthala, pp. 171–88.

Leclerc-Olive, M. 2003b. Arènes sahéliennes: communautaires, civiles ou publiques? D. Cefaï and D. Pasquier (eds) *Les sens du public. Publics politiques, publics médiatiques*. Paris: PUF (coll. Curapp).

Lorrain, D. (ed.) 1999. *Retour d'expériences (6 cas de gestion déléguée à l'étranger)*. Paris, Ministère de l'Equipement, May (research report).

Lorrain, D. (ed.) 2001. *Retours d'expériences (7 cas de gestion déléguée à l'étranger)*. Paris, CEMS/EHESS (research report).

Manor, J. 2002. Partnerships Between Governments and Civil Society for Service Delivery in Less Developed Countries: Cause for Concern. Presented at the *World Development Report 2003/04 Workshop: Making Services Work for Poor People*, Oxford, 4–5 November.

Meï, L. 2008. La gestion de l'eau dans des villages périurbains de Ouagadougou, Burkina Faso. Étude sociogéographique comparative. Bordeaux: Université Bordeaux 3 (doctorate thesis).

Olivier de Sardan, J.-P. 2000. *La gestion des points d'eau dans le secteur de l'hydraulique villageoise au Niger et en Guinée*. Paris: AFD, May (Etudes de l'AFD).

Ombogo, P.L. 2006. Strategy for access to water and sanitation services in informal settlements (the Kisumu experience). Paper at Africities Conference. http://www.pseau.org/event/africites_4/session_ss3_fr.htm.

Otayek, R. 2005. La décentralisation comme mode de redéfinition de la domination autoritaire ? Quelques réflexions à partir de situations africianes. Report to the VIII congress of the French Association of Political Science, Lyon: 14–16 September 2005. www.afsp.mshparis.fr/archives/congreslyon2005/communications/tr4/otayek.pdf.

Ostrom, E. 1992. *Crafting Institutions for Self-governing Irrigation Systems.* San Francisco (CA): Institute for Contemporary Studies.

Ostrom, E. 1996. Crossing the great divide: coproduction, synergy, and development. *World Development*, 24(6): 1073–87.

Page, B. 2003. Communities as the agents of commodification: the Kumbo Water Authority in Northwest Cameroon. *Geoforum*, 34: 483–98.

Pauw, J. 2006. Metered to death: how a water experiment caused riots and a cholera epidemic. The Center for Public Integrity (The Water Barons), www.publicintegrity.org/ water/report.aspx?aid=49.

Piveteau, A. 2005. Décentralisation et développement local au Sénégal. Chronique d'un couple hypothétique. *Revue Tiers monde*, 181: 71–93, January–March.

Ribot, J. 2007. *Dans l'attente de la démocratie: la politique des choix dans la décentralisation de la gestion des ressources naturelles.* Washington: World Bank.

Robinson, J. 2002. Global and world cities: a view from off the map. *International Journal of Urban and Regional Research*, 26(3): 531–54.

Robinson, J. 2005. Urban geography: world cities, or a world of cities. *Progress in Human Geography*, 29(6): 757–65.

Shirley, M. (ed.) 2002. *Thirsting for Efficiency: The Economics and Politics of Urban Water System Reform.* Washington: World Bank.

Smith, L. 2002. The murky waters of the second wave of neoliberalism: corporatization as a service delivery model in Cape Town. *Geoforum*, 35(4), 375–93.

Sonkin, E. 2006. Élaboration d'une stratégie municipale concertée dans la commune de Dschang pour l'accès à l'eau et l'assainissement. Paper at Africities Conference. http://www.pseau.org/event/africites_4/session_ss3_fr.htm.

Subramanian, A., Jagannathan, N.V., and Meinzen-Dick, R. 1997. *User Organizations for Sustainable Water Services.* Washington: World Bank (World Bank Technical Paper, 354).

Tendler, J. 1997. *Good Government in the Tropics.* (Johns Hopkins studies in development) Baltimore: Johns Hopkins University Press.

Tremolet, S. 2006. Un point sur les privatisations de l'eau en Afrique. *Responsabilité & Environnement*, 42: 59–68, April.

UNDP. 2006. *Human Development Report 2006. Beyond Scarcity: power, poverty and the global water crisis.* New York: Palgrave Macmillan.

Valfrey, B. 2005. *Livre bleu. L'eau, la vie et le développement humain. Etat des lieux et perspectives de l'atteinte des objectifs de développement du millénaire dans les secteurs de l'eau et de l'assainissement. Burkina Faso, Mali, Niger, Note de synthèse de trois études de cas.* Montreal: Secrétariat International de l'Eau.

Wegner, L. 2005. Privatisation: a challenge for sub-Saharan Africa. *Policy Insights* (OECD), no. 14: 1–3, October.

Chapter 12

Governance failure: Urban water and conflict in Jakarta, Indonesia[1]

Karen Bakker[1] and Michelle Kooy[2]

[1]Department of Geography, University of British Columbia, Vancouver, Canada
[2]Mercy Corps, Jakarta, Indonesia

12.1 INTRODUCTION: WATERING JAKARTA

In contrast to many other Asian cities, Jakarta has relatively minimal water supply and wastewater disposal infrastructure, and is characterized by lower rates of urban service provision than other national capitals. The World Bank characterizes Jakarta's water and sanitation sector as one of the weakest in Asia (Brennan and Richardson, 1989; Leitmann, 1995; McGranahan et al, 2001; World Bank, 2004b). Between 50 and 60% of the city's residents are connected to the network (Jakarta Water Supply Regulatory Body, 2004).[2] The water delivered through the network is not potable; medical studies repeatedly find faecal coliform contamination, and residents are advised to boil their water.

As discussed below, a number of conflicts have arisen over water supply in Jakarta: social, economic, and environmental. The private sector participation contract signed in 1998 has been one source of conflict, actively contested by NGOs and consumers. Conflict has arisen between the private company, the municipal government and the regulator over water prices and tariffs. Labour unrest, due to conflicts between water industry workers and the two private concessionaires, has also flared up. Underlying these conflicts is a long-standing pattern of inequitable access to clean, potable water for the city's residents; a pattern that has persisted for more than a century. Recent reforms to the national water law, which were intended to address some of these issues, have also generated conflict.

---

[1] This article was originally submitted to the editors in 2007, based on fieldwork conducted over the period 2001 to 2004.

[2] Figure calculated using 2002 data from annual reports of the two private concessionaires operating in Jakarta. This was cross-referenced with figures from ADB (2003b), which reports a figure of 51.2%. Coverage ratios should be understood as rough estimates; their calculation is dependent upon a number of variables that are only imprecisely measured, such as urban population and average size of household. Reported figures vary significantly, and do not indicate the number of households that have a connection but which rely primarily on other sources (e.g. groundwater) due to quality or service concerns (e.g. low pressure).

12.2   SPLINTERED URBANISM: FRAGMENTED ACCESS TO URBAN WATER SUPPLY

As documented below, the production of fragmented networks (or 'archipelagos') has been a consistent pattern in Jakarta during the past century, where inequitable access is literally hardwired into water supply networks. Scarcity, in other words, is mediated by human action; it is socially constructed – but nonetheless very real.[3] This is not to say, however, that fragmentation of the network and exclusion from water supply access are deliberate in a simplistic sense (although this is indeed sometimes the case). Rather, fragmented networks are perhaps better viewed as 'unintended consequences of development' (to use James Ferguson's language):[4] unintentional but nonetheless coherent outcomes of technical practices, political commitments and cultural norms.

The resulting conflicts over water use must be placed in context. Domestic per capita water consumption within the city is estimated to be, on average, between 70 and 80 litres (one of the lowest of the eighteen large Asian cities surveyed) (ADB, 2003b). Jakarta has almost no sewer system (with less than 2% of households connected to a sewerage system) (ADB, 2003b); the vast majority of wastewater is deposited directly into rivers, canals or to (often poorly functioning) septic tanks (Crane, 1994; McIntosh, 2003; Surjadi, 2002). Rivers and canals are sometimes too polluted to use even for washing clothing. Contamination by wastewater and industrial effluent, as well as seawater infiltration due to over-pumping have, in turn, polluted the shallow aquifer in many areas of the city. Reliance on groundwater as one of the multiple sources of water is common in Jakarta, and provides a de facto alternative to networked water. But while wealthy residents rely on cleaner, but more expensive deep wells, poor residents rely on shallow aquifers (Table 12.1). In some areas, depletion and salinization of the latter has rendered water unfit for drinking and cooking (Braadbaart and Braadbaart, 1997). The public health impacts of this situation are predictable, and have been well-documented: high rates of water-related diseases, including gastrointestinal illness due to contaminated water and parasite-related illnesses due to poor drainage, particularly in poorer areas (Agtini et al, 2005; Leitmann, 1995; McGranahan et al, 2001; Simanjuntak et al, 2001; Surjadi, 2003). Echoing the concerns of survey respondents, general public mistrust of the quality of network water in Jakarta is demonstrated by the rapid growth of bottled water consumption by poor households over the past five years, particularly via 'non-name' *air isi ulang* (refilled bottled water) supplied by private (and unregulated) water treatment microplants throughout the city (Forkami, 2006; Weimer, 2006). This has deepened the reliance of poor households on water vendors in North Jakarta – the poorest part of the city, and the area with the greatest salt intrusion into the aquifers, further entrenching the parallel and informal networks of vended water suppliers.

As a result, most Jakarta residents (as in many mega-cities) use multiple sources of water in the home (Table 12.1). Due to poor quality, low pressure and incomplete network coverage, most residents depend on a variety of water sources, including deep and shallow wells, water vendors and bottled water (Berry, 1982; Gilbert and James, 1994; Lovei and Whittington, 1993; McGranahan et al 2001). Residents of Jakarta

---

[3] See Kooy and Bakker (2008a, 2008b). See Xenos (1989) for a discussion of scarcity under modernity.
[4] Ferguson (1994).

Table 12.1 Household sources of water supply in six Jakarta *kampungs* (2005)

Water source	# houses	%
a) Groundwater	39	37
Groundwater with bottled water/vended water/public hydrant	41	39
Network water	10	9
Network water with groundwater	2	2
Public hydrant/vended water with rainwater	11	13
Total	106	100
Total households using at least two sources	65	61
b) DW	3	3
bottled water	12	11
groundwater	70	64
vended water	34	31
HU	7	6
PAM	32	29
other public toilet	4	4
public hydrant	8	7
TA	13	12
Total	183	166
Total % exceeds 100 because some households use multiple water sources.		

Source: household survey by authors (2005, n = 110).

obtain their water supply through a complex, heterogeneous set of sources, techniques and modes of delivery. Few residents rely on one source, using a combination of household piped network water connections, shallow and deep wells, public hydrants, and water vendors for their water supply needs (Surjadi, 2002, 2003). According to our survey of 110 households in six Jakarta neighbourhoods in 2005, 61% of households surveyed used multiple sources (the three most frequent combinations being network and vended water, network and groundwater, and groundwater and vended water)[5] (Bakker, et al 2008).

Use of different water sources varies temporally and seasonally, due to quality and pressure concerns. Low pressure in the piped network means that households prefer to have a backup source – often a well. In some areas of the city, however, shallow groundwater cannot be used for drinking due to salinization and pollution of the shallow aquifer (due to pumping, sea-level intrusion and surface wastewater disposal in the absence of a sewerage system).

This heterogeneity of use is further complicated by Jakarta's spatial pattern of urban development and urban service provision. Within the city, blocks of commercial properties and colonial-era mansions fronting on broad avenues are intermixed with dense 'illegal' settlements of poorly serviced houses and self-built dwellings among the inner

---

[5] These findings are similar to the results of surveys conducted by Surjadi (1994, 2002, 2003) and McGranahan et al (2001).

blocks (Cowherd, 2002; Ford, 1993; Leaf, 1996; Porter, 1996). The latter are located on empty lots and along any streets wide enough to accommodate built structures while still permitting the passage of traffic. This pattern has intensified following the informalization of much of the city's economy following currency devaluation in 1998. Many neighbourhoods do not have access to piped water, as the water network is concentrated in wealthier areas of the city (Martijn, 2005). The resulting spatial differentiation of land use and income has created a form of 'urban dualism', with middle-class houses abutting informal housing in a highly variable urban micro-geography in which multiple water sources will be in use simultaneously.

Even in those areas with networked water supply, many homes do not have individual household connections. Susantono finds, in an extensive survey, that informal water services 'thrive' in neighbourhoods where formal services are available, with households relying on water vendors even when they have the option of house connections with the municipal water utility (Susantono, 2001). The reasons for this are explained below.

12.3 URBAN GOVERNANCE: THE PRODUCTION OF THIRST

As explored above, physical proximity of the network (as indicated by the distribution of a tertiary pipe network in the neighbourhood) is not always associated with residential network connections. Why would this be the case? The answer is that the choice of which source to use is influenced by factors other than physical availability of a network. One important factor is the total cost of water supply (as distinct from the cost per unit volume of water). In a pattern typical of third world cities (Cairncross et al, 1990; Gulyani, 2005; Swyngedouw, 1997), piped water supply costs less per unit volume in Jakarta than other modes of water supply, particularly vended water. In comparisons of prices of vendor water *versus* networked water supply, the price per unit volume was found to be from ten times to thirty-two times more expensive in the case of vendor water.[6] Poor households typically rely on vendor water, whereas wealthier households have access to the networked water supply system; as a result, many poor households pay more per unit volume of water than wealthier residents of the city. Given their lower incomes, many poor households spend a much higher proportion of their income on water than wealthier households. In our survey of 110 households, 43% of households spent more than 5% of their income on water bills (often cited as appropriate threshold by international aid organizations).[7] Wealthier households with a networked connection, in other words, receive water at a lower cost per unit volume, spending lower proportions of income for much greater quantities of water. Unsurprisingly, levels of water consumption are positively correlated with wealth in Jakarta (McGranahan et al, 2001).

---

[6] ADB (2003b); McGranahan et al (2001); and a survey conducted by the author in the neighbourhood of Sunter Agung in January 2001. ADB (2003b) gives a maximum figure of US$4.17/m^3.

[7] A study of 1,000 households in Jakarta which examined the different prices paid by different wealth groups found that, overall, the poor pay on average twice as much per metre cubed as the wealthy (McGranahan et al 2001), and that water expenditure represents, on average, 10% of income in poor households.

On the basis of cost per unit volume alone, then, it would seem counter-intuitive for poor households not to connect to the water supply network where possible. However, the disincentive for connection becomes more obvious when we consider the *total cost* of connecting to the water supply system (as opposed to price per unit volume of water supply). Monthly bills include more than charges per unit volumes of water consumed. Fixed charges (such as the meter fee and the annual charge) are also added on to the bill. For a poor household whose residents consume 50l/person/day (the World Health Organization's recommended minimum), the fixed charges will be anywhere from five to ten times as high as the volumetric consumption charge; the effective cost per unit volume will thus be higher than that of vended water for the poorest consumers. Moreover, a networked water supply implies additional infrastructure costs to be borne by the consumer, in the form of a water tank or holding device, made necessary because of the intermittent nature of water supply through the piped network (with cut-offs of several hours occurring daily in some areas). Transaction costs are also significant; long waiting times at water utility offices to pay bills and clear up meter mis-readings raise transaction costs compared to the ease of complaint handling and convenience of home visits by vendors to collect bill payments. Connection fees are also significant (ranging from Rp 200,000 to Rp 350,000 in the households surveyed), relative to average incomes of poor households (which averaged Rp 1.4 million in the households surveyed), and must usually be provided as a lump sum. This may pose significant barriers to households with small, irregular incomes. Connection fees also vary depending on distance from the network; poor households are more likely to live in areas of lower network density (Figure 12.1), and thus pay higher fees for connection. For all of these reasons, overall costs to poor households of vended water may be lower than networked water supply, even though the latter has a lower price per unit volume. Given these cost barriers, the payment flexibility permitted by vendors (some of whom even allow customers to buy water on credit) is a significant incentive for poor households, who may have limited budgeting capacity, to choose vended water over networked water (Susantono, 2001; Shofiani, 2003).

Another important factor is land tenure. Deep wells are expensive and have higher maintenance costs; this effectively prohibits development by those without permanent tenure. A significant proportion of the city's population lives in temporary (often self-built) accommodation without secure tenure. In these instances, public hydrants and vended water become the sole or primary source of supply. Surjadi et al found that over 20% of the city's residents regularly buy drinking water from vendors (Surjadi et al, 1994). The most recent academic survey found that approximately one-third of Jakarta's households purchase water from street vendors (Crane and Daniere, 1996). These figures correspond with the results of our household survey, which found that 31% of respondents regularly bought vended water (Table 12.1 above).

Another disincentive is perceived quality of water provided through the network. In our survey, networked water was perceived to be of lower quality than other sources of water (especially groundwater), particularly by more educated respondents. Residents of Jakarta perceive groundwater to be of higher quality than either vended or network water. Indeed, the most comprehensive comparative survey of water quality of different sources in poor neighbourhoods in Jakarta to date found that samples of drinking water from the network were more contaminated with fecal coliform than

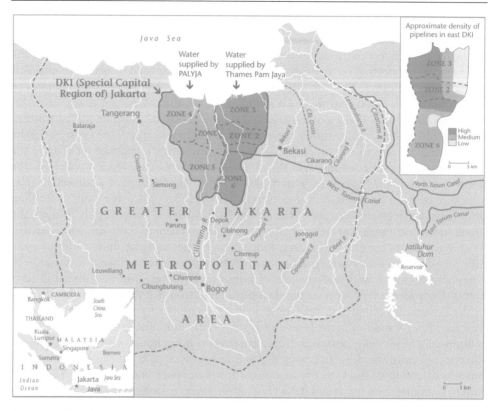

Figure 12.1 **The Capital District of Jakarta, showing the two Private Sector Participation concessions for water supply**

Source: Author.

groundwater (Surjadi et al, 1994). In some cases, vended water was perceived to be of higher quality than networked water supply. The fact that vendors check water quality and may strain the water or let it settle before delivering explains why perceptions of vended water quality may be higher, despite the fact that vendor water often originates in hydrants connected to the networked water supply system.

International financial institutions often claim that one of the most important factors contributing towards low levels of water service provision in Jakarta is low levels of infrastructure finance (see, for example, Akhtar, 2005; World Bank, 2004b), exacerbated by the Asian financial crisis and currency devaluation. Initiatives such as the Indonesia Infrastructure Summit (held in Jakarta in early 2005) explicitly targeted foreign direct investment. The government has identified a significant shortfall in financing requirements for rehabilitation and extension of urban infrastructure. Jakarta's infrastructure 'crisis' is not, however, a new phenomenon. Indeed, the current lack of funding for services is compounded by the legacy of deliberate under-investment during the 1960s and 1970s. This period of extremely rapid population growth saw policymakers actively discouraging rural-urban migration,[8] while articulating a policy stance that sewerage was a 'private concern' (Argo, 1999; Cowherd, 2002).

The lack of finance to deal with the demands created by a rapidly growing urban population only partially explains this situation. Another set of issues pertains to governance. Jurisdictional fragmentation has reduced the ability of any one level of government in Jakarta to effectively govern water resources within a watershed, or even within urban boundaries.[9] Municipal governance structures are another factor. In Jakarta (as with other cities in Indonesia and indeed around the world), water utility budgets were not ring-fenced from that of the municipality. Rather, a small tax base and the relative lack of alternative revenue-generating activities for the municipal government encouraged the use of water utility revenues for non-water related expenditure by municipal politicians and managers. For many water supply utilities in Indonesia, this had the effect of reducing the available revenue for operating costs and funding capital expenditure (notably infrastructure rehabilitation and improvement), a situation exacerbated by relatively low-cost recovery rates. Like many water supply utilities worldwide, Indonesian water providers are often caught in a vicious cycle: low-cost recovery, low revenue, low investment and low levels of service (Bakker, 2003; Cross and Morel, 2005; Nunan and Satterthwaite, 2001).

Two more issues merit consideration. First, a culture of urban governance in Indonesia shapes urban planning policy in ways that discourage government initiatives for systematic, structured water supply initiatives. As Benedict Anderson and Abidin Kusno have argued (Anderson, 1990; Kusno, 1997), the idea of 'power' in Javanese culture can be understood via the spatial metaphor of the 'beacon', in which concentration and affirmation of power at the centre (e.g. the palace, or the capital city) enables an extension of power over the periphery. Power, in other words, is enacted through projecting a strong and unified source of authority. In the contemporary context, state governance, and the patron-client hierarchies which embody it, are directed towards affirming the centrality of state power. In the context of urban planning, this means that state activities may be geared towards the reaffirmation of prestige and reinforcement of networks of patronage, rather than public welfare *per se* (Cowherd, 2002; Kusno, 1998, 2000). In the post-Independence era, for example, high priority

---

[8] Jakarta City alone grew from 1.8 million people in 1950 to 6.5 million in 1980, with equally rapid population growth in the surrounding metropolitan areas (total population of the greater metropolitan area is now estimated at 18 million). This implies rapid increases in population density and significant changes in land use (Chifos, 2000; Firman, 1997, 1998, 2000; Lo and Yeung, 1996).

[9] In the Jakarta region, for example, the majority of the JMA is politically constituted as an independent territory with a status of a province – 'DKI Jakarta' (Special Capital Region of Jakarta). The city governor is independent from West Java province, and (together with the municipal government) controls the city's water supply company: PAM Jaya. The province of West Java is responsible for the urban areas which fall outside of DKI Jakarta, and for the watershed in which the main Jatiluhur reservoir for Jakarta's water supply is sited (well upstream from the city). Environmental and urban planning regulations are not systematically applied within the watershed, and the open canals which act as conduits for Jatiluhur water are polluted by residential and industrial effluent, posing serious water quality challenges to municipal water supply utility engineers. Meanwhile, within the city, the tackling of groundwater pollution from urban effluent is complicated by the division of responsibility amongst the sewerage authority, the municipal water utility (which controls networked water supply), and the national government's Ministry of Mines, which bears responsibility for regulating deep (i.e. drilled) wells, from which a substantial proportion of the city's residents draw water.

was given to an urban redevelopment agenda focused on highly visible infrastructure (avenues, highways, parks, monuments and sculptures). This 'monumentalist' infrastructure served a dual purpose of urban development and source of pride for the new nation, whereas urban service provision and the urban environment were given relatively low priority (Chifos and Hendropranoto, 2000; Firman and Dharmapatni, 1994; Ford, 1993; Kusno, 1997; World Bank, 2004b). This occurred in spite of sporadic national government-led development plans to accelerate service delivery (Silas, 1992). The 'monumentalist' architecture of Jakarta's central corridor under Presidents Sukarno and Suharto (whose reigns lasted successively from Independence until 1998), together with the relative neglect of infrastructure and public services of all kinds for the urban periphery, are illustrative of the implications of this model of urban governance. Where 'pro-poor' water supply initiatives do occur, they are often sporadic, intermittent and emphasize state largesse (and recipient gratitude), rather than an integrated and long-term approach to network extension.

Second, the non-networked systems of water supply in the city, particularly standpipes and water vending, are controlled by vested interests that see little advantage in network extension. For example, water vendors typically operate as spatial monopolists; vendors do not compete, but rather collude – via a network of middlemen – to establish monopoly supply zones and a captive clientele (Lovei and Whittington, 1993; Susantono, 2001). Indeed, the potential profitability of extracting rent from the captive market of water consumers is recognized through the practice of selling informal 'licenses' amongst water vendors (Susantono, 2001). In Jakarta, as in other cities, water-vending is sometimes linked with organized crime, and at times characterized by intimidation (if not outright violence) of customers, competing vendors and police and water company employees who attempt to eradicate informal water-vending practices. Some observers argue that mafia-like control of water vending in poor areas of the city is a significant barrier to network expansion. This is significant, as surveys have found that approximately one-third of Jakarta's households purchase water from street vendors (World Bank, 1993; Crane and Daniere, 1996; Crane et al, 1997 and survey by author in June/July 2005) (Table 12.1). An even more startling form of illegal activity is the existence of illegal network builders (some of whom even operate with business cards), who install illegal tertiary pipes and household connections without the authorization (and often the knowledge) of the water supply concessionaires. Palyja/Suez estimated 30 *kilometres* of illegal pipeline in their concession area alone in 2005.

12.4 GOING PRIVATE: CONFLICT OVER THE WATER SUPPLY CONCESSION CONTRACT FOR THE CITY OF JAKARTA

Proponents of private sector participation (PSP) in water supply have argued that PSP is a means of improving service delivery to the poor (see, for example, Cross and Morel, 2005; Nickson and Franceys, 2003; Franceys and Weitz, 2003).[10] Specifically,

---

[10] See, for example, the Global Water Partnership (www.gwpforum.org/) and the World Water Council (www. worldwatercouncil.org/) – two influential networks of private water companies, governments and lending agencies. The Business Partners for Development links the World Bank with private water companies and governments, and 'aims to produce solid evidence of the positive impact' of PSPs (www.bpdweb.org).

proponents argue that private companies improve performance (including cost recovery rates) and increase access by extending networks and providing new connections to previously 'unserved' customers. This is made possible by a combination of efficiency gains, improved management, and better access to finance than public utilities. Access to network connections benefits the poor, particularly in urban areas, who are often served by a variety of informal arrangements such as water vendors, and typically pay much higher prices per unit volume for poorer quality water than wealthier consumers (Johnstone and Wood, 2001; Shirley, 2002; World Bank, 1994, 1997, 2004a).

Opponents of private sector participation argue that PSPs are not effective or appropriate mechanisms for supplying water services to the poor. Some refer to the relatively limited proportion of 'pro-poor' household connections provided under PSPs (Marin, 2009). Others argue that the profit imperative will foreclose private sector involvement in poor regions, and point to the withdrawal of the private sector from contracts and various regions of the world, in light of risk-return ratios, which have remained unacceptably high in part because of the low 'ability to pay' of poor consumers (Castro, 2008; Hukka and Katko, 2003; Prasad, 2006; Smith, 2002).[11] Other critics of PSP rest their objections on ethical grounds (such as the argument that water is a human right) (Morgan, 2005).

This ongoing debate has resulted, in some cases, in increased political risks for private water companies. Political risks were also high, in part, stemming from political protest against private sector participation, across countries of all income levels (Hall, Lobina and de la Motte, 2005). These included a variety of mechanisms: court cases, campaigns to pressure governments into cancelling contracts and street protests (in rare cases leading to contract cancellation or expropriation). These protests brought together a broad range of groups, representing a range of interests, including organized labour, consumers, environmentalists, women's groups and religious organizations (Bennett et al, 2005; Morgan, 2005; Olivera and Lewis, 2004). In some instances, protests were framed in terms of an outright rejection of private sector participation; in other instances, specific issues were disputed, such as tariff increases. By 2005, the World Bank acknowledged the fact that 'the frequent (not inevitable) result [of water privatization] was popular protests, dissatisfied governments, and unhappy investors' (World Bank, 2005). This had the effect of raising the costs of private sector participation and, according to some observers, contributed to diminishing private sector interest in long-term concessions (given the greater exposure of this type of contract to political risks).

12.4.1 The private sector participation contract in Jakarta

Debates between proponents and opponents of private sector involvement have been acute in Jakarta. The private sector participation contract signed in 1998 with two international operators was intended to improve water quality, mobilize international finance for network expansion, and thereby improve and increase access to water

---

[11] For academic studies critical of the privatizsation process, with a focus on developing countries, see the Municipal Services Project website (http://qsilver.queensu.ca/~mspadmin). For an international public sector union perspective, see the very comprehensive PSIRU website (www.psiru.org). For a campaigning NGO perspective, see the Council of Canadians Blue Planet Project (www.canadians.org/blueplanet/index2. html) and the US-based Public Citizen's campaign on water supply (http://www.citizen.org/cmep/Water/).

supply for Jakarta residents. As documented in this and the following section, however, key original performance targets have been dramatically scaled back, and pro-
poor efforts on the part of both the public and private sectors have not fulfilled
expectations. Tariff pricing (with lower tariff bands below marginal costs), decided by
the municipal government in negotiation with concessionaires, is implicitly 'anti-
poor', providing a disincentive to both the municipality and the private concessionaires to connect the poor. The physical layout of the network, which is spatially
concentrated in wealthier areas of the city – a legacy of public sector management –
constitutes an additional barrier to supply poor households.

Moreover, poor households have multiple disincentives to connect to the network.
The total costs of networked water supply may be higher than alternative sources
(such as groundwater or vended water). Other disincentives include insecure tenure,
the need for flexibility of payment, convenience, status, and the high 'transaction
costs' associated with dealing with formal water utilities. Other disincentives include
transaction costs: infrastructure costs to build storage to compensate for intermittent
networked water supply; line-ups and time off work to pay bills (for those without
bank accounts and regular income); and concerns over meter misreading and bill over-
charging, and the consequent loss of time.

Jakarta's government exhibited a heightened interest in the urban environment and
service provision in the 1990s, as typified by the then-governor's preferred slogan for
the city: 'Bersih, Manusiawi, Wibawa' (Clean, Humane, Powerful) (Leaf, 1996).
Concerns about the poor level of service in the water sector had persisted for decades,
and water shortages and water quality problems were perceived to be increasingly
acute (Berry, 1982; Gilbert and James, 1994; *Indonesia Times*, 1996; Lovei and
Whittington, 1993). One response (in Jakarta as in other Indonesian cities) was limited private sector participation: out-sourcing of routine repairs, billing and payment
collection by Jakarta's water supply utility, PAM Jaya (Mandaung, 2001). Water supply was one of many PSP initiatives ongoing in the country; the Indonesian government had passed legislation enabling private sector participation and privatization for
most public sector utilities in the mid-1990s, and had embarked on private ventures in
various sectors over the past decade, for example, privately funded toll highways
throughout the greater Jakarta area.[12]

Discussions regarding a long-term PSP concession contract with foreign firms began
in the mid-1990s. International water companies were keenly interested in entering the
water services market in Indonesia. The country was a large, middle-income state with
an expanding middle class and relatively low penetration of networked water supply
services. After protracted negotiations, 'cooperation agreements' for the management
and expansion of Jakarta's water supply system were awarded in late 1997 to two of
the largest water services companies in the world:[13] the British company Thames
Water International and the French company Ondeo (Suez-Lyonnaise des Eaux). The
process of awarding the contract for Jakarta's water supply was characterized by what
the political science literature defines as 'collusive corruption' (where government and

---

[12] Private sector participation contracts in water supply have been signed in several other Indonesian cities:
Bali, Batam, Medan, Lhok Seumawe and Sidoarjoand Pekanbaru (Baye, 1997; ADB, 2003b).
[13] Sanitation services were not included in the contract and remain the responsibility of the various municipalities that make up the greater Jakarta area.

private sector officials collude to deprive the government of revenues) (Bardhan, 1997; Shleifer and Vishny, 1993). This took the place of a public tendering process (where international water companies tender unsolicited proposals directly to the government). Under then-President Suharto, partnership with an Indonesian firm was a pre-requisite for international corporations hoping to take over the operations of a utility network. This is not unusual in the international water supply sector, in which private sector consortia typically have local minority shareholders. In the case of Indonesia, however, these private sector consortia were frequently linked directly to the President. By the early 1990s the large Indonesian conglomerates had 'already [become] active within other public service areas, and these groups expected to benefit from the privatization of water services' (Baye, 1997: 201). The two international firms were part-nered with two local private firms, run by individuals closely associated with President Suharto.

In January 1998, each consortium signed a twenty-five-year contract with PAM Jaya, the municipal water supplier in Jakarta, which retained ownership of the water supply assets. That the contracts were awarded despite national laws prohibiting foreign investment in drinking water delivery (Law No. 1/1967; Ministry of Home Affairs Decision No. 3/1990) and local regulations (No. 11/1992 and No. 11/1993) precluding private sector involvement in community drinking water supply was to be a source of conflict in the early years of the contract (Argo and Firman, 2001). The terms of the contracts also became the source of future conflict. The private consortia were to be responsible for operation of the water supply system, including administration of the customer database and billing. Thames' contract allocated the local partnership, Thames PAM Jaya (TPJ), the exclusive right to operate and manage the existing water supply system in the eastern half of the city,[14] supplying 2 million people connected to the supply system out of a potential customer base of 5 million. Simultaneously, Lyonnaise des Eaux's subsidiary, Palyja, was given a contract to supply the western half of the city (Figure 12.1), covering a slightly larger number of potential customers. Ambitious targets were set: the private companies committed to supplying potable water to consumer by 2007 and to reaching universal coverage by 2023.

The contracts were expected to be lucrative for both the local and international partners. Under the terms of the contract, this profit was not linked directly to the revenues of the municipal water supply system. Instead, each consortium was to receive a fee on the basis of volume of water supplied and billed, *not* on the basis of the water tariff (set by the municipality), or the percentage of cost recovery. With no direct equity stake, and with profit de-linked from cost-recovery rates, the international water companies thus sought to minimize the risk inherent in cost-recovery. An additional safeguard was built into the payment mechanism: an indexation formula, linked to the rupiah-US dollar exchange rate and the (Indonesian) inflation rate, was built into the 'water charge' formula used to determine payments made to the private operators. The operators were paid according to unit volume of water delivered to the distribution network rather than billing revenue. Cost recovery and currency risks, in other words, were to be borne by the government, rather than the private companies.

---

[14] *Indonesia Times*, Privatised water supply begins soon, 16 January 1998, p. 3.

12.4.2 Re-regulation: tariffs, profits, and conflictual re-negotiation of the contract

The political and economic turmoil that unfolded in Indonesia in 1998 vitiated these strategies. Riots, the resignation of Suharto, and the abrupt and dramatic devaluation of the Indonesian rupiah[15] threw the country into a period of chaos. After a tense interlude in which senior expatriate managers of the private concessionaires fled the country, local managers cancelled the PSP contracts, and senior British and French executives and diplomats pressured the federal government to have the contracts rein-stated, the private concessionaires resumed operations (having discreetly abandoned their Indonesian partners, now tainted by their association with ex-President Suharto) (Harsono, 2005).

Confronted with public protest over rising prices of staple food items and gasoline, the municipal government refused to raise tariffs to compensate for the devaluation of the rupiah. This delay in tariff increases should not, in theory, have posed difficulties for the private water companies, as revenues are determined by a 'water charge' paid per unit volume of water delivered into the network. This means that revenue of the private operators is not linked to amounts billed or collected from consumers. In other words, the revenue of the private concessionaires is, in theory, independent of cost-recovery as well as tariffs. Indexing the water charge to the rupiah-USD exchange rate provided protection against currency devaluation; should the rupiah fall in value, the water charge (expressed in rupiah), would rise accordingly.

The limitations of this strategy were revealed when receipts in dollar terms plummeted from 1998 onwards. Given the political unrest in Jakarta, the Governor was unwilling to implement agreed-upon tariff increases. The gap between the water charge required for compensating the private companies and the average water tariff increased dramatically. Whereas, the water charge paid to the private operators was 11% below the average tariff in 1997, it rose to over 60% above the average tariff in early 2001. Subsequent tariff increases did not raise the tariff above the water charge until early 2004 (Jakarta Water Supply Regulatory Body, 2004).

As a result, the amount charged by the private concessionaires – via the water charge – to the government increased dramatically, while revenue fell just as dramatically. PAM Jaya (and thus the local government) bore the sole risk for the revenue shortfall, and became increasingly indebted to the private companies. The cumulative deficit by the end of 2001 was Rp 469 billion (approximately US$46 million) and had reached Rp 990 billion (approximately US$97 million) by September 2003 – excluding late payment interest and retroactive tariff increases (Jakarta Water Supply Regulatory Body, 2005).

The time period for repayment of this debt by PAM Jaya is likely to be protracted. With the fall in value of the rupiah, its operating revenues fell approximately four-fold in dollar terms. At the time, PAM Jaya's revenue was forecast to be on the order of Rp 400 billion per year (approximately one-twentieth of the outstanding 'debt'). The negotiated tariff increases are likely to be less than 10% per year (Pam Jaya, 2003). Thus, although tariffs were raised and will continue to increase, these increases will not generate sufficient revenue to quickly repay the 'shortfall'. By mid-2001, the

---

[15] From approximately Rp/US$2,300 in 1997 to Rp/US$10,000 in 1998.

Table 12.2 Original and renegotiated household connection targets (1998–2008)

	Number of connections Unit		Service coverage ratio %	
	TPJ	Palyja	TPJ	Palyja
Baseline (before privatization)	231,607	176,980	52.00	38.00
Original Targets (1997)	361,607	395,522	70.00	70.00
Revised Targets (2002)	335,413	301,048	62,00	45.00
Realization (2002)	336,550	312,879	62,17	44.17
Revised Targets (2008)	403,030	391,980	75.50	61.00
Realization (2008)	379,480	398,507	65.28	61.58

Sources: PAM Jaya 2004; JWSRB (2008); TPJ and Palyja annual reports, Shofiani (2003).

prospect of slow repayment of the still increasing 'debt' provoked a renegotiation of the contract, transforming it into a management contract – with a guaranteed internal rate of return of 22% – rather than the original concession agreement.[16] Technical targets were dramatically scaled back (Table 12.2): most notably, the commitment to provide potable water supply at the point of consumption was dropped, and targets for connection rates were reduced. The large debt and prolonged and conflictual tariff negotiations were two reasons which likely influenced Thames Water in its decision to withdraw from the contract in late 2006.

12.4.3 Conflicts with water utility workers: labour-led protests and unrest

Prior to the initiation of the concession contract, employment at PAM Jaya fluctuated between 1,100 and 1,200 full-time (equivalent) employees. The cooperation agreement signed in 1997 between PAM Jaya and the private operators allowed for the secondment of employees to help TPJ and Palyja run water facilities on the production and distribution side. Of these, between 800 and 900 employees were permanent and on direct contract, with the remaining 200 and 300 on yearly contracts. This secondment arrangement is quite common in PSP contracts in developing countries. Local employees retain their jobs while the (often foreign) concessionaire obtains required local expertise and language skills.

The secondment arrangement has given rise to concerns on the part of employees with respect to their status and benefits. PAM Jaya is a municipally owned company,[17] subject to the relatively restrictive labour legislation applying to the public sector. Confronted with perceived surplus of employees (relatively frequent in public utilities in developing countries), and prevented from firing those who remained nominally

---

[16] As with many such contracts, profits are 'backloaded'. The Internal Rate of Return is calculated over the lifetime of the contract, and is, to date, negative.

[17] The formal category is BUMD (Badam Usaha Milik Derah) which loosely translates as a 'state-owned company', the owner in this case being the municipality.

civil service employees, TPJ and Palyja re-assigned staff to PAM Jaya, while complaining to the municipality of the rigidity of the labour laws and weak sanctions available to the private concessionaires against poor performance. The civil service employees, on the other hand, resented being excluded from day-to-day operations, particularly as new private sector employees – with higher salaries, and sometimes benefits such as company cars – were hired directly by the private concessionaires.

As labour relations worsened, the political situation in the country also deteriorated, with widespread riots leading up to President Suharto's eventual departure. Widespread political unrest led the expatriate staff to flee the country, and the municipality declared that the contract had been cancelled. Against the backdrop of political unrest and the generalized backlash against foreigners, protests by the local water company staff grew more vehement. These peaked in 2000, with union members demanding cancellation of the contract, staging public protests, physically closing neighbourhood kiosks and company offices, and even welding the doors of company offices shut. By this point, labour relations had degenerated significantly. Company managers accused workers of acting to protect lucrative and corrupt practices entrenched in PAM Jaya's billing and subcontracting practices. These accusations were countered by accusations of corruption on the part of the union, directed against the foreign water companies in tendering and, of course, the original bidding process. In the context of endemic corruption in Indonesia with recognized effects on urban management (see, for example, Server, 1996), these unproven allegations were mutually damaging yet unsurprising. The municipal government's adoption of a middle-ground position – refusing to raise water tariffs (despite a contractual obligation to do so) and also refusing to cancel the contract – was a key factor in dampening union protests. Although work slowdowns were still in place in 2001, strike action had petered out.

12.4.4 Connecting the poor? Conflict over tariffs and pricing

Implicit in the original technical target of 100% service coverage, and explicit in public justifications of the PSP contracts, was the belief that private sector participation in water supply would lead to a higher rate of connection of poor households. Service coverage has increased since 1998, but the distribution of new connections has not been 'pro-poor', if this is defined as a rate of connection equal or greater to the percentage of poor in the urban population. This has led to an increase in consumer protest, including court challenges brought against the municipal water utility (highly unusual in a society that was characterized, until the fall of Suharto, by an authoritarian style of governance in which consumers' rights were not articulated through the legal system).

An important goal of the original concession agreement was extension of the network and increase in coverage, for which targets were specified in the original contracts. By 2002, however, service coverage level for both concession areas remained just above 50%, well below the 70% target specified for 2002 in the initial contract (Global Water Report 2002) (Table 12.1). New connections have occurred, but these have not targeted poor customers in proportion to their representation in the urban population. Figure 12.2 illustrates the disproportionate weighting of consumer connections in middle-income tariff bands in 2003. Whereas the majority of residents in Jakarta would fall into the 'lower-middle' and 'low-income' categories, 87% of

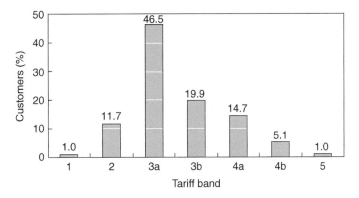

Figure 12.2 Customers per tariff band (%) (2003 data)

Source: PAM Jaya (2003).

Table 12.3 New connections, East Jakarta 1998–2004, by tariff band

	Tariff Group	*# New connections*	*% increase*
I	Social institutions and public hydrants	1,101	1
II	Public hospitals, poor and very poor households	21,898	24
IIIa	Middle-income households and small-scale businesses	51,847	58
IIIb	Upper middle income households and government offices	11,150	12
IVa	Large hotels, high rise buildings, banks and factories	2,323	3
IVb	Harbour/port	1,849	2
Total		90,167	100

Source: Thames Pam Jaya, personal communication (May 2005).

networked connections are provided to tariffs for middle-income households or above. This is, to some extent, the legacy of public-sector management, attributable to unwillingness by the municipally managed utility to extend the network into poor areas due to fears about low cost-recovery (Taylor, 1983), and to a tariff pricing policy in which water rates for public hydrants (used by poor households and water vendors) were higher per unit volume than water rates for individual households – implying a reduction in revenue when a poor household was connected to the network (Crane, 1994).

However, the legacy of public sector under-provision of poor households was not redressed by the private concessionaires. Table 12.3 provides data on the numbers of new consumers connected in each tariff band by one of the two partnerships (TPJ) over the period 1998 to 2004. Only 25% of new connections were targeted in the two lowest tariff bands (public hydrants, intended to serve those without household connections, and 'poor' and 'very poor' households). In other words, three-quarters of new connections were for middle-income and upper-income households, government enterprises and commercial enterprises.

Given that private concessionaires are paid via a 'water charge', linked to volumes of water delivered into the water supply system but independent of revenues and tariffs, this bias towards wealthier consumers might seem surprising. There is no apparent direct disincentive to the private concessionaires to connect low-income households. Why, then, were customers in the lowest tariff bands less likely to be connected? An important part of the explanation lies in the pricing levels of the tariff bands (Table 12.4). The lowest tariff (May 2005 data) is 550 rupiah/m^3, well below the production cost (of approximately 3,000 m^3).[18] Increasing the number of connections in the lowest tariff band thus decreases the average revenue per cubic metre supplied. Reducing the average revenue per cubic metre by connecting poor households would lower the municipality's revenues, in turn reducing their ability to pay the water charge, and repay the debt shortfall owed to the private operators. A secondary disincentive is the higher average cost per connection in poor neighbourhoods, which raises installation costs: given the lack of land-use planning in informal settlements, the highly dense and disordered distribution of homes means that installing connections may be more time-consuming (if conventional below-ground infrastructure is used).

The municipality thus has two direct incentives not to target poorer neighbourhoods for new connections. This is an important explanation for why the utility, when under public management, did not connect poor customers. In turn, this produces an indirect disincentive for the private operators to connect poor customers: the revenue received by the municipality is the source of funds from which the private operators are paid, and decreasing revenues implies a greater chance of debt, longer repayment periods, and increased possibility of municipal default.

The perverse disincentives built into Jakarta's water supply tariff structure are an example of how pricing strategies intended to increase access may have counterproductive goals (Whittington, 1992). The remedy, as most commonly prescribed by international financial institutions, is to increase tariffs (a seemingly counterintuitive strategy), thereby removing the disincentive for connecting poor consumers and providing more capital to finance new connections (e.g. Azdan, 2001; Yepes, 1999). This recommendation is supported by studies, which assert that 'willingness-to-pay' and 'ability-to-pay' of poor customers is higher than previously thought. Frequently, the higher rates per unit volume paid by poor customers relying on water vendors are cited as evidence for this argument (e.g. Soto Montes de Oca et al, 2003; Winpenny, 1994). Indeed, the response to the problem of low tariffs in Jakarta has been a series of negotiated tariff increases, which have disproportionately raised tariffs for poorer and middle-income groups (Table 12.4).

This, in turn, raises a more general point about water pricing, pertaining to the limits of cross-subsidization within a water-pricing regime in cities, such as Jakarta, with a large proportion of poor residents. With a ratio of domestic to industrial customers of 80/20 and with relatively few users in higher tariff bands, possibilities for cross-subsidies in Jakarta are relatively limited. This suggests that in the absence of subsidies external to the water supply pricing regime – as used in Chile and previously in OECD countries such as the UK (Bakker, 2004; Gomez-Lobo, 2001) – raising tariffs will not necessarily have the desired goal of increasing revenues and increasing

---

[18] Interview with Alizar Anwar, Advisor to the Jakarta Regulatory Body, May 2005.

Table 12.4 Tariffs, fixed charges, and tariff increases (2003–2005)

	Average Tariffs per Tariff Band Rp/M3 (2005)	*Monthly fixed charges (Rp) (2005)*	*Tariff Group Description* (2003–2005)	*% customers per tariff band (2003 data)*	*% increase in tariffs (2003–2005)*
I	550	4,695	Social institutions (e.g. religious facilities) and public hydrants	1.0	47
IIa	550	5,060	Public hospitals and very poor households	11.7	47
IIb	2,450	10,440	Low income households	46.5	44
IIIa	3,500	11,950	Middle income households and small-scale businesses	19.9	59
IIIb	5,100	19,390	Upper middle income households and government offices	14.7	32
IVa	9,750	19,390	Large hotels, highrise buildings, banks and factories	5.1	48
IVb	11,500	27,665	Harbour/port	1.0	31

Sources: Pam Jaya (2003); DKI Jakarta (2005).

rates of connection of the poor. Additional measures will also be required to reduce or remove disincentives for poor customers to choose network connections. This is discussed further in the following section.

The slow pace of connections for poor customers, and the renegotiation of contractual targets for connections, have been the target of sustained critique by Jakarta-based NGOs such as WALHI, Kruha, the Indonesian Consumer's Association and the Urban Poor Consortium. These groups have launched campaigns employing the media, direct action protests, testimonials from poor communities, and a civil suit against the Jakarta municipality and water companies contesting the legality of tariff increases. The emergence of water as a conflictual issue in this regard is striking. Interviews during a research visit in 2001 indicated that few Jakarta NGOs thought water supply was a major issue, whereas it had become the most high-profile campaign issue for most of these NGOs by 2005.

12.4.5 Pro-poor initiatives

Having recognized some of these barriers to connecting the poor, both private concessionaires have since undertaken limited initiatives to improve access for poorer households. To render in-house connections more affordable, Palyja introduced a policy allowing poorer households (on the lowest tariff bands) to pay the connection fee in twelve monthly installments, included in the monthly water bill.[19] Partly as a result of

---

[19] The pro-rated monthly connection fee of US$0.71, included in the monthly water bill. A household consuming 20 m^3 a month will thus have a monthly bill of about US$1.50 (US$0.0375 × 20 + US$0.71).

this the number of poor people served in West Jakarta increased from 72,816 in February 1998 to 177,164 in December 2000 (ADB, 2003a). However, the monthly bills remain at a level above what many households can afford. In the eastern concession area, the community of Marunda was targeted by TPJ, which used a grant from its British parent company to subsidize the provision of in-house connections. Fewer than 500 households were connected. To facilitate payment, connection fees were waived and households were required to pay a deposit instead (of approximately US$2.50) (ADB, 2003a). Levels of water consumption have reportedly increased dramatically, while water bills have fallen substantially (ADB, 2003a). Prior to the concession contract, households in Marunda District generally received their water from private vendors who purchased water from tankers. Households used to spend, on average, US$7.50 a month for $3\,m^3$ of water (five 20-litre containers per day, at US$0.05 a container). Now they pay approximately US$1.125 for $30\,m^3$ of water (at US$0.0375 per $metre^3$ – most customers being on a low tariff, reflecting the small size of their dwellings). In short, these households are consuming ten times as much water but paying approximately one-seventh of their previous monthly bills. This is partly the reason for high levels of cost recovery from the newly connected households (BPD, 2003). Recognizing the limited penetration of the water supply network into poor neighbourhoods, the federal government launched a water supply programme in some of the poorest *kampungs* in 2004.[20] However, the provision of household connections was severely limited by the disincentives discussed above, compounded by the unwillingness of private partners to extend the network in conjunction with the government, and by suspicion on the part of some public sector managers that publicly provided infrastructure would end up providing implicit subsidies to the private sector (Shofiani, 2005).

Given the high level of indebtedness on the part of the municipal water utility towards private concessionaires, little interest has been shown by private companies for extending what are essentially charitable, loss-making initiatives. Accordingly, these 'pro-poor' initiatives have remained limited in scope, and have not been duplicated elsewhere in the city. Without an explicit 'pro-poor' policy on the part of the government, and in the absence of specific pro-poor targets in the contract, new connections in poorer areas are likely to lag in proportion to the overall increase in new connections for the reasons discussed above.

Recognizing this, donors have begun re-funding community water supplies in Jakarta. The (American bilateral aid donor) USAID, through its Environmental Services Program (with a budget of US$40 million over five years), is funding decentralized small-scale community water supply systems in West Java, including Jakarta. These community systems will not connect users to the network, but will rely on alternative technologies. In addition, the World Bank approved a US$5 million 'output-based aid' concessional loan for expanding network coverage in Jakarta.[21] With an explicitly 'pro-poor' focus, this latter project provides cheap capital to the two concessionaires to connect the poor.

---

[20] Under the auspices of the Kimpraswil Fuel Subsidy Reduction Compensation programme, created to offset the impacts of a reduction in fuel subsidies on poor households.

[21] The USAID project was underway and the World Bank project was in the tendering stage.

The difficulty of extending water supply to low-income households is demonstrated by the outcomes of the World Bank project, initiated in 2006. The project encountered a stumbling block early on: most households considered to be sufficiently poor to receive a subsidized connection do not have legal land-tenure status, so cannot receive a subsidy. Despite government lobbying by private sector operators to change this tenure policy, the original target was scaled back to fewer than 4,000 households.[22] Even direct subsidization of private sector operators by the World Bank had failed to enable water supply network access to be extended to the city's poorest households.[23]

Meanwhile, one of the two private operators (Palyja) reported an increase in the numbers of low-income households served: from 11,659 in 1998 to 69,581 households in 2007 – a total increase of 57,922 households (PALYJA, 2008). This figure should be contrasted however, with a total increase of 103,154 for middle and upper-class households over the same time period (PALYJA, 2008).[24] Despite constituting a small proportion of total households, wealthier households continue to account for the majority of in-home connections, and have benefited disproportionately from new connections. Moreover, these figures pertain to only the western half of the city (and are likely to be much worse in the eastern, poorer half). Moreover, by 2009, a decade after the start of the contract, government interest in rolling out connections to lower-income households remained a low priority, for two reasons. First, political pressure to keep tariffs low limited the potential for cross-subsidization. Second, a new emphasis on network connections for commercial and high-end properties using deep groundwater (largely in the central business district) meant that new investments would be focused largely on increasing supply capacity via network rehabilitation, rather than expanding the network into unserviced areas or improving supply in low-income areas.

By this time, the civil-society groups that had flourished following Suharto's fall from power had launched a concerted campaign against water privatization. Consumers and environmental groups joined forces to criticize the record of the private companies. Although court cases launched against the municipal and federal governments failed, the public attention drawn to water issues in the capital increased political pressure on Jakarta's governor to rein in tariff increases and to push the companies to improve their lagging performance. Relations grew progressively more strained. By 2006, one of the private companies had withdrawn, and the other had substantially modified its role in running the water supply system, thereby reducing its risks.[25]

[22] Forkami (2006).

[23] Here, I am referring to an 'output-based aid' project, initiated in Jakarta through the World Bank's Global Partnership on Output-Based Aid. Intended to 'accelerate' the connection of poor households (given that the two private firms had failed to meet the original contract targets for new connections), the project had resulted in late 2008 in only 4,000 of an original target of 40,000 new connections. For an analysis of PSP performance in general, see (Marin, 2009).

[24] The precise figures are: from 108,159 to 159,493 'upper class households and small business', and from 64,829 to 116,649 'middle-class' households (PALYJA, 2008).

[25] In 2006, Thames Water sold its stake in its Jakarta water services subsidiary to an Indonesian-led consortium (AERTRA) in 2006, which now manages the eastern half of the city. PalyjaOndeo succeeded in salvaging a proportion of its investments through floating a domestic bond issue for water investment, although it continues to operate the management contract for the western half of the city.

12.5 CONCLUSIONS: GOVERNANCE FAILURE

Our analysis of the history of Jakarta's water supply network suggests that the efforts of government and the private sector with respect to water supply have generated conflicts. We emphasized the fact that the situation is more complex than might be implied by a simple public-private binary relationship. It is, rather, a situation in which market failure, state failure and governance failure have all combined to fragment access to urban water supply. In response, poor households play an active role in finding alternative solutions, some of which are neither public nor private, and are sometimes expressive of resistance to the remedies prescribed by development experts.

This provides an alternative viewpoint from which to consider water supply access in a city like Jakarta. Both public and private operators of the water supply system are subject to constraints that relate to questions of culture (such as the colonial discourse of the 'ungovernable' *kampungs*) economics, and to norms of urban governance as well as utility performance. The seemingly paradoxical resistance to connecting the urban poor living in illegal settlements only becomes comprehensible if this debate is framed in terms of Jakarta's culture of urban governance and the Indonesian experience of modernity in which this culture of governance is embedded. The 'governance failures' evident in Jakarta encompass, on the one hand, practical issues (such as targeting, internal incentives and the capacity to provide services in a manner amenable to the needs of the poor) and, on the other, issues of power and politics (such as elite capture of urban infrastructure). Unless these underlying governance failures are addressed, urban water conflicts are likely to persist in Jakarta, and indeed in megacities around the world.

ACKNOWLEDGEMENTS

Numerous individuals (water company employees, consumers, government officials, NGO representatives) gave generously of their time during fieldwork research in Jakarta in 2001 and 2005. Fieldwork research support by Michelle Kooy, Nur Endah Shofiani and Ernst-Jan Martijn is gratefully acknowledged. Helpful comments were received from Michelle Kooy, Graciela Madanes-Schneier, Philippe Le Billon, Olivier Coutard, Bernard Barraqué and Sylvy Jaglin.

REFERENCES

ADB. 2003a. *Beyond boundaries: extending services to the urban poor*. Manila: Asian Development Bank.

ADB. 2003b. *Water in Asian Cities: summary of findings of the study and a regional consultation workshop*. Manila: Asian Development Bank.

Agtini, M., Rooswanti, S., Lesmana, M., Punjabi, N., Simanjuntak, C., Wangsasaputra, F., Nurdin, D., Pulungsih, P., Rofiq, A., Santoso, H., Pujarwoto, H., Sjahrurachman, A., Sudarmono, P., von Seidlein, L., Deen, J., Ali, M., Lee, H., Kim, D., Han, O., Park, J., Suwandono, A., Oyofo, B., Campbell, J., Beecham, H., Corwin, A., Clemens, J. 2005. The burden of diarrhoea, shigellosis, and cholera in North Jakarta, Indonesia: findings from 24 months surveillance. *Infectious Disease* 5(89).

Akhtar, S. 2005. Mobilizing finance for infrastructure development. Speech (Director General, Southeast Asia Department, Asian Development Bank), Indonesia Infrastructure Summit.

17 January, Jakarta. Accessed on 14 October 2005: http://www.adb.org/Documents/Speeches/2005/sp2005004.asp.

Anderson, B. 1990. *The idea of power in Javanese culture. Language and Power: exploring political cultures in Indonesia.* Ithaca: Cornell University Press, 152–93.

Argo, T. 1999. Thirsty downstream: the provision of clean water in Jakarta, Indonesia. Unpublished doctoral thesis, School of Regional and Community Planning, University of British Columbia.

Argo, T. and Firman, T. 2001. To privatize or not to privatize? Reform of urban water supply services in Jabotabek, Indonesia. *Built Environment* 27(2), 146–55.

Azdan, M. 2001. Water policy reform in Jakarta, Indonesia: a CGE analysis. Unpublished doctoral thesis. Department of Agricultural, Environmental and Development Economics: Ohio State University.

Bakker, K. 2004. *An Uncooperative Commodity: Privatizing Water in England and Wales.* Oxford: Oxford University Press.

Bakker, K. 2003. From archipelago to network: urbanisation and privatisation in cities in the south. *The Geographical Journal*, 169(4): 328–41.

Bakker, K., Kooy, M., Shofiani, E. and Martijn, E.J. 2008. Governance failure: rethinking the institutional dimensions of urban water supply to poor households. *World Development*, 36(10): 1891–915.

Bardhan, P. 1997. Corruption and development: a review of the issues. *Journal of Economic Literature*, 35(3): 1320–46.

Barlow, M. and Clarke, T. 2002. *Blue Gold: the fight to stop the corporate theft of the world's water.* New York: New Press.

Baye, E. 1997. Experimenting with the privatization of water services in Indonesia – from Surabaya to Djakarta. D. Lorrain (ed.) *Urban Waste Management.* Levallois: Hydrocom, pp. 195–211.

Bennett, V., Dávila-Poblete, S. and Nieves Rico, M. (eds) 2005. *Opposing Currents: the politics of water and gender in Latin America.* Pittsburgh: University of Pittsburgh Press.

Berry, B. 1982. Clean water for all: equity-based urban water supply alternatives for Indonesia's cities. *Urban Geography*, 3(4): 281–99.

BPD. 2003. *Jakarta, Indonesia: Water Supply Improvements to Marunda District.* London: Building Partnerships for Development in Water and Sanitation. Accessed 3 January, 2004: http://www.bpd-waterandsanitation. org/english/projects/jakarta.asp.

Braadbaart, O. and Braadbaart, F. 1997. Policing the urban pumping race: industrial groundwater overexploitation in Jakarta. *World Development*, 25(2), 199–210.

Brennan, E.M. and Richardson, H.W. 1989. Asian megacity characteristics, problems and policies. *International Regional Science Review*, 12(2): 117–29.

Castro, J. 2008. Neoliberal water and sanitation policies as a failed development strategy: Lessons from developing countries. *Progress in Development Studies.* 8(1), 63–83.

Cairncross, S., Hardoy, J. and Satterthwaite, D. (eds) 1990. *The Poor Die Young.* London: Earthscan.

Chifos, C. 2000. Meeting unmet expectations: Initiatives in urban environmental service provision in Jakarta. C. Chifos and R. Yabes (eds) *Southeast Asian Urban Environments.* Program for SEA Studies, Arizona State University. Arizona: Temple University Press, pp. 29–72.

Chifos, C. and Hendropranoto, S. 2000. Thirty years of urban infrastructure development in Indonesia. C. Chifos and R. Yabes (eds) *Southeast Asian Urban Environments.* Program for SEA Studies, Arizona State University. Arizona: Temple University Press, pp. 153–81.

Cowherd, R. 2002. Cultural construction of Jakarta: design, planning and development in Jabotabek, 1980–1997. Unpublished PhD dissertation, Department of Architecture: MIT.

Crane, R. 1994. Water markets, market reform and the urban poor: results from Jakarta, Indonesia. *World Development*, 22(1): 71–83.

Crane, R. and Daniere, A. 1996. Measuring access to basic services in global cities: descriptive and behavioural approaches. *Journal of the American Planning Association*, 62(2): 203–21.

Crane, R., Daniere, A. and Harwood, S. 1997. The contribution of environmental amenities to low-income housing: a comparative study of Bangkok and Jakarta. *Urban Studies*, 34(9): 1495–512.

Cross, P. and Morel, A. 2005. Pro-poor strategies for urban water supply and sanitation services delivery in Africa. *Water Science & Technology*, 51(8): 51–57.

DKI Jakarta. 2005. *Sosialisasi Pelaksanaan PKPS-BBM*. Jakarta, Departemen Permukiman dan Prasarana Wilayah; Bidang Infrastruktur Pedesaan.

Ferguson, J. 1994. *The Anti-Politics Machine: "Development", depoliticization, and bureaucratic power in Lesotho*. Minneapolis: University of Minnesota Press.

Firman, T. 1997. Land conversion and urban development in the northern region of West Java, Indonesia. *Urban Studies*, 34: 1027–46.

Firman, T. 1998. The restructuring of Jakarta Metropolitan Area: a 'global city' in Asia. *Cities*, 15(4): 229–43.

Firman, T. 2000. Rural to urban land conversion in Indonesia during boom and bust periods. *Land Use Policy*, 17: 13–20.

Firman, T. and Dharmapatni, I. 1994. The challenges to sustainable development in Jakarta metropolitan region. *Habitat International*, 18(3): 79–94.

FORKAMI. 2006. Social Assessment for the Jakarta Output-Based Aid Project. Jakarta: Forum Komunikasi Pengelolaan Kualitas Air Minum Indonesia.

Ford, L. 1993. A model of Indonesian city structure. *Geographical Review*, 83(4): 374–96.

Franceys, R. and Weitz, A. 2003. Public-private community partnerships in infrastructure for the poor. *Journal of International Development*, 15(8): 1083–98.

Gilbert, A. and James, D. 1994. Water pollution in Jakarta Bay. D. James (ed.) *The Application of Economic Techniques in Environmental Impact Assessment*. Kluwer: Dordrecht, pp. 111–42.

Gomez-Lobo, A. 2001. Incentive based subsidies: designing output-based subsidies for water consumption. *Public Policy for the Private Sector*, 22. Washington DC: World Bank.

Gulyani, S., Talukdar, D. and Mukami Kariuki, R. 2005. Universal (Non)service? Water markets, household demand and the poor in urban Kenya. *Urban Studies*, 42(8): 1247–74.

Hall, D., Lobina, E. and de la Motte, R. 2005. Public resistance to privatization in water and energy. *Development in Practice*, 15(3/4): 286–301.

Harsono, A. 2005. When water and political power intersect: a journalist probes the story of water privatization in Jakarta, Indonesia. *Nieman Reports*, 59(1): 45–47.

Hukka, J.J. and Katko, T. 2003. Refuting the Paradigm of Water Services Privatization. *Natural Resource Forum*, 27: 142–55.

Indonesia Times. 1996. Water shortage maybe come out of control in year 2000. *Indonesia Times*, 19 August.

Jakarta Water Supply Regulatory Body. 2004. *Regulatory Approach to the Jakarta Water Supply Concession Contracts*. Presentation by A. Lanti (Member, Regulatory Body), Stockholm Water Symposium, August 16–20.

Jakarta Water Supply Regulatory Body. 2005. *Urban Water in Jakarta*. Presentation by Urban Water and Citizenship workshop, University of Indonesia, 1 May.

Johnstone, N. and Wood, L. 2001. *Private Firms and Public Water: realising social and environmental objectives in developing countries*. Cheltenham: Edward Elgar.

Kooy, M. and Bakker, K. 2008a. Splintered networks? Urban water governance in Jakarta. *Geoforum*, 39(6): 1843–58.

Kooy, M. and Bakker, K. 2008b. Technologies of government: constituting subjectivities, Spaces, and Infrastructures in Colonial and Contemporary Jakarta *International Journal of Urban and Regional Research*, 32(2): 375–91.

Kusno, A. 2000. *Behind the Postcolonial: architecture, urban space and political cultures in Indonesia*. London: Routledge.

Kusno, A. 1998. Custodians of transnationality: metropolitan Jakarta, middle-class prestige, and the Chinese. H. Dandekar (ed.) *City Space and Globalization: an international perspective*. College of Architecture and Urban Planning, University of Michigan pp. 161–70.

Kusno, A. 1997. Modern beacon and the traditional polity: Jakarta in the time of Sukarno. *Journal of Southeast Asian Architecture*, 1: 25–45.

Leaf, M. 1996. Building the road for BMW: culture, vision and the extended metropolitan area of Jakarta. *Environment & Planning A*, 28(9): 1617–35.

Leitmann, J. 1995. Urban environmental profile: A global synthesis of seven urban environmental profiles. *Cities*, 12(1): 23–39.

Lo, F.C. and Yeung, Y.M. 1996. Global restructuring and emerging urban corridors in Pacific Asia. Y.M. Yeung and F.C. Lo (eds) *Emerging World Cities in Pacific Asia*. Tokyo: United Nations University Press.

Lovei, L. and Whittington, D. 1993. Rent-extracting behaviour by multiple agents in the provision of municipal water supply: a study of Jakarta, Indonesia. *Water Resources Research*, 29(7): 1965–75.

Mandaung, M.E. 2001. Improving the efficiency of water supply management: a case study of the Jakarta public water supply enterprise, Indonesia. Canberra: Centre for Developing Cities, University of Canberra. Accessed 12 January 2001 and 24 June 2002: cities.Canberra. edu.as/publications/Policypaper/Melva.htm.

Marin, P. 2009. *Public-Private Partnerships for Urban Water Utilities: a review of experiences in developing countries*. Washington DC: World Bank.

Martijn, E.J. 2005. *Hydraulic* Histories of Jakarta, Indonesia (1949–1997): development of piped drinking water provision. Vancouver: Unpublished research report, Department of Geography, University of British Columbia.

McGranahan, G., Jacobi, P., Songsore, J., Surjadi, C. and Kjellén, M. 2001. *The Citizens at Risk: from urban sanitation to sustainable cities*. London: Earthscan.

McIntosh, A., 2003. *Asian Water Supplies: reaching the urban poor*. Manila: Asian Development Bank and International Water Association. Accessed 13 October 2005: http:// www.adb.org/Documents/Books/Asian_Water_Supplies/

Morgan, B. 2005. Social protest against privatization of water: forging cosmopolitan citizenship? M-C. Cordonier Seggier and C.G. Weeramantry (eds) *Sustainable Justice: reconciling international economic, environmental and social law*. Boston: Martinus Nijhoff Publishers.

Nickson, A. and Franceys, R. 2003. *Tapping the Market: the challenge of institutional reform in the urban water sector*. London: Palgrave MacMillan.

Nunan, F. and Satterthwaite, D. 2001. The influence of governance on the provision of urban environmental infrastructure and services for low-income groups. *International Planning Studies*, 6(4): 409–26.

Olivera, O. and Lewis, T. 2004. *Cochabamba! Water War in Bolivia*. Boston: South End Press.

PALYJA. 2008. Implementation of the GPOBA Project in Jakarta West Zone. Jakarta, Indonesia.

PAM Jaya. 2003. Laporan Evaluasi Kinerja Tahunan. Annual Report Evaluation 2003. Jakarta: Perusahaan Daerah Air Mimun.

Porter, R. 1996. *The Economics of Water and Waste: A case study of Jakarta*. Indonesia. Aldershot: Avebury.

Prasad, N. 2006. Privatisation results: private sector participation in water services after 15 years. *Development Policy Review*, 24(6): 669–92.

Server, O.B. 1996. Corruption: A Major Problem for Urban Management. *Habitat International* 20(1): 23–41.

Shirley, M. 2002. *Thirsting for Efficiency: the economics and politics of urban water system reform.* London: Elsevier.

Shleifer, A. and Vishny, R. 1993. Corruption. *Quarterly Journal of Economics.* August, 599–617.

Shofiani, N.E. 2003. Reconstruction of Indonesia's drinking water utilities: assessment and stakeholders perspective of private sector participation in the capital province of Jakarta, Indonesia. Thesis LWR-EX-03-30, Stockholm: Department of Land and Water Resources Management, Royal Institute of Technology.

Shofiani, N.E. 2005. Assessment of the Kimpraswil water supply project: fuel subsidy reduction compensation program for water supply. Vancouver: Unpublished report, Department of Geography, University of British Columbia.

Silas, J. 1992. Government-community partnerships in Kampung improvement programmes in Surabaya, *Environment and Urbanization,* 4(2), 35–36.

Simanjuntak, C.H., Larasati, W., Arjoso, S., Putri, M., Lesmana, M., Oyofo, B.A., Sukri, N., Nurdin, D., Kusumaningrum, R.P., Punjabi, N.H., Subekti, D., Djelantik, S., Sukarma, Sriwati, Muzahar, Lubis, A., Siregar, H., Mas'ud, B., Abdi, M., Sumardiati, A., Wibisana, S., Hendarwanto, B., Setiawan, B., Santoso, W., Putra, E., Sarumpaet, S., Ma'ani, H., Lebron, C., Soeparmanto, S.A., Campbell, J.R. and Corwin, A.L. 2001. Cholera in Indonesia in 1993–1999. *American Journal of Tropical Medicine and Hygiene,* 65(6): 788–97.

Smith, L. 2002. The murky waters of the second wave of neoliberalism: corporatization as a service delivery model in Cape Town. *Geoforum,* 35(4): 375–93.

Soto Montes de Oca, G., Bateman, I.J., Tinch, R. and Moffatt, P.G. 2003. Assessing the willingness to pay for maintained and improved water supplies in Mexico City. CSERGE Working Paper ECM-2003-11.

Surjadi, C. 2002. *Public private partnerships and the poor: drinking water concessions. Case study – Jakarta, Indonesia.* London: International Institute for Environment and Development.

Surjadi, C. 2003. *Public private partnerships and the poor: drinking water concessions, a study for better understanding public-private partnerships and water provision in low-income settlements.* Loughborough: Water, Engineering and Development Centre.

Surjadi, C., Padhmasutra, L., Wahyuningsih, D., McGranahan, G. and Kjellén, M. 1994. *Households Environmental Problems in Jakarta.* Stockholm: Stockholm Environment Institute.

Susantono, B. 2001. Informal water services in metropolitan cities of the developing world: the case of Jakarta, Indonesia. Berkeley: Doctoral thesis, Department of City and Regional Planning, University of California.

Swyngedouw, E. 1997. Power, nature, and the city. The conquest of water and the political ecology of urbanization in Guayaquil, Ecuador: 1880–1990. *Environment and Planning A,* 29(2): 311–32.

Taylor, J. 1983. An evaluation of selected impacts of Jakarta's Kampung Improvement Program. Unpublished phd thesis, Department of Urban Planning, UCLA.

Weimer, M. 2006. *Action Research on Point of Use Drinking Water Treatment Alternatives as appropriate for underprivileged households in Jakarta.* Jakarta: Environmental Services Program, August.

Whittington, D. 1992. Possible adverse effects of increasing block water tariffs in developing countries. *Economic Development and Cultural Change,* 41(1): 75–87.

WHO. 2000. *Global Water Supply and Sanitation Assessment. 2000 Report.* Geneva: World Health Organization, UNICEF and Water Supply & Sanitation Collaborative Council.

Winpenny, J. 1994. *Managing Water as an Economic Resource.* London: Routledge.

World Bank. 1993. *Indonesia: urban public infrastructure services.* Washington DC: World Bank, East Asia Region.

World Bank. 1994. *Infrastructure for Development. World Development Report 1994.* Washington DC: World Bank.

World Bank. 1997. *The State in a Changing World. World Development Report 1997.* Washington DC: World Bank Group.

World Bank. 2004a. *Making Services Work for Poor People. World Development Report 2004.* Washington DC: World Bank Group.

World Bank. 2004b. *Indonesia: averting an infrastructure crisis: a framework for policy and action.* Indonesia Infrastructure Department, East Asia and Pacific Region, World Bank.

World Bank. 2005. *Infrastructure Development: the roles of the public and private sectors: World Bank Group's approach to supporting investments in infrastructure.* World Bank Guidance Note. Washington: World Bank.

Xenos, N. 1989. *Scarcity and Modernity.* London: Routledge.

Yepes, G. 1999. Do cross subsidies help the poor to benefit from water and wastewater services? Lessons from Guayaquil. UNDP – WB Water and Sanitation Program Working Paper.

Chapter 13

Man-made scarcity, unsustainability and urban water conflicts in Indian cities

Marie-Hélène Zérah[1], S. Janakarajan[2] and Marie Llorente[3]

[1]Centre de Sciences Humaines, New Delhi, India, and Institute of Research for Development, Paris, France
[2]Madras Institute of Development Studies, Chennai, India
[3]Centre Scientifique et Technique du Bâtiment, Paris, France

I am not the minister of water resources but the minister of water conflicts[1]

13.1 INTRODUCTION

Most cities in India are facing a form of water crisis, be it related to water resource scarcity or water access. Problems and concerns pertain to quantity and quality, equity across different sections of the population, poor sanitation, ineffective and obsolete wastewater management practices and lack of long-term vision, planning and motivation. Poor regulation and overuse of resources, and the development of large hydraulic projects feeding into obsolete and inefficient distribution systems, within a supply-driven approach, contribute to the actual water crisis (in some cases) and the perceived one (in other cases). The ongoing process of urban population growth, which predates the network extension, also critically weighs upon the access issue with the inability of operators (municipalities or state public agencies) to cope with the demographic and spatial growth of cities. Even though the urban transition in India is limited when compared with other countries (around 30% of the population is officially urban), the added yearly urban population is significant in absolute numbers. Levels of service vary from state to state and according to the size of cities. The level of access to piped water and sanitation in secondary cities and small towns is much below that of large cities (World Bank, 2006). The reinforced top-heavy urban hierarchy also aggravates inequities in service delivery. In terms of water service, this leads to problems such as reliance on untreated groundwater, the critical role of which in the process of urban development is given very little attention (Janakarajan, 2004a); the large range of compensatory strategies of users to ensure a minimum level of services (Zérah, 2000; Llorente and Zérah, 2002); and the withering of the current model of governance.

Both resource and service-related issues are of great concern and deserve to be put in perspective in an attempt to present the type of ongoing debates in India. The chapter is based on two case studies with the assumption that, despite their peculiarity, the cases of Chennai and Delhi embody a degree of representativeness. The first section

---

[1] The Minister for Water Resources, Mr. Priyaranjan Das Munsi, in a 2005 World Bank report.

aims to provide a definition of urban water conflicts as well as a general overview of the dimensions of the water crisis in India. The case of Delhi further highlights the limits of the current model of service provision and looks into proposed reforms that led to diverging views around the introduction of private sector participation, the role of international funding agencies and the reform path itself. It reveals the conflicting visions on the 'governance regime' for water provision (see second section). Facing acute water scarcity, Chennai,[2] is emblematic of the looming threat of serious conflicts in resource-sharing between cities and their peri-urban and rural areas. It raises the issue of finding a smooth and acceptable way to transfer water-use rights from farmers to urbanites as the city's footprint expands.[3] This is compounded by the environmental issue of unplanned and unregulated use of groundwater, which leads to a situation like the 'tragedy of the commons' (see third section). The two case studies highlight distinctive processes and issues, and yet both illustrate deficiencies in the existing institutional set up that relate to addressing long-term planning and conflicting concerns. This, we argue, stresses the need for mechanisms, at the least, to create dialogue platforms, and, at the best, to evolve efficient and sustainable resolution mechanisms.

13.2 OUR UNDERSTANDING OF URBAN WATER CONFLICTS

A conflict always implies, irrespective of its origins, objectives or progress, an opposition between at least two categories of actors, whose interests are temporarily or fundamentally divergent. Conflict differs from a crisis situation or even a tension. It acts as a signal that a 'given state' is becoming unstable. We shift from a tension to a conflict when one of the parties implements a credible threat. There are several indicators. For instance, the use of the media, one party bringing the other before the courts, or the production of signs (such as a notice); finally, both parties may enter into a direct confrontation (verbal or physical). Shifts also occur from a crisis to a conflict situation when existing arrangements fail due to varying factors, such as a breach of trust or the realignment of stakeholders. Conflicts can generate debate and fights, but they can also lead to new arrangements. It is important to underline the point that a conflict does not necessarily constitute the last stage in the degradation of a relation, or a market failure. It simply crystallizes divergent particular interests that will take more or less time to be channelled and debated in constructive and democratic ways. Furthermore, controversies can also reveal new aspects of a problem, while uncovering the position of excluded actors, previously without a say in policy-making or implementation (Lascoumes, 2004). As such, conflicts can also be perceived as a signal and a force for competition and change. They constitute a modality of coordination and negotiation, like any other, and can thus contribute to change, which can be both 'positive' and 'negative'; to reorienting a debate; and even to a solid project around actors and issues raised and mobilized by the conflict (Lascoumes, 2004).

In India, there is a growing consensus about the water crisis and worsening tensions. A primary element of this crisis is the availability of water resources (for a recent

---

[2] Chennai (ex-Madras) is the capital of Tamil Nadu.

[3] This question is of growing importance in the context of Indian urbanization, but few articles examine the specific dynamics of peri-urban zones (see Kundu et al, 2002 and Shaw, 2005).

update, see Briscoe and Malik, 2006) both in terms of quantity and quality. As demands from all sectors increase, decline in water quality generates conflicts between uses. A second tension is the decline of the Hydraulic State model, where solutions usually seek to increase the availability of water resources, despite inefficient networks. The escalating costs of this approach delay investment and thereby network improvement, due to expectation of freshwater resources from new and further away dams. Substitutive modes of provision and reliance on groundwater by users are a signal of the limits of the supply-based model, which de facto increases scarcity and worsens the current governance regime. Indeed, the literature has dealt at length with the various deficiencies in the management of urban networks and service-delivery mechanisms, such as the unsustainable financial situation of providers (high costs of operation and maintenance, low tariff, high debt), inefficient organizations and the lack of accountability among others (Davis, 2004; Zérah, 2009; World Bank, 2006 for an overview of the governance issues).

Scarcity, in the Indian urban context, is a mostly man-made phenomenon that is conducive to the flaring-up of localized conflicts. We assume that a situation becomes conflictive when the existing conventional mode of supply does not enable the system to sustain itself. The situation becomes socially unstable (with user complaints, press coverage and mostly political consequences) and also unsustainable from an environmental resource point of view. The following sections bring to light the shift from crisis to conflict (either open or latent), which indicates both a critical state as well as the potential development of a new alignment to redefine water policies. In both cases, we shall try to highlight the conditions of the crisis that crystallize in a conflict, the ins and outs of the question at hand, the stages of the conflict, an analysis of the winners and losers, the links with the actual stakes, and the existence or absence of resolution mechanisms that prevent or make possible a credible and stable way out of the conflict and provide a new equilibrium.

13.3 THE ISSUE OF WATER ACCESS IN DELHI

In Delhi, the public undertaking, the Delhi Jal Board (DJB),[4] is unable to meet the water and wastewater needs of the nation's capital. This is a typical case of man-made scarcity with a high demand-supply gap, despite potentially available resources (Maria, 2005). The DJB provides its citizens with an erratic[5] and unequally distributed water supply that is well below international standards. The Board's financial situation is dismal with a considerable debt, high operation and maintenance costs, and limited streams of revenues due to very low tariffs. The urban poor, in particular, are more affected with lower access rates to piped water and little water available overall (27 litres/capita/day in slums as per Llorente, 2002). Water access conflicts emerge as a result and are in a way displaced through alternative provision modes (Zérah, 2000).

---

[4] In the case of mega-cities, supply is either the responsibility of a municipal department (Mumbai and Calcutta are two examples) or is under the control of a separate water supply and sewerage board (Delhi and Chennai). Boards are usually considered more autonomous from a financial and an organization point of view but relatively well-run corporations can perform as well as water boards, and eventually better (Ruet, Saravanan and Zérah, 2002). Urban water supply is thus largely dominated by the public sector, yet restrictions on access (complete or partial) are no less real (Zérah, 2000; Llorente, 2002).

[5] Around four hours a day.

Limitation of access is less severe for better-off households, which to a certain extent worsens the already existing inequities.

13.3.1 User strategies and cross-bred networks

Substitutive strategies, which we call 'decentralized governance structures' in contrast to centralized water provision through the network, can be broadly divided into two categories: formal and informal strategies.

Formal strategies develop along market lines and are answerable to existing legislation. In Delhi, a diversity of markets, or rather niche markets, co-exist. The first type is developed by private water tankers, who sell water in large quantities for specific events, cooperative housing societies or industrial needs. The second type concerns packaged water, either in bottles or jars. Both these options are costly and affordable only to high-income households. Another problem is the lack of guarantee on water quality. Llorente and Zérah (2002) showed how regulation weaknesses enabled the emergence of small companies with short-term speculative strategies, which simply resell public water or sell untreated groundwater. However, some companies set up sophisticated production lines with a view to establishing themselves in the market on a long-term basis. Such companies have not come up with innovative solutions to provide services at affordable prices. Instead, these private ventures build their strength on the inefficiency of the public sector and only offer peripheral, temporary and localized solutions.

Informal strategies are external to any regulated market structure and can be individual, privately led, or result from a form of collective action. On an individual basis, the poorest people 'free-ride' on public water via illegal connections onto which they install cheap devices to pump water from the network. Higher-income households adopt more expensive strategies: electric pumps fixed on the network to obtain better pressure; storage in rooftop tanks; and tube-wells relying on groundwater.

In poor settlements, recent years have witnessed a rise in private or collective innovation. Ragupathi (2003) describes the setting up of 'small local entrepreneurs' who dig their own well, install a powerful motor and lay a simple system of pipes through nearby alleys. This 'network' can service up to about 200 households. The cost of an individual connection amounts to the expenditure incurred on the necessary plumbing. The household also pays a monthly subscription, six or seven times higher than the cost of municipal water, but in return benefits from a home connection and neighbourhood service. For the 'entrepreneur', investments are recovered in two years. Other initiatives owe more to genuine collective action, for example, the *gali* or alley taps described by Tovey (2002). Under this system, users contribute jointly for the installation and maintenance charges. Often, the process is triggered by political patronage and/or local leaders. Once the mechanism is operational, informal relationships come into play that sanction and maintain these connections. In each case, they involve different players (the police, local elected representatives, parliamentarians and eventually employees of the DJB). Regulation of these alley taps also involve collectively accepted rules and norms, which define a hierarchy in the allocation of rights. Households that contributed financially to the *gali* taps have prior right to the service. Residual rights are then granted to the tenants, and then to households situated in the vicinity of the water points, even if they have not contributed financially.

This ultimately complex system demonstrates the use value of water but also acknowledges its social value.

In more well-off (middle-class) areas, especially in new urban extensions, cooperative group housing societies, instead of relying on individual storage tanks and wells, have developed what Maria and Levasseur (2004) consider to be 'water systems'. These systems comprise three subsystems of treatment, intelligent storage and supply. They can be more or less innovative, based on reverse osmosis or ion exchange technology for treatment, dual system for supply, and elaborate storage devices, especially if water recycling is implemented. Under this last arrangement, a prerequisite is access to technologies in an emerging market of firms able to provide tailor-made equipment and services. A second condition is reliance on technical expertise provided by consultants (often architecture companies) to the cooperative society members.

Indeed, most of these informal strategies are linked to the formal economy (through credit, expertise and consultancy). They are also mostly characterized by their ability to evolve adequate and short-term solutions for users, and are embedded in the centralized piped network (of individual connections and public fountains), creating a crossbred network (Coing, 1996), whose management is increasingly complex.

13.3.2 The question of sustainability

From an economic point of view, all these strategies generate direct investment as well as time opportunity costs. In 1995, Zérah (2000) estimated these costs for connected households. The cumulated expense of households for such strategies was 6.5 times higher than the bills paid to the public undertaking. The aggregate cost of water unreliability at city level was equivalent to almost twice the annual expenditure incurred by the water board.[6] Over ten years, the level of service in Delhi has further declined and despite investment, only 1% of connected households receive a twenty-four hour water supply (World Bank, 2006). Consequently, the ratio of substitutive strategy costs per year to yearly water bills has gone up to 8.0 instead of 6.5 (World Bank, 2006: 19). The cost of these strategies can vary greatly but in almost all cases, they are not economically sustainable. The detailed work of Maria and Levasseur on 'water systems', which could be assumed to present a proper alternative (in terms of quality of service) to public piped water, shows that the cost of these systems can vary according to the amount of water available from the municipality. As public supply is characterized by uncertainty, optimizing the planning and management of these systems remains a distant objective, although they could be economically competitive in the future. Nevertheless, the economic sustainability of collective planned strategies can only be estimated if such systems are integrated into the supply system rather than being buffer solutions.

A major issue is the environmental impact of stopgap solutions. Unauthorized connections and sucking pumps contribute to the deterioration of existing infrastructure. During service interruptions, untreated or contaminated water leaks into the network, thereby increasing the risk of waterborne diseases. Second, the multiplication of tubewells and the emergence of water systems aggravate the already critical situation

---

[6] In 1995, the public undertaking was the Delhi Water Supply and Sewerage Disposal Undertaking. The Delhi Water Board was constituted in 1998.

of the groundwater table. In some areas, the table has gone down by a few meters in less than twenty years, notwithstanding a decline in water quality. In other words, a system of negative externalities becomes self-sustaining with a harmful impact on the environment and users' health. Furthermore, in the context of poor service provision, the availability of groundwater simply postpones strong and real conflicts among users, but this apparent functioning system cannot last long term. Thus, the situation gives rise to intergenerational conflicts between present and future users, whose interests are not taken into account.

From a social equity perspective, improving access for lower-income sections is a priority. Institutionalization of community participation mechanisms is desirable for at least three reasons. First, this would allow the additional costs of substitution strategies to be internalized and a more equitable redistribution system set up. Second, households would be provided with an effective means for ensuring that the infrastructure is properly maintained. Lastly, water resources would be more effectively managed, thanks to a demand-oriented approach and by facilitating leak detection. Thus, access rights to water would be secured. However, this would require major institutional change and, in particular, the democratic representation of all interests and the setting up of agreed-upon negotiation procedures (Haider, 1997; Llorente, 2002).

13.3.3 A chaotic reform process and unexpected outcomes

Current strategies are a response to an inefficient service run under an institutional set up unable to provide suitable incentives. They are affected by the absence of formal rules and do not permit a sound allocation of the resource. The existence of such arrangements suggests that reform of the sector should be analysed in a systemic way,[7] and that consideration should be given to the opportunities offered by decentralized governance structures. Until now, and despite a consensus around the notion of 'crisis', the main solutions have remained in the supply approach framework with the construction and management of new water treatment plants. This has just resulted in the addition of more capacity to a leaky network.[8]

However, in the last few years, the Government of the National Capital Territory of Delhi has recognized the urgent need for reform. Encouraged by the apparently successful privatization of the electricity supply public undertaking, the Delhi government has prepared a project with the World Bank's technical and financial support. Its main objective is to gradually improve service management, extend the infrastructure to underserved parts of the city, and financially strengthen the water utility through O&M cost recovery. The project does not deal with regulation of groundwater or with institutional mechanisms to foster decentralized governance structures. It can be described as a conventional attempt to enhance network services and demonstrate the

---

[7] By a systemic approach, we mean analysing all interaction between the agents, the resource and the institutional environment. In the case of water, this analysis reveals huge differences between developed and developing countries that preclude the mere transposition of a contractual model without any other kind of consideration.

[8] One such project involved private sector participation and led to a movement opposing it. However, we cannot deal at length with this specific instance due to lack of space.

feasibility of providing twenty-four hour supply in an Indian city. Technical features of the project include secured monitoring of produced and supplied volumes, and the reduction of non-accounted for water. For this purpose, two zones were selected as pilot projects. The proposal was to introduce a private operator in both of these zones, through a management contract to operate the distribution network twenty-four hour a day. One significant impact of this would be an increase in tariff for the whole city. Yet, one should recall that, with the exception of short-duration contracts for the construction of water treatment plants, attempts to initiate large-scale projects with the international private sector have failed in India, and that tariff increases are always a contentious issue.[9]

In order to operationalize the reform project, a consultant was appointed (financed through the World Bank) to carry out a feasibility study. 'However, the disclosure of documents related to this project by a Delhi-based group called Parivartan, which obtained them under the Delhi Right to Information Act, *forced the government to retreat* [emphasis added], and put it on hold' (Bhaduri and Kejriwal, 2005). During a period of around six months (from July to December 2005), the reform process generated an open conflict with the emergence of powerful actors, gathered around the Parivartan group.

One major point of controversy related to the issue of introducing private sector participation. The Parivartan group developed a number of arguments substantiated by detailed analysis of the project itself (Bhaduri and Kejriwal, 2005). Most of these arguments related to the cost of the project, the inability of institutional mechanisms to ensure performance, the absence of accountability as well as the increase in tariff. The notion of equity, explicitly of services to the slums, was at the core of the mobilization movement. Interestingly, to back up its stand, this group built its strength on a second argument, related to the procedures of the project themselves. By using the recently passed Right to Information Act, Parivartan secured access to the correspondence between the Delhi Water Board and World Bank officials (Parivartan, 2005). With some of these documents, they demonstrated that the award procedure for the consultant was not transparent: the World Bank had requested the opportunity to revise the selection criteria and marks given by the selection committee to favour one of the bidders. This 'arm-twisting of a democratic government by the Bank' seriously damaged its credibility (Bhaduri and Kejriwal, 2005). Despite refutations issued by the World Bank Country Director, this scandal led the Delhi Government to reject World Bank support for the reform.

There is no doubt that the controversial consultant selection, compounded by a general distrust towards private sector participation, led to the withdrawal of the Bank. However, the opposition movement that Parivartan created went far beyond this. Indeed, within the span of a few months, a large number of meetings had been organized to explain the reform process and counter it with the support of resident welfare associations (many from the pilot zones themselves). Other organized groups,

---

[9] Private operators are not wholly absent, but are small in size. They undertake local contracts, involving a limited number of operations and no investment. Various factors explain the absence of concession contracts: weak political will, civil society opposition, very low tariffs – which make it impossible to achieve economic equilibrium, insufficient return on investment, and lack of guarantees from the government (Zérah, 2003).

such as the Centre for Civil Society, also supported Parivartan. This action was not the result of a large mobilization movement; it mostly relied on a number of active and resident associations from wealthy and middle-class colonies in the south of the city. Nevertheless, this emerging Indian middle class, which was itself not receiving adequate services, has been increasingly able to create a balance of power and compel the government to revise its policies.[10] This strength (disproportionate to its numbers) is further reinforced by a process of ever-increasing expertise.[11]

The proponents of the project argued that opposition would hamper improvement of the service and leave most users as losers in the process, especially 'the poor who unfortunately pay the highest price as a consequence' (Jagannathan, 2006). Interestingly, both positions argued for more accountability. Indeed, one of the arguments presented by the World Bank for involving the private sector was enhancing accountability: at present, the Delhi Water Board is at once the provider, the regulator and is driven by political considerations. The Chief Minister is also the Chairman of the Water Board. The same quest for transparency and accountability is the driving force behind this civil society movement. In fact, we argue here that water was chosen as an emblematic case because social values are attached to it. But the main battle of the middle-class is the enforcement of greater transparency, more accountability and better democratic procedures: 'Successive governments so far have tried their best to hide than to reveal. It does not speak well for our democracy, and the people's campaign over the water sector reform is a small step in our long march towards changing this state of affairs' (Bhaduri and Kejriwal, 2005).

Thus, the case of Delhi illustrates how politically sensitive reform is in the water sector. In a way, this opposition reflects the vested interests of the middle class as well as its will to conquer political space and influence politicians.[12] Alternatively, this hostility to the ongoing reform process also stems from aspirations for a new governance regime based on accountability and reform of the public sector, rather than reliance on imposed models. The priority is to focus on public action reform and to redefine the role of the different institutions involved in the governance of water services. The key question is thus to find out the right incentives to persuade the government and public agencies to perform new roles and become accountable. This is the central issue underlying the largely misleading public vs. private debate.

13.4 CHENNAI: EXPANDING NEEDS AND GROWING CONFLICTS WITH PERI-URBAN USERS

The Chennai Metropolitan Water Supply and Sewerage Board (also called the Metro Water Board or MWB), set up in 1978, has the sole responsibility of augmenting water supply and sewerage in the city, as well as ensuring service provision. Chennai embodies an acute scarcity crisis. When the Metro Water Board was created, an imbalance between demand and supply already existed. Today, the situation has deteriorated further: the Board can supply only 50% to 65% of the requirements of the city's population, for

---

[10] This can also be seen in the case of Mumbai for other urban services (Zérah, 2009).

[11] See, for instance, the convincing critique of the project proposed by Bhaduri and Kejriwal (2005).

[12] In India, poor people vote in large numbers in comparison to middle-class people.

just a few hours a day, and not on a daily basis (World Bank, 2006; Ruet et al, 2002). Consequently, groundwater plays a critical role in filling this gap. In order to preserve and regulate this resource, the Chennai Metropolitan Groundwater (Regulation) Act (1987) entrusted MWB with the task of controlling abstractions and prioritizing public water supply. But in solving the problem of water availability, political choices remained confined to traditional approaches, in other words, increasing supply from far away sources through mega projects, which also involve inter-basin transfers. These projects are very expensive and riparian states are reluctant to release water for Tamil Nadu. Desalinization is another option, but this is also very expensive. Locally available solutions (such as restoration of ancient tanks) could be explored as the city's rainfall is quite substantial (over 1,200mm),[13] but to date these have not been seriously considered. Therefore, the short-term strategy of the city is to pump increasing amounts of groundwater further and further to the periphery, as it expands. This chapter argues that this does not account for a sustainable policy, being based on ad-hoc contractual agreements, informal mechanisms and a cyclical needs-based approach.

13.4.1 The central role of peripheral groundwater

Chennai has undergone a series of changes in recent years, with spatial expansion of the city being matched by an increased willingness to pay for water on the part of industrial and urban users, and a shift in allocation of resources, with water use rights being transferred from farmers to urbanites. This process induces economic changes as well as socio-economic tensions. First and foremost, it is not environmentally sustainable. Groundwater extraction is partly the result of private appropriation with the rise of the water tanker industry. In the particularly dry summer of 2004, around 6,000 private tankers plied the roads to Chennai. The municipal supply was limited to 20lpcd (World Bank, 2006: 18), and a number of apartment buildings residents were willing to pay very high prices for water. Ensured of profitable returns, a number of fly-by-night operators began selling water – most with just one lorry. A 12 m^3 tanker of water would cost as much as Rs 800–1,000[14], that is, Rs 66 to Rs 83 per m^3. This is at least twice the average tariff of around Rs 35 per m^3, but it represents much more for domestic users, as Chennai is one of the few Indian cities to charge its industrial users at almost full cost (O&M plus capital).[15] Furthermore, these private water tankers aside, the MWB has a contractual agreement with an association of tankers and provides them with treated water (Ruet, Saravanan and Zérah, 2002).

This weighs on the finances of the MWB. It is estimated that the MWB spends around Rs. 500 million to buy 3.7 million m^3 of water each month. Most of this comes from public appropriation of peripheral groundwater. Water extraction from village common lands is not new and over the past two decades, MWB has relied heavily on transport of water from public wells and agricultural wells located in peri-urban villages. The MWB also compelled farmers of many villages to sell water from their

---

[13] Despite observed cyclical fluctuations, as anywhere else, no declining trend in rainfall could be observed over the period of the past 100 years.

[14] US$1 = 50 INR.

[15] As in the case of Delhi, the cost is even higher if one relies on purified water (Rs. 2000 per m^3 of water) or polythene water sachets (Rs. 8,000 per m^3 of water).

own irrigation wells. Many agreed as it was a profitable venture. However, this second arrangement yields little water compared to the current supply of about 103,000 m³/day (Janakarajan, 2005).

From a resource perspective, the existing system looks like a stopgap policy. Groundwater is under threat and the consensus is that Tamil Nadu is one of the worst states in terms of underground resource degradation (Briscoe and Malik, 2006). Clearly, meeting present urban water needs conflicts with environmental sustainability. Interestingly, these tensions occur despite a stringent regulatory system: while most other Indian states have not passed groundwater regulations, Delhi and Chennai have. However, in a sharply polarized political arena, effective implementation of such rules is minimal. Indeed, the Chennai Metropolitan Area Ground Water (Regulation) Act of 1987 has several important features:

1. MWB is the authority entitled to grant/not grant permits to sink wells in the designated area, and to grant/not grant licenses for extraction, use or transport of groundwater
2. a database of existing wells has to be kept
3. no person shall extract or use groundwater in the planned area for any purpose other than domestic, and no person shall transport groundwater by means of lorry, trailer or any other goods vehicle.

However, after almost two decades, the MWB is chiefly responsible for the groundwater overdraft in its peri-urban zone and as far away as 50km from the city limits. Furthermore, its own lorries and private tankers operate without licenses. Some private operators complain that, despite having applied for a permit, they have received no response to their requests. The procedure for new well/borewell licensing is mere eyewash and remains on paper. Many industries are not only drawing groundwater in violation of the Act, but also degrade the quality of the groundwater. None have paid a penalty, and no stringent action has been taken against them by MWB. A more recent act has also been passed, as well as government resolutions. However, loopholes and weakness in implementation remain (Geetalakshmi and Janakarajan, 2005), highlighting a more general Indian situation, where powerful sets of legislation exist, but are not enforced, undermining this potential tool of conflict resolution.

13.4.2 Short-term winners and losers: a transition towards conflicts?

The water resource (its quantity and quality) is an obvious 'loser' in the process of unplanned infrastructure for urban expansion. In addition, this creates increasing socio-economic tensions and polarization of interests. The higher value of 'urban water' leads to a general shift in water allocation criteria, demonstrated in some villages through analysis of the strategies of farmers who are willing to sell water.

On the urban demand side, a large share of this water is provided to a small number of industrial users, who effectively constitute the 'winners'. To convince reluctant farmers, MWB made the case for thirsty urban dwellers. In fact, through a rapid cost assessment based upon a study of two villages, Gambiez and Lacour (2003) made the point that MWB was making profits by selling groundwater to these industrial users at a higher rate. Indeed, these industries, especially a cluster located in the north of Chennai, make up a considerable share of MWB's revenues. As such, it is critical for

MWB to supply them to ensure that they do not exit the system[16] (Ruet, Saravanan and Zérah, 2002).

In villages, the situation is complex with socio-economic consequences varying among different types of farmers and according to the geographic location of the villages. Using the same field approach, Gambiez and Lacour (2003) calculated the economic gains/losses of three types of farmers in different villages, where some had signed agreements to sell water from wells they owned to MWB. These constitute the first type. The second type does own wells and depends on the former to buy water and irrigate its fields. The third type, 'independent farmers', does not sell its water to MWB and is not affected by the agreement. The authors assessed the evolution of agricultural practices and the consequences in terms of income. The results show that the independent farmers, who serve as a reference group, suffered a slight fall in their income due to the reduction in cultivated areas. This trend is likely to be explained by the growing influence of the city and the steady decline of agricultural employment due to urbanization. On the contrary, for the two other categories, the evolution of both cultivated areas and incomes is very marked. Between 2000 and 2001, well-owners contracting with MWB reduced their cultivated areas by 43%. Out of the approximately thirty farmers selling water, only three, owners of several borings, maintained their previous level of activity. The others reduced their cultivated area and their revenue increased by 80% between 1999 and 2002. Dependent farmers were the losers: whereas fifteen farmers supplied water to them before 2001, only two kept this relationship after. This has resulted in a considerable reduction in the irrigated area and, in consequence, a substantial drop in income for the dependent farmers. In-depth research carried out by Janakarajan (2005) substantiates and reinforces these results (see Box 13.1). A comparison of present occupations with the situation twenty years ago in the selected villages, shows clearly that there has been a huge shift from agricultural employment to non-agricultural employment. The research also highlights inter-village variations. Villages with a location advantage (proximity to Chennai) and higher connectivity are better equipped to make the transition towards an urban-based economy.

Some villagers are eager to sell water, but the tension levels between rural and urban interests as well as within rural interests is strong. In some cases, this has induced direct conflicts between diverging interest groups. The case of violent and direct conflict in the village Velliyur (see Box 13.1 for details) illustrates the continual degradation of a situation where no resolution mechanisms (local negotiation, legal action, or agreed-upon commitment) succeeded in implementing a credible solution. In this case, there is clearly an asymmetry of bargaining power among actors. Local opposition raised by some villagers is not credible enough to stop powerful actors – backed by the priority accorded to drinking water by the National Water Policy – from supplementing their water requirements with short measures. Under pressure, dependent farmers quit and move to the city which continues to expand in an unplanned manner – a process that goes on indefinitely (Janakarajan, 2005). Following the example of other metropolises (Kundu et al, 2002), the city of Chennai is expanding and developing by imposing new social and environmental costs, even to some extent on urban dwellers, who have to pay a high premium for water.

---

[16] This highlights another distorted allocative process, further leading to intra-urban conflicts not developed in this chapter.

Box 13.1 Example of an acute water conflict in Velliyur, a peri-urban village of Chennai

Situation
The village of Velliyur is located at a distance of 50 km from Chennai and has a population of around 4,300. In 1980, there were 280 wells with depths in the range of 50–80ft. Today, there are 220 wells with depths ranging from 130–160ft. Quality of water has deteriorated over the last decade. Quantity of drinking water declined rapidly from a round-the-clock supply from 4 bore wells in 2000 to two hours a day from 12 bore wells four years later.

Background to the conflict
In 1969, 11 bore wells were installed by MWB to pump water from the common village land to supply additional water to Chennai and nearby industries. In 2000, 9 out of the 11 bore wells failed; since then, water has been purchased from farmers. As a result, groundwater availability has shrunk considerably with agriculture being badly hit. Water sales from 75 irrigation wells belonging to individual farmers made things worse, affecting landless labourers.

Narration of the conflict
The situation prompted a NGO working in the area to motivate Self Help Groups (SHG) and other members of the landless population. In April 1995, SHGs began to oppose water sales, insisting that the Panchayat should pass a resolution banning water sales. The Panchayat (the local self-government unit for rural areas) did not do so as groundwater then was pumped only from government land. After 2000, as water was purchased from farmers, the same group pushed again for a Panchayat resolution. This was denied on the grounds that water is sold by individual farmers from their own land. Some village residents then filed a case in the court to ban water sales from the village. They were successful in getting a standing, but it was soon vacated through an appeal petition filed by a water-seller supported by the MWB.

Meanwhile, the groundwater table was becoming further degraded as a result of sand-mining by the government. The farmers selling water took the sand-mining issue to Metro water, threatening to stop sales if sand-mining was allowed. Metro water took the issue to the government, which acted to stop sand-mining. Labourers working in sand-mining stood to lose and joined the opposition to water sales. Open conflict between sellers and non-sellers broke out in August, 2004. The entire village apart from the sellers asked the Panchayat to pass a resolution to ban water sales and resorted to road blockages. A Peace Committee was formed consisting of water-sellers, non-sellers, SHGs and officials (including Metro Water Board officials). A decision was taken to stop the sales from farmers to MWB from mid-September 2004. However, when the date arrived, MWB officials refused to discontinue water purchase, stating that their own higher authorities had rejected the agreement of the Peace Committee. In addition, water-sellers resorted to the court and obtained a stay order against the decision of the Peace Committee. Non-sellers also resorted to the court to get a stay on water sales. This proved an unsuccessful move for both sellers and non-sellers. In parallel, water pumping continued. In September 2004, the entire village blocked the road, and some people from agitating groups damaged the MWB pipeline. Following this violent outbreak, forty-seven people arrested by the police and booked under the Public Property Damaging Act. MWB filed an order requesting Rs. 30,000 damages towards compensation.

Status
The case is pending in the courts and water-selling was stopped. However, the MWB has circulated a notice among farmers stating that whoever is willing to sell water can approach Metro Water for a one-year agreement. The tender should have been submitted before 22 February 2005. But to date water sales have not restarted.

Source: Janakarajan (2005).

13.5 CONCLUSION: A COMMON FRAMEWORK OF WEAK AND INEFFECTIVE CONFLICT RESOLUTION MECHANISMS

Both case studies underscore the consequences of a water crisis, whether related to the resource or management. They also illustrate distinct, but partially interrelated, processes relating to urban-rural tensions, intra-urban inequities and the unsustainable long-term situation, and debates surrounding the choice of governance regime. Nevertheless, both cases display a similar inability among public utilities to develop long-term solutions to their water resource problems, and to take into account the central role of groundwater. As the case of Chennai points out, regulation is not a credible tool and remains a weak institutional mechanism. In other words, as stated by the opponents of the Delhi reform: 'The Bank would be less than intelligent to expect that the same inefficient and corrupt government (or some of its officials designated as 'regulator'), which failed to make its water utility perform, would make the private company perform' (Bhaduri and Kejriwal, 2005).

This further confirms the growing lack of confidence in the credibility of public utilities to evolve solutions or carry out reforms. As a consequentce, litigation becomes increasingly prevalent, either to obtain stay orders as in the case of Chennai, or access to government correspondence as in the case of Delhi. In Chennai, delays and counter-cases seem to have annulled the potential of this mechanism, despite its increasing use across the country. In Delhi, the revelation of unpleasant information led to the formation of a strong group, which proved able to block a policy high on the agenda of the Delhi government. This example also demonstrates the strengthened bargaining power of the organized middle-class in matters related to urban affairs.

However, despite the organization of interest-groups, the critical question remains unresolved: how can a shift be made towards a win-win situation – from conflict scenarios to cooperation, or at least to a situation where negative outcomes are minimized or compensated? Attempts have been initiated in Chennai to create a multi-stakeholder platform and dialogue (Janakarajan, 1999, 2003 and 2004b). Several local meetings were held, which were attended by researchers, NGOs, farmers from peri-urban villages and certain government officials, and were followed by two regional stakeholder workshops. This type of cooperative endeavour – in the sense that stakeholders agree to a minimum agenda concerning issues – enables expertise and knowledge to be exchanged and conflicting views to share a common platform.

Finally, it should be particularly stressed that a rationed water supply and an often inefficient service in the cities, together with the disregard of formal rules (that are moreover vague), lead various users to a 'pumping race', and over-exploitation of the resource, through either individual or joint initiatives. Today, all these decentralized solutions have a fairly high cost (which can only continue to increase under the 'business as usual' scenario), despite water being apparently free. Although presently unsustainable, these solutions have the potential for improvement, subject to several conditions.

Private markets for the resale of water are not sustainable in the long term given the present state of affairs. The lax regulatory framework offers private operators the possibility to supply a private commodity at an excessive price, but at a quality which is not guaranteed. Only the most affluent households can take advantage of this service, which ultimately contributes to the segmentation of different categories of the population.

These are provisional solutions that do not meet the overall requirements of urban management, or of the resource. They reduce the scope for territorial equalization or any other unifying mechanism, specific to a public service monopoly.

These modes of organization reveal the incapacity of the institutional environment to prevent agents from carrying on unsustainable practices, as most rules can simply be bypassed. In return, the institutional environment cannot evolve much, being tuned to major malfunction and growing discontent. Each of the system's factors is governed by its internal dynamics, without clear interaction with others, thus emphasizing the magnitude of the institutional deadlock. Thus, we have a situation of tacit laisser-faire, which contributes to depletion of the resource and the degradation of the infrastructure.

The role of the institutional environment is, among others, to lay down the rules enabling transactions to take place, at a lower cost; in other words, to ensure the transfer of rights that accompanies these transactions. In most Indian cities, several problems combine to exacerbate the bad management of water and infrastructure: poor coordination between various agencies, both vertically and horizontally, which results in erratic planning; political instability, which constitutes a permanent threat; the problem of corruption; and the limits of the judicial system.

All these problems are, of course, very difficult to solve and we can only indicate the goals that should be kept in view. This confers a very normative character to our propositions. The first goal should be simplification of the institutional framework by redefining responsibilities to better coordinate the various decision levels. Overlapping of tasks should be avoided and the intervention capacity of discretionary powers should be limited, thus increasing the legitimacy of each of the actors. The second goal stresses the concept of a democratic decision-making process in which all the interest groups at stake are represented (from the infra-local level to that of the whole area). This would act like a broad-based regulatory framework. Lastly, it is essential to redefine the constituents of the public service and its articulation in operational terms. This implies a reversal of perspective, in the sense that the service should not be conceived in a technocratic top-down manner by imposing arbitrary norms, but in terms of the fundamental needs that should be met, taking into account the different systemic effects.

REFERENCES

Bhaduri, A. and Kejriwal, A. 2005. Urban water supply. Reforming the reformers. *Economic and Political Weekly XL*, 52.

Briscoe, J. and Malik R.P.S. 2006. *India's Water Economy: bracing for a turbulent future*. New Delhi: Oxford University Press/World Bank.

Coing, H. 1996. Mimétisme ou métissage? *Annales des Ponts et Chaussées*, 80: 58–63.

Davis, J. 2004. Corruption in service public delivery: experience from South Asia's water and sanitation sector. *World Development*, 32(1): 53–71.

Gambiez, M. and Lacour, E. 2003. *Rural impact of farmers selling water to Chennai Metropolitan Water Board: a case study of Magaral Panchayat*, Research Report. New Dehli: INA of Paris-Grignon and Centre de Sciences Humaines.

Geetalakshmi and Janakarajan, S. 2005. Intricacies of Chennai Metropolitan Water Laws. Draft paper presented in a mid-term workshop organized by MIDS, Chennai.

Haider, S. 1997. Community participation in basic services and environmental protection: case study of a Jhuggi-Jhompri cluster. *Man & Development*, 19(4): 158–88.

Janakarajan, S. 1999. Conflicts over the invisible resource: is there a way out? M. Moench, E. Caspari and A. Dixit (eds) *Rethinking the Mosaic: investigations into local water management*. NWCF and ISET (USA), 1999.

Janakarajan, S. 2003. Multi-stakeholders' dialogue as a tool for natural resource management. Paper presented in a National Seminar on Water, organized by CESS, Hyderabad, 30–31 July.

Janakarajan, S. 2004a. Trading in groundwater: a source of power and accumulation. M. Moench (ed.) *Selling Water: conceptual and policy debates over groundwater markets in India*. Ahmedabad; VIKSAT/Pacific Institute/Natural.

Janakarajan, S. 2004b. A snake in the grass!! Unequal power, unequal contracts and unexplained conflicts: facilitating negotiations over water conflicts in peri-urban catchments. Paper presented in a workshop on *Urban and peri-urban conflicts*, organized by MIDS, Chennai.

Janakarajan, S. 2005. Dying agriculture, weakening environment and fading institutions: declining livelihood options and capacity to adaptation for livelihood resilience in peri-urban villages of Chennai. Draft paper presented in a mid-term workshop, organized by MIDS, Chennai.

Jagannathan, V.N. 2006. Story of a hiss. Water scarcity and the need for reform. *Economic and Political Weekly XLI* (11 March).

Kundu, A., Pradhan, B.K. and Subramanian, A. 2002. Dichotomy of continuum: analysis of impact of urban centres on their periphery. *Economic and Political Weekly*, 37(14): 5039–46.

Lascoumes, P. 2004. Controverse. L. Boussaguet and S. Jacquot, P. Ravinet (eds) *Dictionnaire des politiques publiques*. Paris: Presses de Sciences Po, pp. 123–31.

Llorente, M. 2002. Une approche néo-institutionnelle de la gestion urbaine de l'eau à Delhi: quelle régulation pour quel service?, PhD dissertation. Nanterre (France): University of Paris X.

Llorente, M. and Zérah, M.H. 2002. Urban water sector: formal *versus* informal suppliers in India, *Urban India*, XXII(1) (January-June): 35–49.

Maria, A. 2005. La crise de l'approvisionnement en eau à Delhi: réponse des acteurs et scénarios d'évolution. Presented at the IDDRI seminar on Essential services in developing countries, Paris, 22 April.

Maria, A. and Levasseur, J. 2004. Strategies for water supply in a planned urban extension: a case study of the Dwarka sub-city in Delhi. Research Report. New Delhi: Cerna and Centre de Sciences Humaines.

Parivartan. 2005. Letter to Paul Wolfowitz, President, World Bank. Available at: http://siteresources.worldbank.org/INTINDIA/Resources/Parivartan_lettertoPDW.pdf.

Raghupathi, U. 2003. Small private water providers – an alternative solution for the poor. *Shelter*, 6(3), New Delhi.

Ruet, J., Saravanan, V. and Zérah, M.H. 2002. The water and sanitation scenario in Indian Metropolitan Cities: resources and management in Delhi, Calcutta, Chennai, Mumbai. Occasional Paper 6. New Delhi: Centre de Sciences Humaines.

Shaw, A. 2005. Peri-urban interface of Indian cities: growth, governance and local initiatives, *Economic and Political Weekly*, 40(2): 129–36 Mumbai.

Tovey, K. 2002. The institutional responses to the water needs of the urban poor: a study of collective action in Delhi slums, India. Ph. D. dissertation. University of Cambridge.

World Bank. 2006. India, water supply and sanitation: bridging the gap between infrastructure and service. Background Paper, Urban Water Supply and Sanitation. South Asia Region, World Bank.

Zérah, M.H. 2000. *Water: unreliable supply in Delhi*. New Delhi: Manohar Publishers.

Zérah, M.H. 2003. Dix ans de libéralisation de l'économie indienne: les effets limités dans le secteur de l'eau et de l'assainissement. *Autrepart*, 27, no. Variations: 91–106.

Zérah, M.-H. 2009. Reforming solid waste management in Mumbai and Hyderabad: Policy convergence, distinctive processes. In J. Ruet and S. Tawa Lama-Rewal. *Governing India's Metropolises: Four case studies*. Delhi, Routledge: 241–269.

Chapter 14

Urban water conflicts in the western US

Stephanie Pincetl

Institute of the Environment and Sustainability, University of California, Los Angeles, USA

This chapter focuses primarily on issues related to urban water conflicts in the western United States, as they are currently some of the most contentious in the country. However, there is little conflict within cities, for example, about water pricing: the growing issue is conflict between cities and urban development on the one hand, and other water users, in particular irrigated agriculture but also minimal ecological flows, on the other. The western United States is an arid region that has benefited from massive federally built irrigation systems to supply agriculture. Cities in the west are among the fastest growing in the country, requiring more water to keep up with suburban growth. Not only is there competition among cities for water, but the also with agricultural water users for federally developed water. This chapter highlights one particular water conflict as an illustrative example of water conflicts in the west.

The Imperial Valley of California is an irrigated desert in the south-eastern portion of California, bordering Mexico and Arizona. This region has historically been allocated the lion's share of water from the Colorado River, a once wild watercourse that spans seven states, 1,500 miles, and crosses an international border. It produces over a billion dollars worth of agricultural products on 500,000 acres.[1] In 2003, the Imperial Irrigation District – the water wholesaler and distributor in the Imperial Valley (and water contractor to the federal government) – was told by the federal government that it had to reduce its water use. This was the result of complex, interacting socio-environmental, political and economic factors that set in motion a cascade of changes, many of which are still unfolding.

Such changes are increasingly supporting an organizational and/or institutional shift along a continuum of water-management options toward a process of cohabitation, competition and eventual displacement of public policy decision-making allocation principles, based on social equity and economic self-sufficiency via market principles prioritizing economic efficiency and allocation techniques (Bakker, 2002: 769).

There are several important issues that characterize the water conflicts in this case study, although many others could also be investigated. First, is the spatial scale. In this case, an integrated interregional scale involving agriculture and urban areas, coast

---

[1] 1 acre = 0.4 ha.

and interior, demonstrates how there are winners and losers and unintended long-term consequences in such water disputes, when a larger landscape level is used. Secondly, there are the roles of institutions, structures, norms and values which define the rights, constraints and power that have influenced and shaped the development of water and the region. These include the structures of political decision-making, including multiple jurisdictions and single-purpose agencies, as well as private interests. Finally, this case study illustrates how introducing new allocation rules – in our case a water transfer – engenders a new hydrosocial contract between users and their environment, a new exchange value for the resource, and transfers of wealth and power.

14.1 WATER DEVELOPMENT AND THE ENVIRONMENT

In order to contextualize the issue of water conflicts, the critical linkages to explore are those between environmental degradation and the institutions, structures and norms that constitute and create our current social organization(s). Water development materially transforms environments for the maintenance of current lifestyles, and does so through regulatory mechanisms. In order for change to occur in water allocations – as among cities, or between cities and agricultural regions – rules have to change. In other words, the appropriation of natural resources and their transformation into material goods (including water and food), is effectuated through institutions and institutional arrangements that are organized in certain patterns. Each regime of accumulation entails a mode of nature appropriation that produces a dynamic resource landscape in its own image (Bakker, 2002) and has socioeconomic ramifications.

In California, cities and agriculture receive water from federally (and state) – subsidized water projects. The long-term future of these water projects is in question due to potential impacts of climate change on the hydrologic cycle, and increased water intensive urbanization in several of the states receiving water from the Colorado River, among other water sources. As in much of the west, agriculture receives the majority of developed water – approximately 80%. With increased urban growth in the state, there has been pressure to re-allocate some of the water from agriculture to urban areas. In 2003, such a transfer was enacted where the Imperial Valley Irrigation District entered into an agreement to transfer a portion of its water to the San Diego County Water Authority and the Metropolitan Water District of Southern California, that is, an urban water supplier and the regional bulk water transfer system from the Colorado river.

14.2 WATER IN CALIFORNIA AND THE TRANSFER FROM AGRICULTURE TO URBAN AREAS

Water development and agriculture in California grew up together, inextricably bound, for one would not have existed without the other. Historically, water development in the west was predicated on providing inexpensive water for agriculture, with urban areas paying the great bulk of capital costs: there are substantial price differentials between agricultural and urban water.

It is important to note that federal water development in the west constitutes a kind of regulatory regime, the component parts of which are fundamentally intertwined – rule changes for one part of the system have implications for all other water projects

Figure 14.1 **Map of Imperial County, California**
Source: USGS.

developed under the federal 1902 Reclamation Act – the central enabling act under which all federal water projects were built in the west. Currently, federally developed water is considered a common good, regulated by the federal government, but agriculture is the privileged user.

By the 1980s, the conventional appropriation of water began to be challenged by proposals to increase water conveyance to cities in southern California. Environmental organizations were concerned about the environmental effects of this water wheeling on the places of origin and the environmental impacts of agriculture itself, and concerned that additional supply would fuel further urban sprawl.

Water in California is distributed by local water districts. Institutional relationships exist between the federal government and local water districts for federally developed water systems, and between local water districts and the State Department of Water Resources for state-developed water. There are no California state-level water districts. Southern California receives water from the federal Colorado River delivery system – up to 12% of California's developed water; the rest comes from the State Water Project. (The City of Los Angeles receives water from the Owens Valley through its own conveyance system.)[2] The water is then provided to several irrigation districts – including the Imperial Irrigation District – as well as to the Southern California Metropolitan Water District (MWD). MWD is a water wholesaler to 95% of the

---

[2] It is worth recalling that the capture of the Owens valley water by the city of Los Angeles triggered a quasi-civil war with the residents and irrigators of the valley in the 1930s (Kahrl, 1982).

South Coast region. The agency includes fourteen cities, twelve municipal water districts and the San Diego County Water Authority (SDCWA). Decisions are made on the basis of weighted votes by the board, which consists of the member agencies.[3] The SDCWA provides water to the second largest metropolitan area of the state. The MWD and the SDCWA have been historical antagonists, due to SDCWA being a member agency of the MWD, yet not having control over its own water policy or water allocation. The MWD and the SDCWA are key actors in the Colorado River delivery system and are strategic players in the entire state, and indeed the whole west.

14.3 INTRODUCTION TO THE IMPERIAL VALLEY CASE

By 1905 there were already 120,000 acres of land under production in the Valley. However, in 1905 a swollen Colorado River washed away the private primitive irrigation canals and control gates and water came flooding into the Salton Sink, refilling the northerly portion of a depression, which was a sea in geological times. The Salton Sea was reborn. Capital costs of irrigation infrastructure led to the creation of the Imperial Irrigation District in 1911 by Imperial Valley businessmen who submitted the creation of the district to the local voters (permissible under California law). A five-member elected board was established to govern the district, each member representing a division within the Imperial Irrigation District (IID)'s service area. The new IID began rebuilding the water delivery system and installing a farm drainage system (tiles under ground) capable of carrying away excess water and salts from the heavy clay soils to the Salton Sink. It also lobbied the federal government to build dams on the Colorado River to regulate its flow. They were successful in doing so, and henceforth, the IID was the agency in charge of water distribution. The district also entered the energy sales and distribution business.

Once the federal government became involved in the development and allocation of Colorado River water, the question was how to divide it as it flows through seven states and into Mexico, emptying into the Sea of Cortez. The solution developed at the time was to divide the states into upper basin states and lower basin states. Water flows for the river were estimated at about 17 million acre feet (MAF) a year.[4] Since the water originated in the upper basin, any arrangement required the upper basin states to guarantee water for the lower basin states: each basin was to receive 7.5 MAF a year, Mexico to receive 1.5 MAF, leaving 1 MAF of surplus. However, the 1922 Colorado River allocation figures were based on unusually high river flows; in fact the average river flow is closer to 12 to 14 MAF a year, leaving potentially significant shortfalls in promised water allocations.

This 1922 arrangement, entitled the Colorado River Compact, allocated 4.4 MAF to California. However, until very recently, other states along the river did not use their entire allocation, and the river's flow over the past half-century was unusually high. California negotiated to receive the surplus under the compact. However, following the construction of the Central Arizona Project, Arizona began to make use of its

---

[3] Water agencies in California come in many different sizes and governance arrangements, including: one person, one vote; property weighted voting; and appointed boards. The Imperial Irrigation District (IID) Board is a district elected board.

[4] 1 acre-foot (AF) = 1,234 m^3; 17 million MAF = 21.2 km^3.

allocation to supply new urban growth, while Nevada too claimed its full entitlement for Las Vegas, thus threatening a reduction in California's water surplus. In turn, this meant that the MWD share would decrease, leading to fears that it would not be able to deliver sufficient water to supply all new urban growth. California had been using 5.2 MAF from the Colorado River, of which the IID received 3.1 MAF (about 60%); the rest (about 2.1 MAF) was shared between MWD, its member cities and two Imperial Valley area water districts. One of the entities concerned had to give up water to get consumption down to the allocated 4.4 MAF.

14.4 THE DECISION-MAKING PROCESS LEADING TO TRADING

Under current water law, to receive water from the Bureau of Reclamation or from the state, it is necessary to demonstrate a 'need' (Water Code 100–100.5). With the decrease in water availability from the Colorado River, demonstrating 'need' has become more important. Due to IID-sponsored drainage policies (to leach salts), the level of the Salton Sea kept rising, flooding adjacent agricultural lands. A lawsuit was initiated by a landowner against the district before the State Water Resources Control Board and the State Department of Water Resources, accusing the IID of wasting water.

In the early 1980s, investigations by the state Water Resources Control Board (WRCB) and the Department of Water Resources (DWR) (State Water Resources Act, 1945), found that the IID could conserve 438,000 AF of water a year by lining earthen canals that seeped, improving imprecise measuring systems and inefficient distribution systems, and reducing spillage on farm use. The practice of flooding fields, using gravity irrigation – and exaggerated salt leaching – and growing high water-use crops, did not meet best management practices. Thus, the existing mode of production – the organization of how water was delivered and used (regulated by IID and federal rules), itself led to waste. Moreover, a further problem for the DWR was the huge amount of non-recoverable losses caused by evaporation.

Initially rejected by the IID, and denied as a practicable possibility by a representative of the MWD in the DWR, a proposal was put forward that water conserved within the IID could be exchanged with the MWD for financial help with implementing conservation methods (Rudy, 1995). Given the impending completion of the Central Arizona Project to provide water to Phoenix – thereby reducing California's access to surplus – water trading began to be seen as a viable option. This first step established the foundations for challenging agricultural water use in the Imperial Valley. The groundwork had been laid for the IID to *sell* agricultural water to growing urban and metropolitan areas in Southern California, as a more efficient use of water. This would also ensure that cities would continue to have sufficient water for their needs, despite California having to reduce its share of Colorado River water. The selling of water to urban areas represents the introduction of market logic to water resources management and allocation, with commercialization of the resources. It carries with it the application of private sector criteria such as efficiency and pricing.

The IID, however, alleged that it could not shoulder the cost of putting conservation measures in place. The Metropolitan Water District, in response, proposed a water-marketing scheme that would pay for conservation measures and liberate hundreds of thousands of AF of water for the southlands (Gottlieb and FitzSimmons, 1991: 81–82). After much infighting and intense negotiation, between the IID and the MWD, and the MWD and the SDCWA, an initial agreement was signed in 1988, under which the MWD would pay for certain conservation projects in exchange for water. However, this agreement crumbled after the Coachella Valley Water Authority (CVWA) to the north of the Imperial Valley declared its water rights violated by the accord.

In 1996, then US Interior Secretary Babbitt told California to come up with a plan to bring its Colorado drawback down to the legal limit – 4.4 MAF a year – within twenty years. MWD and other water agencies came up with a plan to reduce their dependence on the Colorado River, relying on a series of complex water re-allocations and storage plans. By 1997, following federal actions to reduce California's allocation, the Colorado River Board of California drafted an agreement entitled California's Colorado River Water Use Plan. It called for the IID to sell part of its Colorado River allocation (200,000 AF) to San Diego, in support of the plan's goal of using Colorado River water more efficiently. However, the water would be conveyed through MWD facilities: SDCWA, being a member agency, did not have its own pipes for moving water. When this deal was first proposed, SDCWA objected to the price set by MWD for transportation of water through its facilities. The issue was resolved when the state agreed to pay US$200 million to make up the difference.

However, in 2001, environmentalists raised concerns about the effects of the diversion on the shrinking Salton Sea, and the species of fish and birds that rely on it. In December 2002, negotiations for water-trading broke down with the IID claiming that it was being railroaded and objecting to specific parts of the deal. The federal government then reduced water supplies to Imperial County by 11%. Intense last-minute negotiations at the end of 2003 led to the approval of an agreement, transferring water from the IID to San Diego County Water Authority and to the Coachella Valley Water District. The federal government, under pressure from environmentalists, instructed the IID that before it could sell the water to San Diego, it had to somehow mitigate the impact of the loss of drainage water ('wasted water') into the Salton Sea. A polluted man-made sea had become an important bird habitat, constraining decision-making regarding water allocations: a new nature had arisen with its own requirements.

Water wheeling from Imperial currently involves a combination of agricultural land fallowing – more in the initial years – and the lining of the All-American and Coachella canals, which will yield up to 100,000 AF per year for the Metropolitan Water District. The trading will also provide the San Diego region with up to an additional 21.5 million AF of water over the 110-year-life of the agreement (SDCWA Annual Report, 2004). In addition, it will provide up to 150,000 AF a year over fifteen years to the Salton Sea (Revised Fourth Amendment Agreement, 2003).

The re-allocation from the Imperial Irrigation District to San Diego has created a series of policy trade-offs. Fallowing land is the most immediately available way to liberate water for the transfer to San Diego; it does not decrease the 'wastage' of water in agriculture, but ensures that the Salton Sea continues to receive the water it needs to be sustained. Fallowing increases air pollution in the valley and reduces employment. The lining of irrigation canals and other water conservation measures will reduce water seepage upon which a portion of nearby Mexican agriculture depends, as well as valuable wetlands. So, though still emergent, water wheeling creates a new regulatory regime for water, and has important environmental implications in the Imperial Valley, San Diego County, and the adjoining Mexican jurisdiction of the Mexicali Valley.

These water transfers are also taking place concurrent with other shifts in the valley. The county's population is growing rapidly, expanding 30% from 109,303 in 1990 to 174,528 in 2011 (2011 US Census). Cities in the county, like the county itself, are poor, and will receive none of the proceeds of the water sales as the IID is an independent agency. Opportunities for urban growth are seen as a way to increase sales taxes and to become more prosperous.

The question for the Imperial Valley concerns the future of agriculture relative to other forms of land use, such as residential and commercial development, which would potentially yield greater local revenues directly to the cities and the county, and probably higher wages. In essence, there would seem to be parallel gains to be made by both the IID for funds derived from the water sales from fallowed land to San Diego, and for the cities and counties for funds that would be generated by urban growth. Each entity gains from the decline in agriculture.

Yet, agriculture exhibits unique characteristics (Guthman, 2004: 63–64). Food systems are fundamentally dependent on biophysical production to the extent that much of the value is created by biological processes. The rhythmicality and seasonality of many biological processes limit (still) the extent to which food production can be controlled or sped up, and where food can be produced. From this perspective, the Imperial Valley offers exceptional conditions for growing, if water and drainage are available. Crops can be grown year round, and a number of crops can be cropped three to four times in the Imperial Valley. Finally, land is the major medium of agricultural production. It is unique, and unsubstitutable in the absolute sense; once it is developed for other uses, the costs of re-conversion are very high, if it can be done at all.

14.5 DECISION-MAKING IN SAN DIEGO

Over a seventy-five year period, the water transfer from the Imperial Irrigation District to San Diego will provide 200,000 AF per year. This amount of water is sufficient to meet the projected urban growth of approximately 1 million people by 2030 at the current water consumption rate of over 200 gallons per capita per day for all urban uses, and up to 280 gallons per day in the summer. The price of water will start at US$258 per acre-foot and will increase to US$420 by year fifteen. SDCWA as a water retailer delivers water to its member agencies including cities, irrigation districts, municipal water districts, county water districts and the Pendleton Military Reservation. As previously mentioned, 70% to 95% of the county's water supply is currently imported from the Colorado River and Northern California, and SDCWA has relied on MWD for imported water supplies. The total water delivered via the SDCWA system is 720,700 AF. The transfer will also provide MWD with up to 110,000 AF per year and other transfers from agriculture to urban water use are included in this highly complex arrangement, largely among water wholesalers.

The water transfer was based on projections of population growth in the region, and was negotiated by the SDCWA. While SDCWA sets local water rates, direct water consumption management is not part of its administrative authority; it simply sells water wholesale – and the more it sells, the greater the profitability. While it advocates best management practices to obtain water conservation savings, the water transfer from the IID to SDCWA has not entailed any mandatory measures for urban water conservation. Yet, the SDCWA 2005 Urban Water Management Plan Projects a 12% water conservation rate by 2030 (SANDAG, 2004: 2–5).

However, estimates to improve urban water use efficiency and conservation in California show much greater potential. The Pacific Institute, a non-profit think-tank, conducted a peer-reviewed study of the potential for urban water conservation in the state and estimated a potential of 33% greater savings. According to the Pacific Institute, the residential sector is the largest urban water user and offers the largest volume of potential savings. With current technologies and policies, residential water use could be as low as 60 to 65 gpd[5] without any change in the services actually provided by the water. In San Diego County, residential use ranges around 200gpd. Even without improvements in technology, the Pacific Institute estimates that indoor residential water use could be reduced by nearly 40% by replacing remaining inefficient toilets, washing machines, showerheads and dishwashers, and by reducing the level of leaks. With improved management practices and available irrigation technology, another 34% of water savings could be realized on outdoor residential water use. They also found that water savings could be made in industry, adding up to a potential 39% (Pacific Institute, 2003). SDCWA may argue that water pricing will eventually reduce use, as the price of outdoor irrigation may become prohibitive; yet it appears that there has been no real consideration of what an investment in conservation, equivalent to the ultimate additional costs of the transferred water (US$420/AF), might yield in terms of newly available water. However, the implementation of water conservation measures poses a structural problem for the revenue stream of water agencies – the less water sold, the less income received. Thus, a water conservation strategy can be seen as one that undercuts revenue. This is due to the way in which the water districts were organized at the turn of the twentieth century, on the basis of rules encouraged growth.

San Diego County now has about 3 million people, and 62% of the homes in the region were built after 1970. Under current plans and policies more than 90% of the vacant land designated for housing over the next twenty years (a projected additional growth of 37%) is planned for densities of less than one home per acre (many of these are located by wells). Of the remaining vacant land planned for housing in the eighteen incorporated cities, only about 7% is planned for multifamily housing (SANDAG, 2004). Multiple family homes use less water and less land per unit. Much of the growth since 1990 has occurred in the northern part of the county, in areas with high ecosystem diversity, including the endangered coastal sage scrub ecosystem. Land-use decision-makers in the San Diego area are able to ignore water scarcity as the obligation to provide water falls under the responsibility of the SDCWA and there are no substantive policy linkages. The regulatory structure is fragmented and siloed. Water districts are responsible for water, and cities and counties are responsible for their own land-use planning, each to its own.

---

[5] gpd for Gallons per capita per day; 1 US gallon = 3.8 l.

The water-trading between neighboring Imperial and San Diego interweaves their future, although the regions are treated as distinct and unrelated. Thus, it is quite plausible that the transfer will have effects in both places that involve negative impacts on nature and resources, and community development (most of the planned future growth in San Diego County is in the affluent rural areas at low densities), that would not have occurred had there been no, or less, water traded. But urban clout, and the sense that agriculture wastes water, have combined to force the transfer of water from one region to another far more affluent one. SDCWA is the big winner in the transfer, and the claim of water scarcity for future urban growth is unchallenged.

14.6 INSTITUTIONS, SPATIAL SCALE AND INTER-REGIONAL RELATIONS

Much of the water in the west was developed by the federal government in the first half of the twentieth century, and is regulated by federal policy. Existing structures of political decision-making facilitated the process. First, there is the fragmentation of decision-making in the area of water, with local water districts autonomous in their processes from the counties and cities they provide water to, and from state government. Second, cities and counties need not coordinate their decisions, while federal decision-making is also entirely autonomous.

Within the regions themselves, water authorities and districts depend on water sales for revenues, and relations are competitive. All take part in a zero sum game situation. If aggressive conservation takes place, districts will sell less water and reduce their fiscal viability. If one district reduces its water use, another district will receive more water and make more sales. County governments and water retailers have no linkages with the water districts, but each county and city is in competition with the other for growth. Looking towards the future, however, there are several factors that are now less predictable, and notably among them the effects of climate change on Colorado River flows. The amount of water that flows in the river – while it can be regulated and stored[6] – is ultimately a material factor that will impact the future mode of production and regulatory regime built up over the past century. How climate change will affect rainfall in the states through which the river flows is still unknown, although most predictions are for drier conditions. This will further reduce the flow of the Colorado River, and intensify competition over its allocation, both among states and between agriculture and urban uses.

14.7 CONCLUSION

The case of water wheeling from the Colorado River to San Diego County highlights the structures, norms and values that have defined the rights, constraints and power which have influenced development of the regions. The regulatory framework that exists is a historically accreted tangled complexity that operates on several interlinked geographical scales. It is defined by multiple, distinct institutions that regulate at different jurisdictional scales in quasi-autonomous self-reproducing spheres. At the heart of this complexity lies water and land, the material substrate utilized to produce

---

[6] Total storage capacity is five times the mean annual flow, due to exceptional dam sites like Glen Canyon.

human sustenance. We have observed how diffuse, conflicting and de-centered 'authority and power are inscribed across the intermingling domains of nature and culture' (Whiteside, 2002: 114). These set up the dynamics of water conflicts.

The selling of federally developed and subsidized water from the Imperial Valley, where it was historically destined for agriculture, to an adjacent urban area at much higher prices represents a significant shift in the institutions of water management toward the application of private-sector thinking. Bakker (2002) terms this the commercialization or *mercantilización* of water. It is premised on a new universal condition of water scarcity, simultaneously natural – justifying cost-reflective pricing – and social, the result of flawed public management (inefficient water management on the part of both the IID and other users). As Bakker notes (idem: 772), this is a process initiated not by the market, but by the state, and should thus be understood as a process of re-regulation rather than deregulation, characterized by an emergent new form of governance of natural resource allocation. This chapter has illustrated how this process of re-regulation of the social metabolism of nature is highly contingent on local variables, including the accidental Salton Sea and its environmental advocates, a highly concentrated corporate and absentee landlord form of agriculture, and a fragmented governmental structure with little or no coordination or collaboration. Water conflicts between urban water wholesalers (the MWD and the SDCWA) and between agricultural and urban water uses will intensify over the next century as water becomes scarce. Ironically, conservation is not the outcome, rather outcomes are a result of strategic manoeuvering.

REFERENCES

Bakker, K. 2002. From state to market? Water *mercantilización* in Spain. *Environment and Planning A*, 34: 767–90.

Gottlieb, R. and FitzSimmons, M. 1991. *Thirst for Growth: water agencies and hidden government in California*. Tucson: The University of Arizona Press.

Guthman, J. 2004. *Agrarian Dreams, the Paradox of Organic Farming in California*. Berkeley: University of California Press.

Kahrl, W. 1982. *Water and Power: the conflict of Los Angeles' water supply in the Owens Valley*. Berkley: University of California Press.

Pacific Institute 2003. *Waste Not, Want Not: the potential for urban water conservation in California*. www.pacinst.org/reports/urban_usage/ (Last accessed 23 April, 2005).

Revised Fourth Amendment Agreement Between Imperial Irrigation District and San Diego County Water Authority for Transfer of Conserved Water, October 2003. www.Sdcwa.org/manage/mwd-QSAdocs.phtml (Last accessed: 24 April, 2005).

Rudy, A.P. 1995. Environmental conditions, negotiations and crises: the political economy of agriculture in the Imperial Valley of California, *1850–1993*, PhD Dissertation in Sociology, University of California, Santa Cruz.

SANDAG. 2004. *Regional Comprehensive Plan for the San Diego Region*, July. Urban Water Management Plan www.sdcwa.org/manage/pdf/2005UWMP/Final2005UWMP.pdf (Last accessed 9 November 2006).

State of California, Department of Water Resources (SCDWA). 2004. *Colorado River QSA & SS Ecosystem Restoration Legislation*, March 4. Available at: http://www.saltonsea.water.ca.gov/legis.indes.cfm

Whiteside, K.H. 2002. *Divided Natures, French Contributions to Political Ecology*, Cambridge: MIT Press.

Chapter 15

Urban water reform in Italy: A live bomb behind outward unanimity[1]

Antonio Massarutto

University of Udine, and Centre for Research on Energy and Environmental Economics and Policy (IEFE), Bocconi University, Milano, Italy

15.1 THE WEAVING OF PENELOPE

At the end of the 1980s, Italian water and waste water systems were showing clear signs of unsustainability: poor performance records, increasing difficulty in guaranteeing quality and continuity of supply, an inability to achieve adequate environmental quality, and rapid decline of investment levels and subsequent degradation of infrastructure.

During the last twenty years, a far-reaching reform has changed the Italian water management system and consequently the organization of water and sanitation services (WSS). The pre-existing organization, based on direct public operation of networks, regional planning of infrastructure, and a high level of subsidies for investment, had to adopt a completely new philosophy: industrial asset management and operation, integrated management of water supply and sanitation at a district scale, delegation to professional water companies (either publicly or privately owned), financial self-sufficiency and full-cost recovery, with regional and basin institutions retaining strategic stance and regulatory functions.

The reform started in 1994 with a national framework law (l.36/94, 'Galli' from the name of its spokesman). This law needed to be transposed and adapted by regional laws and later implemented by local authorities, with the latter retaining responsibility, but under a regime of compulsory association within each new management unit, the ATO.[2] L.36/94 was 'quasi'-unanimously approved.

The rhetoric employed at the time presented the reform as a way to end with infrastructural deficits and obtain more effective and less vulnerable water supplies, a better water environment and a more efficient and rational management system.

---

[1] The support of the FIRB research programme 'Evoluzione delle forme di mercato e delle modalità di regolazione dei servizi locali di pubblica utilità in Europa' and the EU-5FP Euromarket project is gratefully acknowledged. The author wishes to thank Vania Paccagnan, Elisabetta Linares and Alessandro de Carli for their collaboration on both projects. Special thanks to Bernard Barraqué and Esteban Castro for their useful comments on a previous version of this chapter.
[2] ATO is the acronym for *ambito territoriale ottimale* (optimal territorial units).

The law asserts universal principles and rights and claims public ownership over all water resources. Full-cost recovery was intended as a cornerstone of the policy, with the aim of guaranteeing financial self-sufficiency of WSS undertakings.

Nearly all observers seemed to agree that the goals set were appropriate, and the solutions adopted the most effective (for a survey, see Massarutto, 1993 and 1995). The timing initially imagined was short and straightforward: six months to create regional laws, and another six months to establish inter-municipal agencies, develop an inventory of existing assets and a management plan, and identify a suitable operator. No obstacles would seemingly delay a rapid implementation of such an unanimously shared reform.

However, more than fifteen years later, the reform is far behind schedule. Investments ceased for more than a decade, while the procedures for implementing the law took nearly ten years to be completed – ten times longer than expected.

Today, implementation is slow: the building of the institutional regime and the parallel evolution of the water industry have become a never-ending task: the painful achievement of a temporary solution is suddenly contradicted by new legislation, delegating decisions to local authorities then subsequently forcing them towards compulsory tendering; imperious deadlines are established then postponed with ad hoc ordinances; one day favours concentration and partial privatization, the next obliges privatized companies to terminate their contracts.

To further complicate matters, during the same period Italy was undergoing a regional decentralization reform. Water (and environment in general) became a regional responsibility, while matters regarding market and competition remained with the central state, thereby creating the potential for inter-institutional clashes. Many regions began introducing their own regulations for water services, sometimes integrating, often contradicting the national dictate. Disputes concerning respective attributions and powers were often referred to the Supreme Court, where many are still pending today.

The parallel evolution of European policy concerning services of general economic interest (SGEI) added further opportunities for legislative and regulatory changes, the actual outcome of which resulted from the relative power of lobbies and pressure groups: national vs. local, private industry, former municipal companies, construction companies, non-globalist movements and so on. This endless legislative bradyseism has resulted in a confused, contradictory, fuzzy and unpredictable regulatory framework, with new rules inconsistent with previous norms, and new laws just about covering implementation as defined, only for a new dictate to alter the situation once more.

The net outcome is thus not open conflict, but rather endless debate, permanent stagnation and delayed action. In 2009, a Law Decree (DL 135/09) imposed compulsory competitive tendering within a short timespan; all in-house delegations[3] would be terminated in 2011. Once again, water companies will wait until the complicated procedures and subsequent predictable litigations and complaints are finished. But one can

---

[3] According to the EU law, in-house delegation is made by public authorities to own companies, entirely controlled, without competitive tendering. Tendering is a competitive process, where many candidates compete for the contract; in-house delegation is made to publicly owned companies in a non competitive way.

be certain that the deadline will be postponed and an escape route for the majority of public water companies will be introduced in the next round.[4]

The net losers at the moment are clearly the next generation, who will have to confront with a new investment cycle; and water quality, which is still lagging behind the targets imposed by the 1991 Wastewater Directive even before the WFD.

The aim of the present chapter is not to discuss this ongoing situation, but rather to single out the most important critical aspects of the Italian reform that block the transition towards the new system and prevent modernization; and at the same time to unravel the maze of trade-offs that have remained mostly silent so far, but have nonetheless impeded scheduled implementation. Open conflicts might arise in the future, unless these critical issues are appropriately addressed by a coherent set of institutional regimes, cultural approaches and political synthesis.

Now, the main question is: how is it possible that a Western country, which claims to be seventh or eighth in the world ranking of GDP, is unable to mobilize the financial resources, in the ordrer of a few tens of euros per capita/year, for a purpose that everyone agrees is a fundamental necessity? How can it be that every solution attempted so far has met more or less explicit opposition, resulting in delayed public action, endless paperwork and vain abstract discussions?

We argue that three main issues have to be dealt with: first, the setting up of intermunicipal agencies to share resources yet, it may generate tensions since in each municipality the infrastructure may be in a different condition and degree of completeness. The second issue concerns the creation of a professional water industry and the necessary regulatory checks and balances. Regulation is still dominated by traditional command and control mechanisms, which prevent the effective delegation of decision-making power to water companies. The third issue concerns the implementation of full-cost recovery and the polluter-pays principle, as the implications for water prices and income distribution have been largely undervalued.

These issues can be followed in three steps: we first describe the state of the art prior to the reform. We then describe the management system framework introduced by L. 36/94. After, we analyse the implementation phase in more depth, as well as the many stop-and-gos that have characterized it to date. In the last section we examine how these unresolved issues have rendered unmanageable the achievement of targets imposed by the EU Water framework directive.

15.2 PATH TOWARDS THE REFORM: THE CRISIS OF THE TRADITIONAL MODEL

Italy is overall a water-rich country, where supply for human consumption has never represented a major problem (Irsa-CNR, 1999). This is especially true in the north and along the peninsula, with the few areas suffering from water shortage being concentrated

---

[4] The situation dramatically changed after the writing of this chapter. A popular referendum was organized in June 2011 with the aim of abolishing both the 2009 decree and the mandatory rule of "adequate remuneration of capital." An overwhelming majority voted the double abolition. This was intended to stop the trend towards privatization, but post-voting scenarios are very difficult to build. More information available at: www.lavoce.info.

in the big islands and some southern districts. Water transfers exceeding the intermunicipal or district scale occur only in exceptional cases – among which the best known is the Acquedotto Pugliese, created at the beginning of the twentieth century.

Local availability of good quality, easily distributed resources historically allowed urban water management systems to retain a local dimension and a relatively low level of industrial and technical complexity. Until the 1990s, it was impossible to obtain a precise figure on management units, estimated in the reach of 13,000 for the whole country. Local responsibility for urban water systems was accompanied by state – and, later, increasingly regional – responsibility to provide municipalities with the means to fulfil this duty.

Management of local water supply and sanitation (WSS) was thus increasingly incorporated into planning systems for all infrastructure required to meet local needs. The need for water supply plans started to be felt in the 1960s, as a consequence of rapid urban development growth in the post-war period. This was accomplished through regional plans assigning to each urban area the right to obtain definite amounts of water, calculated on the basis of estimated needs. Where these quantities were not available in the same territory, the state (and later on regions) were compelled to implement and finance transfer schemes, aqueducts and supply facilities.

The same approach was adopted once sewage collection and sanitation became mandatory and required urban areas to be endowed with appropriate treatment facilities. This occurred in particular after the passing of law 319/1976, which initiated policies for water quality protection in Italy.

Regional planning can therefore be seen as the counterpart to urban water systems managed by local authorities, taking care of decisions concerning infrastructural development and finance.

In turn, urban WSS systems remained in municipal hands. In a number of cases – and especially in urbanized areas in the north – municipalities were able to create their own management systems capable of dealing with local problems and to develop urban networks. The model of the municipal company (similar to the German *Stadtwerke*), often organized as a multi-utility also dealing with electricity, gas distribution, transport, waste management and other local services), proved to be quite a successful management system. As in the German case, it allowed cross-subsidies and provided municipalities with a powerful engine for driving infrastructural policies. This model remained for the most part confined to cities, although voluntary associations and joint boards of small municipalities were also created in rural or semi-urban areas, especially in the north.

The *ancien régime* was characterized by a dualism between regional planning and municipal management, dominance of a supply-side approach (estimated 'water needs' as the basis of regional planning), financing through the public budget of all new facilities, and the physical fragmentation of management systems.

Regional planning of facilities revealed a powerful and effective way of starting the process and ensuring a baseline of treatment facilities. Yet its bureaucratic approach, dominated by a culture of public works, construction and engineering – easily exposed to the 'capturing forces' of construction industry and suppliers of technology – quite soon revealed itself as inconsistent with the need to ensure efficient design and appropriate management of the newly built facilities. Infrastructure development was often considered as an end in itself rather than as a way to improve management systems.

Regional planning was quite effective when the local counterpart was a municipal water company with adequate managerial capabilities and enough contractual power to impose its investment agenda on the regional planner. Much worse results were obtained when regional planning met with small, under-staffed and under-resourced direct management organizations. Poor performance records, overall lack of professional capabilities, inefficiency and financial unsustainability all became apparent in the second half of the 1980s (Massarutto, 1993).

Pressure for change arose as a result of both external and internal reasons.

The former included the mounting wave of liberalization of public services, mostly concerning energy and telecoms, but also affecting water companies especially when organized as local multi-utilities. The debate on liberalization in Italy had, at that time, some peculiar focuses, chief of which was political favour – a result of contemporary scandals on corruption and the enormous distrust and collapse of credibility of whatever belonged to 'the public'. The enormous public debt and the need to recover national accounts in order to join the European Monetary Union forced the government towards a massive sellout of assets. On the other hand, evidence that at least a handful of municipal companies had achieved good or even excellent performances inspired the idea that a more open internal market could allow them to grow and evolve as players on the national scene.

Among the latter reasons was increasing evidence of overall underperformance of water services, despite some areas of excellence. The main factors rendering utilities systems unsustainable can be summarized as follows (IRSA-CNR, 1999).

Decreased availability of resources for drinking water supply as a result of qualitative factors in the north and quantitative factors in the south. Water supplies in the north rely substantially on underground resources that are increasingly contaminated by agriculture and threatened by the inheritance of unplanned and poorly regulated urban and industrial development during the past decades. In the south, the main challenges concern the need to share storage facilities between drinking water and other uses – which also gave rise to inter-regional disputes[5] – and the still insufficient supply of many areas, particularly in Sicily. Peak summer demand for tourism also creates temporary scarcity problems.

Surface water quality is overall poor. The torrential nature of most Italian rivers makes comparison with northern European rivers difficult, if not unfair. However, the few water quality protection plans issued by regions according to WFD requirements show that the present situation remains far from the target of 'good ecological status'. This problem is largely due to urban waste water discharges. The EU Urban Wastewater Directive (CE 91/271) is yet to be fully implemented and is currently driving capital expenditure. Over the last thirty years, treatment capacity has rapidly reached 60% of pollution loads, but in later years investment ceased and existing facilities suffered from poor maintenance and management difficulties. The outcome in terms of pollution control remains quite insufficient: while stopping (and in some cases reverting)

---

[5] This happens, for example, in the case of the upstream reservoirs located in the Basilicata region that supply drinking water and irrigation to the downstream Puglia region. A complex institutional agreement has finally been reached between both regions and the national government (Mazzola, 2005).

deterioration trends in the bigger rivers, there is also evidence of worsening conditions in smaller watercourses, even in rural or semi-urban areas.

Rainwater management is experiencing growing difficulties, especially in large suburban areas, characterized by urban sprawl and chaotic development.

Low water prices guaranteed affordability, but in the context of a drastic reduction of available public funds this simply led to severe underinvestment and incapacity to face new challenges, while EU directives are calling for dramatic investment.

Leopard-skin is a good description of the management system, with many areas of excellence but also very poor performance in other areas. Reproduction of 'best practices' is required as well as cross-fertilization to export good management solutions to under-developed areas. As a result of the centralized financial system, most operators suffered from limited decision autonomy and limited capacity to begin the search for new solutions.

Box 15.1 Acquedotto Pugliese: when social rights and economic sustainability clash

The case of Acquedotto Pugliese (AQP) illustrates the current debate in Italy, and also shows how misplaced solutions based on ideological arguments can easily complicate the problem rather than solving it (Massarutto and Linares, 2006).

Formerly established as a state-owned undertaking under public law, AQP was created at the beginning of the twentieth century with the purpose of managing a region-wide system supplying Puglia – possibly the driest and most water-stressed area in the country – with long-distance transfers originating from nearby regions. It was also involved in the management of urban services (distribution and sanitation). AQP remains the largest transfer scheme for public water supply in Europe, and was regarded at the time as a technological wonder and a worthy successor to the traditional Roman aqueducts.

However, performance records declined over time, and in the 1980s AQP found itself facing a long-lasting and irremediable crisis, while being possibly the largest employer and generating the highest added-value in the region of Puglia. A popular sentence claimed that 'AQP gave more to eat than to drink', underlying its corrupted relations with stakeholders along the value chain (suppliers of inputs, employees). Service quality was fair – compared to other worse cases in southern Italy – but still unsatisfactory. Water prices were low and far from cost recovery. Charging efficiency was very poor, and substantial illegalities had been tolerated for a long time (delinquent payments, illegal connections, water thefts). Managing AQP and working for it at any level was perceived as 'job-for-the-boys' in terms of local politics. As a result, AQP was close to bankruptcy and permanently in need of central government subsidies, even for operational expenditure.

In the 1990s, the national government initiated a strategy to transform AQP into a private-law equity company, still 100% in public hands, but with the aim of entering into full privatization at a later stage. To prepare for this, a new management strategy imposed a readjustment of accounts, reducing the burden of corruption and recovering unpaid charges. This restored economic viability with respect to operation, but was far from the investment the system needed. A €250 million loan was thus agreed, with some success, under the expectation that privatization would ease an increase in tariffs and allow a margin for sustaining the debt.

(continued)

The strategy of privatization was initially backed by Enel – the former national electricity supplier, newly privatized and striving for diversification. Once this company's interest faded, the government transferred ownership to the region of Puglia. Privatization was still on the agenda, but was increasingly meeting with political difficulties. These were possibly created by non-globalist movements, but in reality were more concretely influenced by many local interests, which would lose out from an improvement of efficiency (workers, suppliers, delinquent customers). Legal actions were initiated on the supposed basis that price increases were not legitimate, while bureaucratic resistance, political skirmishes and even intimidatory actions created obstacles to attempts to initiate a more radical reform.

With the aim of relaunching AQP but avoiding privatization, the newly elected President of Puglia, Mr. Vendola, (belonging to the radical-left party) appointed Mr. Petrella as new President of AQP, a well-known leader of the world campaign against privatization and a supporter of the claim for free and public water as a social right. Conversely, the chief executive, Mr. Scognamiglio, who came from a managerial background, was chosen in keeping with the ongoing strategy of readjusting accounts and achieving financial equilibrium. Petrella's stakes, however, prevailed: one of the first decisions of the new board concerned the introduction of a large (50 l/d) free allowance of water for all customers. The next decision announced the end of the private-law system, reverting to the highly politicized model. As it was impossible to correspondingly raise the charges of larger consumption units, this decision was immediately reflected in a worsening of AQP's financial position. Scognamiglio resigned, given the impossibility of maintaining an economically sustainable management strategy. At the same time, Petrella was fired. In an interview released to the newspaper *il Manifesto* on 12 December 2006, Vendola accused Petrella of 'abstract rationalism' and of 'being too much fascinated by his own ideas'. Vendola himself is now persuaded that to be sustainable AQP requires management at arm's length from political interference, while he judges the now abandoned proposal to grant a free allowance of 50 l/day as 'unrealistic dogmatism', given the huge financial needs that have to be met. His verdict on the experience is that: 'Before becoming a common good, water should become common sense'.

Petrella was been chosen to demonstrate that public management oriented towards people's needs rather than profit could also be economically sustainable, but his approach turned out to be ideological and demagogic. The new President, Mr. Monteforte, has clearly stated that this challenge can be confronted only if politics remains detached from the process and respects economic and financial constraints. Regardless of ownership, a sustainable AQP means that no further delays can be tolerated in modernizing the system and investing in it. This will undoubtedly require a drastic increase in prices and a substantial cut in privileges and corrupt relations. It also means that an over-staffed company will have to lose jobs, and possibly that lower qualified workers will have to be replaced by more skilled ones: not a popular policy in a region with very high unemployment rates. Water prices are still quite low compared to both national and international averages, and the impact on family income low enough to imagine that even a substantial increase might still be tolerable, even though it will hardly match long-term financial needs. In the event of a significant number of poor people being negatively affected, there would be a need for 'smart' redistributive mechanisms. However, the biggest step backwards under Petrella's presidency was the idea that free water was still conceivable and desirable (at least in Europe), and that the need for economic viability could be postponed or left behind. Once again, the encroaching conflict appears to lie between the present generation and the next, while a sort of 'tyranny of the status quo' blocks innovative solutions in the name of those who benefit in the short term.

15.3 THE REFORM

Law 36/1994 aimed at a comprehensive reform to tackle these problems. Its basic principles were as follows (Massarutto, 1993):

- *Full intermunicipal responsibility* for public supply and sewerage within geographical areas individuated by regions (ATOs) with the aim of increasing the size of management areas in order to redistribute water resources and costs, minimize the need for inter-area transfers, and achieve economies of scale.
- *Vertical integration* of responsibility across the whole urban water cycle under the same authority (inter-municipal agency for each ATO). Basic quality requirements are provided by legislation and regional environmental planning and external regulation.
- *Professional operation* made compulsory through the phasing out of direct labour management, obliging the ATO to designate a single operator for all the utilities. Accepted management formulas range from publicly owned companies to private sector delegation; mixed-venture solutions were also allowed. In all cases, including public undertakings, a contract is required in order to formalize the obligations of each involved party.
- *Full-cost recovery* through charges (although some margin remains for the state to subsidize large projects judged to be in the general interest). The state fixes rules for water prices concerning tariff structure, maximum increase rates and bench-marks to be reached.

More precisely, the regulatory system devises a model in which investment choices are made by the ATO (through the asset management and development plan (AMDP), which represents the basis of the contract with the operator). Assets are provided as a gratuitous loan for use, while the plan foresees that tariffs increase in order to allow the operator to finance the investment levels to which it is committed by the AMDP. There are margins for the operator to renegotiate the AMDP, in the first place through the tendering process, and secondly through subsequent voluntary agreements during the contract.

The aim of larger-scale utility management and full responsibility over asset planning and operation is to create self-sufficient management units that can realize better oversight over available resources while feeling more responsible for their appropriate management. Individuation of 'reserve zones' – water resources suitable for future drinking water purposes which should be managed according to precautionary principles – is a legal requirement for each ATO. It should be possible to single out such areas quite easily almost everywhere from a technical point of view. However, they must then be subjected to selective development limits in the general interest of the ATO.

The ATO system is also conceived to allow market-based financial mechanisms, leveraged by water tariffs, to carry the burden of investment. Following this design, regional planning would concentrate on environmental regulation and framing long-term investment strategies. Larger scale utility management is also meant to alleviate the impact of water prices by binding together areas with different marginal

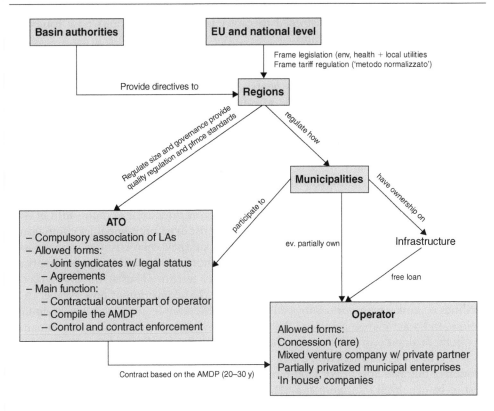

Figure 15.1 Organization of the water management system according to law 36/94

Source: Author.

costs of supply (especially urban and rural areas, assuming that the main explanatory factor for cost differentials is urban density).

Self-sustainability should also guarantee industry viability and allow investment to restart – both new investment required by the EU and the renewal of ageing infrastructure. Privatization and corporatization of water companies is expected to foster efficiency gains and favour economic rationality in the setting up of asset management and development plans.

In the following years, however, the reform faced many problems and difficulties. Many controversial issues set municipalities against regions, with the former fiercely defending the autonomy of the existing system, and the latter affirming integrated management at a larger territorial scale and supporting the creation of super-municipal multi-utilities extending their role to energy, waste management and natural gas. Most regional laws were approved only between 1997 and 1999.

Even more controversial issues set municipalities against one another: high against low average cost; recently built against ageing and incomplete infrastructure; high against low records of cost recovery; those who spent from their own budget against those subsidized by other levels of government levels; and those owning multi-utility companies against those accustomed to direct labour organizations now forced to merge into existing conglomerates.

In addition, the establishment of ATOs and the preliminary tasks (inventory of assets, choice of management model, development of the management plan, identification of the operator) required many years. After fifteen years, implementation of the reform is still on a leopard-skin base, although institutional transformations are consolidating (Muraro, 2004; Utilitatis, 2007). National framework legislation has finally been transposed to all regions, but this required many years. The process is now nearly complete, but in 2006 only 50% of ATOs had signed a contract with the operating water company.

Meanwhile, utility reform was affected by liberalization policies affecting all public utilities. The reform of the gas and electricity industry, in particular, created opportunities for many municipal companies – in many cases organized as multi-utilities – to act as players in the new liberalized market. The most dynamic companies commenced an ongoing repositioning process, either building on their own business capacity and assets or as a result of legislation. Some of the most noticeable trends are transformations into private-law undertakings, mergers (mostly on a territorial basis), quotation on the stock exchange, vertical integration (e.g. through the acquisition of power-generation facilities or bulk gas import contracts) and diversification (e.g. towards telecommunications and other market activities) (Vaccà, 2002). While allowing municipal companies to obtain brilliant economic results, these strategies also imply a reduced commitment to invest in local infrastructure development. The water sector, characterized by low economic margins and heavy long-term risks due to investment, has been one of the first to be penalized (Dorigoni et al, 2005).

At the same time, the national government – under pressure from the EU – has engaged in a reform of local utilities clearly oriented towards liberalization (Massarutto, 1999). Bills on compulsory competitive tendering for all utilities – including water and waste management as the most likely candidates – were almost passed, and were finally blocked or weakened due to the fierce resistance of local governments. Three legislative reforms were passed between 2000 and 2003 (Bognetti and Piacentino, 2005; Robotti, 2002; De Vincenti, 2005).

In a compromise reached in 2002, municipalities retained the possibility of directly running their own undertakings, but only by following strict criteria relating to in-house provision.[6] Otherwise, competitive tendering processes must take place, even if the companies' shares are partially owned by local authorities. Tenders may concern delegation, as well as the choice of private partners in municipally owned companies. This latter mechanism was used, for example, in the first ever tender implementing law 36/94, in the ATO of Arezzo in Tuscany, where the French multinational Suez (with some local partners) finally won the tender and become co-owner of the company (de La Motte, 2005).

---

[6] The definition of 'in-house provision' comes from EU legislation and in particular from certain judgments of the European Court of Justice. It relates to undertakings of any legal form, provided that municipalities retain the same degree of control that they would have with regard to their own offices; and provided that the concerned undertaking delivers a substantial part of its activity on behalf of the owner local authority.

Just a few years later, this provisional equilibrium was disrupted once again. A proposed law concerning compulsory competitive tendering was presented in 2006 by the center-left government of Prodi. After the sudden end of the legislature and the following change of majority in favour of of Berlusconi, a new proposal, similar in its general framework – at least in the appearance, was finally passed (l. 133/08) and modified the following year (DL 135/09). This law introduces compulsory competitive tendering for all public services. All actual undertakings made without tender (not only in-house delegation, but also creation of public companies later partially privatized through quotation) will terminate in two years. Municipalities unwilling to enter the tender process can maintain existing contracts with public companies by selling a share to the market.

The approval of this law has seen unexpressed, long-dormant tensions emerge in the public arena. As often happens when a conflict remains latent over a long period, its explosion generates a surplus of pathos and rhetorical violence, further increased by the inability of most participants in the debate (journalists, bloggers, politicians, militants) to focus on the point at hand. In fact, privatization is the issue of the debate, although the law does not actually impose it. The private sector is targeted as the real cause of tariff increases, and the few true cases of private sector involvement, such as the two concessions awarded to Veolia (Latina) and Suez (Arezzo), are dissected in order to demonstrate evidence of corrupt business dealings.

A flood of discussions quickly emerged at all levels. Television programmes were dedicated to water, for the most part riding the wave of popular discontent about tariff increases. (These increases are limited to date with Italian tariffs still being much lower than European averages, although the rises are sudden enough to create concern.) Fiery debates took place with public opinion stoked by independent blogs and militant campaigns, resulting in the above-mentioned referendum (in *Footnote 4*).

15.4 A NEW MODEL FOR PROVIDING WSS: CHALLENGES AND TRADE-OFFS

The management system before the reform was clearly giving birth to unsustainable trade-offs. It was unable to guarantee adequate appropriation of resources for future urban water supplies and to avoid segregation problems caused by diffused pollution and externalities. In short, it seemed unable to ensure the viability and economic efficiency of the water industry.

The new model aimed to provide a sustainable and economically viable solution, and was welcomed at its inception by an overwhelming consensus. After fifteen years, however, its implementation is still lagging behind.

The reasons for this delay have often been deemed transitory and reduced to simple 'technical' problems. Many observers blamed local authorities, unwilling to give up privileges; municipal companies reluctant to abandon their captive markets and the related monopoly rents; the incoherence of regulatory behaviour (which at least once allowed price increases) and so on. Our point, instead, is that policymakers dramatically undervalued the political and institutional dimension of the governance challenge of designing and implementing reform. This neglect, if not adequately resolved in the future, could seriously hamper the potential benefits of the reform.

We can use the classification of conflicts proposed in this book (see the concluding chapter) to elaborate on this, and then to extrapolate some insights on future sustainability of the reformed system.

First of all, the concentration of management systems into ATOs has raised *status* issues, representing a formidable obstacle that has blocked achievement of local consensus for many years, and still causes tensions and conflicts within the ATOs.

An example of status issues is the above-mentioned designation of reserve zones, where some municipalities are forced to impose restrictions on local development for the sake of safeguarding resources in the interest of the whole ATO. These restrictions cannot be compensated unless through general measures, such as rural development plans. These, however, are not in the hands of the authorities in charge of water policies.

More generally, the convergence into compulsory intermunicipal agencies is a process in which loss of autonomy over service organization and tariff-setting is causing many redistributive effects. The effects concern not only decision-making and tariffs, but also more practical issues such as the need to define common criteria for assessing the economic value of assets that each municipality confers to the system, and then who is responsible for past debt and who is the owner of cash flows generated by operation. Some municipalities had old and ageing assets while others were new; some had spent their own money while others had been largely subsidized, and so on. Governance structures for the new boards require complex balances between larger urban areas and small municipalities, whose weight in terms of population is usually smaller. Regions – in charge of defining ATO's governance rules – have introduced complex mixtures, weighed according to the number of municipalities, surface, population and so on.

The second category of conflicts regards *legitimacy and democracy*, and is directly linked with the setup of the regulatory framework and the choice to move away from public finance systems.

As noted above, the institutional system created by law 36/94 is not fully consistent with the kind of development – both in industry strategy and liberalization policies – occurring in the present phase. The latter calls for a more flexible but firmly adversarial regulatory approach, while the existing one is dominated by rigid ex-ante norms (contracts, tenders, price regulation formulas) and is subject to collaborative discretional renegotiation.

Even if the preferred option adopted by municipalities so far is the mixed venture publicly controlled company, these undertakings are in fact becoming much more autonomous from local strategic control. They grow in size acquiring a much larger scale than that of the original municipality, whose individual weight declines. Entrance of private partners or quotation on the stock exchange is another factor enhancing the business-like nature of former municipal companies.

Privatization, even in this relatively soft version, is often presented as an undue sale of assets of public property aimed at allowing private profit from the provision of essential services (Massarutto, 2011). So far, this has resulted in further active campaigning without significantly strong opposition; yet the greatest challenge will be confronting the people's reaction when the privatized water industry faces its first critical decisions.

In the present phase, while public reaction to privatization remains moderate over-all, the main issues lie in the institutional sphere and the regulatory model. Attention has to be drawn to the weakness and inconsistencies of the regulatory system. There is an evident lack of regulatory devices to enable local authorities to engage in a more adversarial relation with utility operators, especially if this relation is designed to be long-term (twenty to thirty years of concession are the norm) and the operator is charged with investment responsibility. Ex-ante regulation, based on tenders and contracts, was soon shown to be ineffective as a means to renegotiate contract obliga-tions. Risk of regulatory capture is evident, but there is also a corresponding risk that municipalities will not remain fully committed to maintaining business viability after the contract has been released. If this occurs, the water company takes the risk that recovery of its sunk costs will not be permitted if the assumptions of the original AMDP are revealed as wrong.[7]

It can also be questioned whether asking a private company to carry on investment plans that have been designed in detail by local authorities is the best way to 'delegate' investment responsibility. In fact, the risk for the water company is reduced (since investment levels are clearly indicated in the contract and further investment would need renegotiation), but at the same time increased. There is no guarantee that the choice of local authorities was really effective and efficient, while prior to the tender, the company is not in the best position to assess its validity. Renegotiation is allowed and explicitly foreseen, but not regulated; it occurs only on a voluntary base. A detailed study conducted by IEFE through a comparative analysis of selected countries shows that contract incompleteness, fuzzy and fragmented attribution of responsibility, con-fusion and overlap of competences, and failure to explicitly consider contingencies are the most critical features in the economic regulatory regime, and can be blamed for the difficulties met so far by the sector in actually implementing investment plans (Massarutto, 2009).

These difficulties are now creating obstacles and delays. Investment has not started according to plan, and the takeover of responsibilities is slower than expected. Systems that could count on a previously established management structure continue to per-form substantially better, but with the serious risk that formerly public, now semi-privatized water companies, also engaged in development strategies in other sectors, are left free to 'forget' the local water system, while they continue to enjoy an informa-tion monopoly that makes ATO enforcement less effective and credible.

Many observers stress the point that the ongoing evolution is quite contradictory: while promoting market development and competition, it creates strong obstacles to the overall transformation that municipal companies began on their own (Drusiani et al, 2004). In fact, mergers and partial privatization favoured overall development of the local utility industry between 1995 and 2005, but this was also favoured by the fact that municipalities could still maintain a degree of freedom in the choice of

---

[7] An example is provided by water consumption levels. In order to maintain the low price per cubic metre, many AMDPs foresee a significant increase in per capita water consumption. Contracts do not usually guarantee that the total revenue will match costs. In many cases, water companies have not been able to obtain the foreseen revenues since demand did not grow as foreseen. However, municipalities have been reluctant to consider this as a case for renegotiation.

organizational model. The rigid model based on the alternative between 'in-house' and tendering has had the paradoxical effect of bringing back management models that the previous reforms had tried to get rid of. Fearing the risk that tenders would force them to lose control over water utilities, most ATOs are now favouring the creation of brand new 100% public water companies, operating at each ATO level. The future relations between these companies and the former municipally owned companies, now partially privatized, remain unclear.

A third category of issues concerns *patterns of risk allocation* implicit in the chosen regulatory model. In the economic literature about auctions, there is a broad consensus on the idea that the operating company should be relieved of investment responsibilities in order to provide a base for efficient tendering (Massarutto, 2009). This is in fact what ATOs try to do with the detailed AMDPs they develop as a basis for tenders and contracts. However, this reproduces the same dualism between operation and investment planning that law 36/94 aimed to eliminate. On the other hand, as global experience shows, placing investment risk on operators is feasible at acceptable costs only if adequate degrees of flexibility are allowed by the contract and if market risks are substantially reduced (Massarutto, 2009).

This was illustrated empirically in Italy (Drusiani et al, 2004). Immediately after the first tenders and service assignments, most ATOs met unexpected difficulties in the management of the approved AMDPs and relations with water companies. After the contracts were released many water companies discovered that the economic and technical assumptions on which they were based had to be substantially changed, and this gave rise to conflicts, in particular, in cases where the water company had been significantly privatized.

The evolution of regulatory institutions, framed by law 36/94, was dominated by the idea of delegating managing responsibilities, from operation to investment and asset planning, to water companies. The AMDP, first issued by the ATO and later renegotiated with the water company, quite obviously implied a direct relationship between both actors and a collaborative regulatory framework. On the other hand, this was made increasingly difficult both by the strict legal requirements for tendering and adversarial contracting out, and by the increasing distance – in terms of strategy and behaviour – of former publicly owned companies, now clearly oriented towards the open market and unwilling to sacrifice their economic margins in order to satisfy local water policy agendas. Municipal officers are often aware of the inherent contradiction in being at once members of the shareholders' assembly of water companies and on the board of the ATO, with the conflicting pressures of maximizing company returns, consumers' welfare and long-term sustainability.

Such a mechanism could lead local politicians to sacrifice sustainability, as they are simultaneously pressured to avoid service price increases and to maintain the economic viability of the companies. This also allows municipalities to alleviate their budget crisis through dividends, capital gains and eventually company sales (Dorigoni et al, 2005). Long-term investment is thus the best candidate for sacrifice, since the time horizon of public decision-makers is constrained by their electoral mandate.

Many municipalities are now trying to avoid this problem by creating asset ownership companies that rent the municipalities' facilities to water companies in exchange for concession fees that can be calculated on the basis of asset value. This solution is regarded as a practical way out of the dilemma, since it can at least guarantee that cash flows originated by depreciation of assets come under the control of municipalities (Fazioli, 2007). On the other hand, there are obvious counter-indications: if final prices do not change, the net loser is the shareholder of the water company (sometimes the municipality itself, but often also the private partners to whom shares have been sold on the basis of a business plan that did not include a fee for the municipality). Yet the major risk lays in the fact that asset ownership companies could be tempted to play a stronger role, not limited to financial management of assets, but extended to design and investment decisions. This is not to be welcomed, since it would once again reproduce the dualism between asset management and operation that law 36/94 tried to eliminate (Massarutto, 2009).

A fourth and last issue is related to *allocation of costs*, and concerns the actual capacity of the existing mechanism to transfer the financing of utilities onto tariffs without creating substantial tension and conflict. In fact, the lack of evident conflicts in the present phase can also be attributed to a still very limited perception of real change by final customers. What will happen once water prices increase to the cost-recovery level? The next section develops a first exploratory analysis of this issue.

15.5 IMPLEMENTING THE WFD: THE FIRST CHALLENGE FOR THE PRIVATIZED WATER SYSTEM

In order to provide a first exploratory answer to this question, we have created a simulation based on the ongoing experience in two regions – Emilia-Romagna and Lombardia. These regions have been among the first to complete the regional planning documents implementing the EU WFD, and therefore permit a comparison between the actual situation and the system's expected performance.

The study originated in the legal requirement (Dlgs 152/99), which mandated regions to set up an action plan to implement the requirements of previous directives (with a special focus on 91/271 and 91/676), as well as to achieve good ecological status (GES), thus anticipating the WFD, although with some inconsistencies.[8] For this reason, the action plan contains either mandatory actions or discretional ones. Economic assessment of the action plan is also required, although, the approach chosen was far from a real cost-effectiveness analysis (and much further from a cost-benefit analysis), since scenarios and action plans were individuated by regional technicians, with a simulation comprising hydrologic models. Economic analysis could intervene only in the last phase to assess the cost of the chosen actions. No real trade-off of alternative scenarios and actions was therefore possible.

A second limit is that only structural actions (new treatment facilities and connection to sewage systems) could be modelled in detail, thus achieving a reasonable approximation of final costs. Other actions could be assessed only on the base of literature data and very simplified models.

A third limit – acknowledged by both regions' planning documents – is that the envisaged plan is only a starting point towards GES. There is no guarantee that the

[8] See IEFE and Gruppo 183 (2003) for an analysis of the main divergences between Dlgs 152 and the WFD.

Table 15.1 Economic indicators for structural and non structural actions

	Structural actions	Non-structural actions
Total cost (social cost)	Total full cost of existing assets[1] Actual sustainability gap[2] Differential total cost (per year) of new assets, broken down by single actions Total investment in new assets	Identification of main use and non-use EF Quantification of main consumptive uses Total value of water for productive uses Total cost dependent on actions (broken down by economic sector)[3]
Specific cost (cost arising to specific stakeholders)	Water price increase (per m³ and per capita) Irrigation water price increase (per ha)	Estimated number of dwellings, surfaces and working places affected Income reduction for affected firms
Affordability	Total per capita annual expenditure (of average and low-income families)/per capita GDP	
Cost-effectiveness	Cost per unit of target achieved (in terms of reduction of main pollutants concentration)	

Source: Author.

[1] Criteria for cost assessment are based on reconstruction costs of existing assets and a depreciation schedule that is coherent with true economic life. For more details see Tables 15.2 and 15.3.

[2] The sustainability gap is intended as the difference between the actual full cost of existing assets, calculated as in note 7 and direct revenues from water prices.

[3] For agriculture, industry and hydropower costs are estimated on the base of the best available alternative.

foreseen actions are sufficient to achieve it in all cases. This phase of the planning process is designed as a first selection of actions that apply in a general way to all water users, while ad-hoc actions will be singled out in a later phase in basins that do not achieve GES. For this reason, the plan also lacks proper individuation of criteria to decide how to define 'disproportionate costs'.

A fourth limit lies in the fact that models operate in a deterministic way and are based on average situations. The impact of short-lived problems (e.g. temporary water shortages) is not considered, neither are specific actions foreseen to cope with temporary problems and emergency situations.

Within these limitations, the action plan sets out two categories of actions (Table 15.1):

Structural actions: improvements in existing sewage treatment, reduction of leakage in both household supplies and irrigation, and reuse of treated effluents in agriculture and industry.

Non-structural actions: increases in minimum flows, reduction of licensed abstraction from rivers, introduction of cleaner technologies in industry, codes of good practice for agriculture and livestock breeding, and relocation of discharge points.

Table 15.2 **Summary of main economic indicators**

	Scenario 1		Scenario 2		Scenario 3	
	Lombardia	Emilia-Romagna	Lombardia	Emilia-Romagna	Lombardia	Emilia-Romagna
Gap between actual prices and FCR (%)	38	41	58	58	87	78
Average water bill as a % of family income						
Average	1.06	1.51	0.69	1.07	0.46	0.80
Range	0.77–2.83	1.39–2.69	1.52–2.40	0.96–2.04	0.30–1.48	0.70–1.71
Affordability index[1]						
Average	3.70	4.90	2.40	3.46	1.62	2.58
Range	2.37–8.03	3.58–6.85	1.52–5.20	2.48–4.81	0.93–3.29	1.80–3.67

Source: author.
[1] Affordability index = % of water bill on poorer families' income.

The related costs are established both in total absolute terms and with respect to the actors materially required to pay for it. Thus, for example, infrastructure costs for the WSS system are translated into water price increases for consumers. Those concerning irrigation networks are supposed to be partially subsidized and partially paid by farmers through charges. Non-structural actions are normally referred to the income losses of the economic actors affected.

Table 15.2 summarizes the main economic indicators developed.

Scenarios are characterized in terms of institutional arrangements to increase cross-subsidization of costs among areas or to avoid too drastic increases in water prices.

The main results, aggregated at the regional level, are summarized in Table 15.3. They show that the sustainability gap is quite pronounced in both regions, but particularly in Emilia-Romagna (where greater investment was made in the past). Conversely, the additions required to the existing system are comparatively greater in Lombardia. In both regions, the starting point shows a substantial gap between water prices and actual full costs.

Overall, a significant increase in the water price is expected (up to €120 and €150 per capita/year, with significant differences among areas). The most important aspect of this cost can be singled out as piping systems rather than treatment facilities, with a dramatic increase in rural and less concentrated areas. Overall, the total cost of planned actions can be estimated at €101 million/year for Emilia Romagna and €588 million for Lombardia).

These results are highly dependant on assumptions concerning capital cost. In the simulations, two alternative values were considered in relation to public and private sector debt, for new and existing capital assets respectively. In scenario 1 we simulated the impact of a strict full cost-recovery (FCR) requirement, implying the adoption of the market rate for all assets. The result is striking: regional averages jump to 1.06%–1.51% (though remaining below the threshold of 3%), yet average values are problematic in some districts (reaching peaks of 2.8% in some cases). Low-income families

Table 15.3 Cost of non-structural actions, Emilia-Romagna

	ml€/year	€/y per ha irrigated	€/y per ha affected	Range
Irrigation				
Water-saving	9	37		5–281
Reduction = 1.6%	7	27	1.681	309–3.682

Hydropower *Reduction*	*Social cost* ml€/year	*Private cost* ml€/year		
20%	25.7–32.9	18.9–25.7		
30%	38.5–49.4	28.4–38.5		
40%	51.3–65.9	37.8–51.3		

Source: Author.

in some cases would have to pay just under 10% of their annual income for water, which would be very difficult to accept. Both indicators show that inter-district equalization and action to alleviate the condition of low-income families are needed to make the policy acceptable. In any case, the increase in water prices will be quite significant and sudden, and will require careful management on the political side. These results are even more worrying, since the estimate of total costs is probably undervalued, being based on standard costs and estimates. Moreover, it does not take into account further costs required to upgrade the water supply and sewerage system not directly linked to the WFD. Once these considerations are made, the affordability of full-cost recovery is seriously in doubt. This confirms what other authors have noticed in case studies in other EU countries (Barraqué, 2005). Even in England and Wales the acceptability of FCR has reached a critical limit, corresponding to the need for many water companies to engage in high investment to replace existing assets. In some cases this difficulty lies at the origin of the creation of innovative financial solutions substantially based on the take-back of ownership of facilities and responsibility from the public sector, or mixed public-private entities that take over ownership of investment and responsibility for its long-term maintenance (Drusiani et al, 2005).

The cost of non-structural measures has been evaluated taking into consideration alternative scenarios for minimum flow requirements and their likely impacts in terms of reduced irrigated land and power generation. In the first case, the diminution of available irrigation water due to minimum flow is sometimes compensated by actions aimed at using water more efficiently.

Table 15.3 shows the expected reduction of irrigated surface and the related income loss in part of the district considered. Under the chosen rule for minimum flow, the expected reduction is contained at 1–2% as a result of the above-cited compensatory actions.

Overall, the expected cost of minimum flow requirements is of an order of magnitude lower than that for infrastructural actions: €7 million in terms of reduced income for farmers plus €9 million for water-saving actions funded by public contributions in Emilia-Romagna; while in Lombardia the expected impact on farmers is 0, since water savings are expected to compensate fully the reduction in raw abstractions. However, this cost is shared by a small minority of actors; farmers affected by a reduced irrigation capacity would entail an income reduction that is low with respect to total irrigated land (€27/ha on average), but much higher for the farmers that are actually affected (the value of being irrigated ranging between €309 and €3.682/ha depending on the crop). It is also interesting to compare this figure with the cost of water-saving actions (on average €37/ha with a range of €5–281ha). Apparently, reducing water losses is socially beneficial, and would remain acceptable if at least part of this cost is charged to farmers.

A similar deduction can be made for hydropower, for which the total estimated cost of minimum flows is calculated in the range of €20–30 million/year in Emilia Romagna.

Antonioli and Massarutto (2007) argue that these figures show evidence of potentially inefficient allocation of water rights between the civil and agricultural sectors: since the cost of infrastructural actions is much higher, households would be willing to pay farmers to reduce irrigation (and improve minimum flows) instead of paying for further WSS infrastructure. However, the Italian system precludes the possibility of such an arrangement, which is otherwise widely diffused in other EU systems (Heinz et al, 2002).

15.6 CONCLUSIONS

The Italian case is a good illustration of the kind of problems created by the transition from traditional management to a 'modern' WSS – one which is professional, autonomous and financially self-sufficient, if not necessarily private. The legislative reform of 1994 was probably inspired by good principles of water management, but suffered from difficulties in understanding the political, institutional and social dynamics needed to implement it.

Achieving sustainable patterns of urban water management requires new institutional settings. This does not simply mean enlarging the territorial scale: if anything, it refers to the way problems and solutions are shared among the community, costs are allocated, individual rights are traded-off against the general interest, and powers and responsibilities are allocated.

It also requires a new division of labour between water professionals, politicians, regulators and final users, enabling the system to attract the necessary human and financial resources. All this must either accept the general rules of the Common Market, or provide the necessary counterbalances to ensure that the monopoly power of operating companies is not used against the public interest. The delegation of power to professional systems is essential, but cannot occur if people have no trust in them. Trust cannot be simply invoked, it must be earned. The more sensitive the object retained, the stronger the trust required to make delegation of power acceptable.

The map of power has to be redrawn, and such a process cannot be enacted by decree. It requires instead cooperation, goodwill and a fundamentally shared vision of the outcomes and redistribution of costs, benefits and power among the actors involved. Without such a cohesive force, policies will simply remain unimplemented. Theorists have developed the concept of the 'tyranny of the status quo', which arises whenever unanimity is required for a decision. The Italian reform of the water industry shows that such an outcome can also arise when policymakers have the formal power to decide, but do not fully control the implementation chain. In this way, command and control are a largely theoretical exercise, condemned to remain on paper.

So far, the different players in the WSS orchestra – institutions, water companies, professionals, experts, the public opinion – are not playing the same tune. As soon as abstract talks about general principles meet the basic economic and technical reality, the whole theatre collapses. Popular discontent with 'privatization' can be seen, in this perspective, as an indicator of the failure to create this virtuous circle. At present, all the actors seem more concerned with impeding undesired outcomes rather than generating the desired ones.

Many observers have blamed the apparent lack of goodwill among those responsible for implementing the reform, who have instead rowed in the opposite direction (Muraro, 2004). This frustration assumes that the design of the reform was just and optimal, and that unwillingness to implement it could only derive from the defence of rent positions and local selfishness. Yet the claim of the reform to be 'optimal' can be questioned from many different angles, despite the unanimous consent it received at the time of approval.

Our analysis in the last section, for example, shows that the chosen model for allocating economic risks is inconsistent with the expected financial burden. Financial investment may be affordable for a time and can mobilize capital, but this amount is far from the figure actually needed to keep the industry viable over the long term. The delays in the implementation of existing investment plans also reveal that the assumptions made by regulatory agencies from the start were too optimistic. There is no possible match between the investment effort and the maximum tariff increase that regulators at present allow.

Distributive outcomes have also been neglected. Prior to the reform, the fiscal budget operated as a de facto equalizer at the national level. After the reform, equalization occurs only within the ATOs. Even if ATOs are designed to include cities and rural areas, this is hardly sufficient. Our analysis shows huge differences within each region, but far larger differences can be imagined between the north and south. Tariff structure on a pure volumetric base will probably further exacerbate differences, since elasticity of demand to income is typically very low.

As water policy targets become more ambitious and targeted towards the good ecological status required by the WFD, the traditional approach dominated by infrastructure investments reveals its weaknesses. In the past such an approach seemed affordable as most of the finance was subsidized. Yet, once customers have to pay the full cost in their water bill, the relative higher efficiency of alternative actions becomes clear. However, the Italian system still lacks instruments aimed to facilitate cooperation and sharing of costs among different water uses, in particular, agriculture.

The institutional setting has clearly lagged behind. The 1994 legislator believed it was possible to shift overnight from a system relying 100% on public subsidies to one that implied that water companies could rely on financial markets for their needs. Such a transition cannot be improvised; however, one cannot simply imagine that delegating control to the few professional water companies that have adequate capabilities and skills will be straightforward (Anwandter and Rubino, 2006).

It is clear that the final cost will depend on the cost of capital – our analysis shows that 1% of capital remuneration corresponds to 10% of operational cost-saving. However, the system has yet to develop a financial mechanism which could intermediate between the fiscal system and financial markets in order to provide the latter with adequate guarantees (Massarutto, 2009).

Economic regulation has also remained tied to the previous culture in which 'public control' of water utilities meant public planning of facilities and public works. The very idea that water companies should be responsible for deciding investments, instead of politicians, has mostly remained on paper and is still poorly understood (Massarutto, 2009).

Poor economic regulation means that contracts cannot be easily adapted to new requirements. As a result, environmental improvements are difficult to introduce once contracts are signed. This also means that a sort of a vicious circle is created: the weaker the regulatory system, the lower the confidence in delegating power to water companies and the stronger the temptation to solve everything by keeping the water company public. However, this is hardly a solution: even public companies exhibit good performance only when politicians remain at arm's length. More independence, even within the public sector, can also easily mean self-referential. But less independence might be even worse: raiding of funds for the current budget of municipalities, concession fees and improper costs imposed on the water company, forcing debt increases until leverage is unsustainable.

Lastly, an overall contradiction emerges between the general pro-liberalization approach – mostly based on competitive tendering – and the evolutionary path more often followed at the local level, based instead on the evolution of former public monopolies into structures based on private participation. Uncertainty of the regulatory framework and its patent difficulty in managing long-term issues and investment responsibilities represent a serious risk to the whole construct, which despite its sophisticated articulation, will fail to deliver its intended outcomes: modernization of utilities systems, maintenance, replacement and new investment to face the new challenges imposed by EU regulation.

If these trade-offs are not appropriately dealt with in the next phases of reform implementation, modernization of the Italian water industry will probably remain a dream. But this dream could quickly turn into a nightmare, one in which delaying actions are no longer feasible and solutions are found against the backdrop of an emergency. The popular vote in the 2011 referendum has made these contradictions very clear and the emergency even worse.

REFERENCES

Antonioli, B. and Massarutto, A. 2007, Efficienza allocativa e politiche dell'acqua: potenzialità degli strumenti di mercato nel caso italiano, *Economia delle fonti di energia e dell'ambiente*, forthcoming.

Anwandter, L. and Rubino, P. 2006, Perché la finanza privata asseta il settore idrico? *Mercato Concorrenza Regole*, 1: 197–228.

Barraqué, B. 2005. ¿Tienen los europeos los medios para costearse sus servicios públicos de agua y saneamiento? Fundacion Nueva Cultura del Agua (ed.) *Lo público y lo privado en la gestión del agua: experiencias y reflexiones para el siglo XXI*, Madrid, Ediciones del Oriente y del Mediterraneo, pp. 59–89.

Bognetti, G. and Piacentino, D. (eds) 2005. Privatizzazione e riorganizzazione dei servizi di pubblica utilità: Riflessione economica e caso italiano. *Economia Pubblica*, 1–2 (monographic volume).

De Vincenti, C. 2005. Dopo il nuovo art. 113 del TUEL: come creare un'anatra zoppa. P. Polidori (ed.) *Politiche locali e organizzazione dei servizi pubblici economici*. Milan: FrancoAngeli.

Dorigoni, S., Cozzi, G., Massarutto, A., Santoro, L. and Sileo, A. 2005. *La creazione di valore per i diversi stakeholder nel panorama attuale e prospettico delle utilities italiane*. Milan: Quaderni Iefe, Università Bocconi.

Drusiani, R., Fazioli, R., Massarutto, A. and Matino, P. 2004. *Regolazione e ristrutturazione del servizio idrico italiano*. Bologna: Clueb.

Fazioli, R. (ed.) 2007. *Economia e politica della regolazione del servizio idrico: modelli ed esperienze a confronto*. Bologna: Clueb.

Heinz, I., Brouwer, F., Andrews, K. and Zabel T. (ed.) 2002. *Co-operative agreements in agriculture as an instrument to improve the economic efficiency and environmental effectiveness of the EU water policy*, Final report, European Commission Contract ENV4-CT98-0782.

Iefe-Gruppo 183. 2003. Per l'attuazione della direttiva quadro europea sulle acque in Italia. *Economia delle fonti di energia e dell'ambiente*, 3.

IRSA-CNR. 1999. *I problemi dell'acqua in Italia*. Quaderni di ricerca Irsa, Roma, 2001.

Massarutto, A. 1993. *Economia del ciclo dell'acqua*. Milan: FrancoAngeli.

Massarutto, A. 1995. La legge Galli: una rivoluzione per i servizi idrici? *Economia delle fonti di energia e dell'ambiente*, 1.

Massarutto, A. 1999. *La riforma della l.142/90: verso il mercato dei servizi pubblici locali?*, in 'Economia delle fonti di energia e dell'ambiente', vol.XLII, n.3, pp. 121–60.

Massarutto, A. (a cura di). 2009. *La regolazione economica dei servizi idrici nell'esperienza internazionale*, IEFE Research Report, Università Bocconi, www.iefe.unibocconi.it

Massarutto A. (ed.) 2009. La regolazione economica dei servizi idrici nell'esperienza internazionale, Rapporto di ricerca IEFE, Università Bocconi, www.iefe.unibocconi.it

Massarutto, A. 2011. *Privati dell'acqua? Tra bene comune e mercato*. Il Mulino, Bologna.

Massarutto A., Linares E., 2006. 'A difficile apertura al mercato dei servizi idrici: il caso dell'Acquedotto Pugliese', *Economia e Politica Industriale*, 2, 171–89.

Mazzola, R. 2005. Improving drought preparedness through water transfers: the agreement between Puglia and Basilicata regions, *Proceedings of the Training course by Gruppo 183*, Torino, 15–24 November 2005. http://www.gruppo183.org/mdb/file/watertransfers_mazzola.pdf

de la Motte, R. 2005. WaterTime National Context Report – UK.WATERTIME Project. London: University of Greenwich.

Muraro, G. 2004. La riforma dei servizi idrici e il ruolo del Comitato di vigilanza. *Economia delle fonti di energia e dell'ambiente*, 3.

Robotti, L. (ed.) 2002. *Competizione e regole nel mercato dei servizi pubblici locali*. Bologna: il Mulino.

Utilitatis. 2007. BlueBook 07 – I dati sul servizio idrico integrato in Italia, Roma. Available at: http:www.utilitatis.org

Vaccà, S. (ed.) 2002. *Problemi e prospettive dei servizi locali di pubblica utilità in Italia*. Milan: FrancoAngeli.

Chapter 16

Water infrastructures between commercialization and shrinking: The case of Eastern Germany[1]

Matthias Naumann[1] and Markus Wissen[2]

[1]Leibniz Institute for Regional Development and Structural Planning (IRS), Erkner, Germany
[2]Institute of Political Science, University of Vienna, Austria

16.1 INTRODUCTION

The German water market is characterized by debates on liberalization and new forms of regulation, increasing privatization and processes of commercialization, tight public budgets, decreasing demand, the need for investment and new technological solutions. These general transformations create particular dynamics in shrinking cities and regions. Shrinking – understood as a three-dimensional phenomenon of decreasing population, de-industrialization and increasing socio-spatial differentiation – is the main spatial development for most parts of Eastern Germany. While reduced water consumption is an issue in all parts of Germany, shrinking regions are particularly affected by decreasing demand, questioning the functionality of existing infrastructure systems. In order to deal with rising costs for investment and the need to reorganize public finances, many municipalities partly privatize their water companies or commercialize water supply or wastewater disposal. Furthermore, water companies in shrinking regions have to deal with high fixed costs leading to rising fees for water supply and wastewater disposal. High prices endanger regional competitiveness and the quality of living in cities and regions, and tend to deepen existing regional disparities.

Using empirical evidence from research undertaken in the German federal state of Brandenburg and the city of Frankfurt (Oder) this chapter investigates the impact of shrinking processes on water infrastructure and the emerging regional disparities of water supply and wastewater disposal systems. We first summarize the main transformation trends in the German water sector. We then give an overview over the problems of drinking water supply and wastewater disposal in shrinking regions in Eastern Germany. Following this, we investigate the conflicts around the privatization of water infrastructures in Frankfurt (Oder) and Brandenburg. Finally, we interpret our empirical findings in the light of the 'splintering urbanism' approach, and suggest some

---

[1] This chapter largely draws upon the results of research conducted in the years 2002 to 2006 within the spatial analysis part of the project 'Socio-Ecological Regulation of Network-Related Infrastructure Systems (netWORKS)'. The project was funded by the Federal Ministry of Education and Research within the research programme 'Socio-ecological research'. For more detailed results, not only on East Germany but also on the West German cities Munich and Hanover, see Naumann and Wissen (2006).

conceptual modifications to enable this approach to grasp the socio-spatial dimensions of infrastructural change under conditions of shrinkage.

16.2 THE GERMAN WATER MARKET IN TRANSITION: THE MAIN TRANSFORMATIONS IN WATER SUPPLY AND WASTEWATER DISPOSAL

The German water sector is highly decentralized, consisting of about 6,000 mainly public water supply companies and more than 7,000 wastewater disposal companies. For decades, water supply and wastewater disposal in Germany were characterized by stability and very little change. In recent years, however, the German water sector has come under increasing liberalization and privatization pressures. This is due to five interrelated shifts.

First, water supply and wastewater disposal increasingly take place in the context of a neoliberal discourse of efficiency, competitiveness and profitability. According to this, the more efficient, innovative, customer-friendly and profitable a company is, the more it is exposed to competition constraints. The decentralized structure of the German water sector is said to be an obstacle to international competitiveness: water companies are too small to develop the technological and business know-how necessary to play an international role; and thus, are unable to take advantage of scale effects, with the consequence that prices for water and sewage are too high (Deutsche Bank Research, 2000).

Second, the neoliberal discourse corresponds to a severe financial crisis in municipalities. The German municipal budget deficit is estimated to amount to about €5.8 billion for 2005 (Jakubowski, 2006: 238). Decreasing revenues, growing expenditure and high debts, which are largely due to tax reforms and high unemployment rates, and thus part, and an expression, of the neoliberal transformation, make it increasingly difficult to fulfil public tasks. In the water sector, the financial crisis coincides with a high need for investment. Between 2000 and 2009, €29.8 billion had to be invested in German water supply systems in order to maintain networks, adapt them to changing demand patterns, and cope with the standards prescribed by European environmental legislation. The estimated need for investment in the German sewage systems amounts to €62.9 billion (Reidenbach, 2002). Nevertheless, municipal investment in water and wastewater infrastructures was expected to decrease between 2003 and 2007 by 6% (Tegner and Rehberg, 2006: 281). In this context, private investors are said to provide both technological know-how and the capital to help municipalities meet their investment demands. Against this background, one can observe a shift in municipal discourses and power relations with respect to the operation of public utilities, favouring either their (partial) privatization in order to mobilize private capital, or their transformation into publicly run, profit-making companies.

Third, there is the changing scalar organization of the water sector, largely due to the increasing influence of the European Union, however contradictory. On the one hand and with regard in particular, to the Water Framework Directive, the EU fosters 'environmentalization' of water supply by setting higher standards for groundwater protection, and by strengthening the role of environmental NGOs in the management of water resources. On the other hand, there have been several attempts by the European Commission to promote the privatization or even liberalization of the water sector (Herbke, Pielen and Kraemer, 2005). In Germany, these attempts are complemented by initiatives at the national level such as the 'Ewers-Gutachten', a report commissioned by the Federal Ministry of Economics, which argued for the necessity and feasibility of liberalizing the German water sector (Ewers et al, 2001). Even if such initiatives have not affected municipal water politics in a direct material way, up to now, they have brought about important symbolic effects: they have created a climate of insecurity concerning the future legal framework of water supply and sewage disposal and thus made local decision-makers more willing to adjust their utilities to a socio-political environment that may become increasingly competitive.

Fourth, there has been an entrepreneurial shift in public utilities, with a change from provision of public goods to profit-making. This is particularly the case for utilities that are transformed into municipal services and organized as limited liability companies. Even if these are completely or predominantly publicly owned, they become increasingly independent of municipal influence. As major actors in the local and regional economy that contribute to municipality budgets, they are able to reclaim the freedom of entrepreneurial decision-making for themselves. A 'new logic of infrastructure supply' (Marvin and Guy, 1997: 2027) has thus taken possession of public utilities, which today see themselves as 'normal enterprises' (see also Edeling et al, 2004; Kluge and Lux, 2001). Karen Bakker defines these processes of commercialization as the introduction of 'commercial principles (such as efficiency), methods (such as cost benefit assessment) and objectives (such as profit-maximation)' (Bakker, 2005: 542). Commercialization of German water markets thus leads to benchmarking projects, the introduction of new pricing systems, and the reduction of employment.

Fifth, in some parts of Germany there is increasing spatial competition among water providers. The reason for this is the dilemma faced by public water companies. Due to the employment of water-saving technologies in private households as well as the crisis and decline of many Fordist, old industrial branches, water consumption in Germany has dropped significantly since the 1980s. The supply capacities of many water companies therefore exceed the demand of their customers. However, as seen above, public water companies are increasingly being commercialized. They have to make profits in order to survive. A strategy which water companies in the north-west German federal state Lower Saxony apply in order to cope with this dilemma is to expand their provision and service areas – in other words, to compete with other companies for new customers (municipalities and firms). The result is the simultaneous fragmentation and regionalization of provision areas, which puts pressure, in particular, on small and medium-sized companies, some of which are in danger of disappearing (Naumann and Wissen, 2006).

The consequence of these developments, however, is not liberalization of the German water market, as demanded by neoliberal economists and politicians. Liberalizing the water sector would mean opening one and the same network to water from competing providers. Unlike the electricity and telecommunications sectors, which were recently liberalized in this way, the water sector poses severe obstacles to such a step. The most important of these is the physical properties of water itself, which preclude its transportation over long distances and the mixing of water from different sources, as these risk a loss of quality (Kluge et al, 2003: 47). Water turns out to be an 'uncooperative commodity' (Bakker, 2004). Thus, the five developments sketched above have not resulted in a fundamental reform of the water sector, which would be comparable to the liberalization of other infrastructure systems, but in terms of a more or less insidious form of commercialization. A central feature of this development is that the inherited decentralized and predominantly public structure of the German water sector is currently being reorganized in organizational as well as in spatial terms. Global players compete for supply areas or shares of municipal companies with private national companies, municipal services and public, special-purpose organizations.[2] Municipal services themselves try to expand by regionalizing their supply areas or by entering the market for water-related services on a national scale. The municipal services of Dortmund and Bochum together even took over the water company Gelsenwasser from the German energy multinational E.on.[3] The German water market is thus increasingly characterized by a public-private patchwork and a tendency towards larger supply units.

In the following section we shall turn to the situation in Eastern Germany, with special reference to Brandenburg and the city of Frankfurt (Oder), where commercialization has been superimposed and thus intensified by the shrinking process.

16.3 FROM 'FLOURISHING LANDSCAPES' TO SHRINKING REGIONS: POST-REUNIFICATION DEVELOPMENTS IN EASTERN GERMANY

In the first years after reunification, extension of access to water and wastewater networks in Eastern Germany was a major task. In 1990, only 53% of the population of Brandenburg were connected to centralized sewage. Peripheral rural areas with a low population density were particularly disadvantaged in this respect. In infrastructure terms, large disparities existed between city and countryside. Some households still lacked connections to public water supply, and had to obtain water from springs. Furthermore, many water and wastewater facilities had become obsolete and did not meet the technical, ecological and sanitation standards of reunified Germany and the European Union. Leakage rates were up to 20% and in industrialized areas there were problems with the quality of drinking water (Berkner and Spengler, 1991: 586).

The demand for modernization and equalization measures which arose out of this situation met with a widespread support concerning development after reunification. The future of Eastern Germany under capitalist conditions was conceived mainly in terms of growth to catch up with levels of wealth and productivity in the West. Helmut Kohl's dictum of the 'flourishing landscapes', which would replace what was supposed to be a socialist desert, was possibly the most concise expression of the way of thinking shaping the future development of Eastern Germany. Infrastructure systems played a crucial role in this respect: they were considered as preconditions for attracting the expected economic and social activities. Modernization and extension were thus seen as a primary task and were subsidized by national and European structural policy.

---

[2] For example the Oldenburgisch-Ostfriesische Wasserverband (OOWV) in the north-west of Germany successfully took over smaller municipal companies.

[3] This was in 2003 when E.on took over the gas provider Ruhrgas. In order to prevent E.on from becoming a monopolistic enterprise the cartel authority tied its permission for the acquisition to certain conditions, one of which was the sale of Gelsenwasser. The municipal services of Dortmund are a predominantly public company. However, 47% of the shares of its water and energy subsidiary are held by RWE.

Table 16.1 Decrease in water demand in Eastern Germany

	1991	2004	1991–2004
	in 1000 m³		in %
Brandenburg	188.2	108.6	−42.3
Mecklenburg-Vorpommern	142.3	83.6	−41.3
Sachsen	331.7	190.4	−42.6
Sachsen-Anhalt	221.7	108.3	−51.2
Thüringen	185.4	97.2	−47.6
Eastern Germany[1]	1,069.3	588.1	−45.0
Germany	5,747.9	4,733.3	−17.7

[1]Without East-Berlin
Source: Statistisches Bundesamt (1995, 2006).

Consequently, large investments were undertaken in the first half of the 1990s. The federal government introduced an 'urgent drinking water programme' and between 1995 and 2001 up to 3,000 new wastewater treatment facilities were built in Brandenburg (Kreistagsfraktion der PDS Barnim, 2003). The proportion of people having access to centralized water and wastewater networks was thus enhanced.

However, investments were partly miscalculated, insofar as the development assumption did not prove true. Instead of a growing economy in flourishing land-scapes, large parts of Eastern Germany in general, and Brandenburg in particular, faced economic and demographic decline. Today, there are nearly 1.3 million vacant housing units in Eastern Germany (Sigismund, 2006). Eastern German cities are expected to lose up to 25% of their population in the lead up to 2025 (Krautzberger, 2001). As a consequence, water consumption in Eastern Germany has decreased dramatically (see Table 16.1). Thus, the optimistic investments of the years after reunification resulted to a large extent in infrastructural overcapacity with some facilities operating only at half or less capacity (Interview, BGW/DVGW). The creation of over-capacities in some places was also the result of dirty deals. Constructing or moderniz-ing sewage treatment plants was an interesting field for private investors, who sometimes managed to convince a municipality to build (and pay for) a plant the capacity of which exceeded local needs and requirements. The higher the investments, the higher the charge of private building companies, engineering firms and consult-ants. In some places, corrupt local state officials took advantage of the modernization and construction wave in the first years after reunification (see Boecker, 2004; Halbach, 1992; Runge, 1994: 348).

The investments undertaken in the 1990s and the subsequent reduction in water consumption resulted in high indebtedness among the water and wastewater utilities of Brandenburg. It was estimated that the debt of the public special-purpose organiza-tions for water and wastewater, which form only a part of all Brandenburg utilities and mainly address rural areas, amounts to approximately €1 billion. Meanwhile one-third of all water companies in Brandenburg worked under the condition of high deficits (Interview, INFRANEU). Not surprisingly, the fees for wastewater disposal were high. In Frankfurt (Oder), they accounted for €2.66 per cubic metre in 2003, compared to €2.14 on average in Germany. The average fee in Eastern Germany as a whole was €2.50, compared to €2.09 in Western Germany (BGW, 2005). In Johanngeorgenstadt, a shrinking city in Saxony, the costs for gas, energy, district heating and water supply as well as for wastewater disposal are expected to rise by 38% until 2016, if no adaptation measures are undertaken.[4] Rising fees for wastewater disposal and increasing prices for drinking water supply in East Germany are especially dramatic, if one considers the fact that private house owners in the German Democratic Republic (GDR) had to pay only €0.20 per cubic metre drinking water or wastewater disposal. Tenants of public houses – the majority of the population – did not have to pay at all for water supply and wastewater disposal (Runge, 1994: 346).

The massive reduction in demand has led to technical, economic, health, environmental and institutional problems: the increasing residence time of drinking water in supply networks leads to the danger of bacterial contamination. Wastewater canals are affected by problems of smell, plugging and corrosion (Koziol, 2004). In order to maintain technical and health standards, additional measures like chlorination of drinking water, flushing of pipes or reduction of cross-sections of pipes have become necessary. These measures mean extra costs for the water companies. Thus, while the returns of the latter have been diminishing due to decreasing demand, the fixed costs are increasing. Investments for the adaptation of over-dimensioned infrastructures have to be borne by a decreasing number of customers (Schiller and Siedentop, 2005). Rising prices and fees for drinking water supply and wastewater disposal could lead to negative spirals: less customers demand less water, causing rising fixed costs, causing higher prices, leading to reduced demand among remaining customers, and so on.

Companies and municipalities often do not have the financial or personal capacities to deal with the problems caused by shrinking processes. The traditional logic of 'build and supply' of continuously expanding infrastructure systems is still persistent, and technological or organizational innovations are predominantly developed by companies in growing regions instead of by companies suffering from decreasing demand and rising fixed costs. Furthermore, in contrast to housing enterprises, until recently water companies did not receive any support within the framework of the urban renewal programme, by which the national state and federal states try to cope with the shrinking process. Even worse, the large-scale destruction of buildings in shrinking cities, subsidized by this programme, confronts water companies with the parallel necessity to deconstruct their networks, or, in cases where abandoned residential areas are situated not at the end of a service area but in-between places that still need to be served, raise the expenditure for the maintenance of networks (Bernt and Naumann, 2006).

Technical, economic and institutional problems cannot be seen separated, but must instead be understood as a complex network of problems (see Figure 16.1). The rising costs of utilities and the high demand for further modernization and adaptation of water and wastewater facilities, impossible to meet within the status quo, pave the way for the further commercialization or privatization of infrastructure provision. In the following section, we take a closer look at these processes in the federal state of Brandenburg, with special reference to the city of Frankfurt (Oder).

---

[4] Investing in the adaptation of existing systems could significantly lower the increase to 14% (Bose and Wirth, 2006: 20).

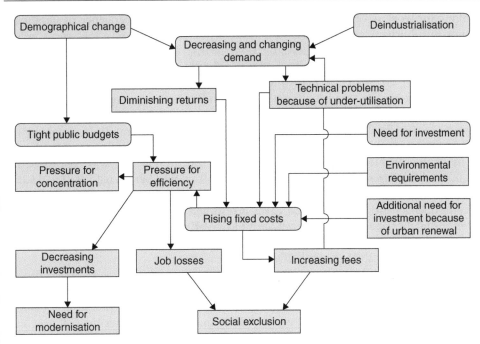

Figure 16.1 Interrelation of problems for water infrastructures

Source: IRS.

16.4 CONFLICTS ABOUT INFRASTRUCTURE SUPPLY AND PRIVATIZATION: THE CASE OF BRANDENBURG AND FRANKFURT (ODER)

Situated on the River Oder, which marks the regional border between Poland and Germany, Frankfurt is Brandenburg's fourth-largest city. In the 1970s and 1980s Frankfurt was the centre of the GDR's semiconductor industry. As such it received major investments and subsidies from the central government. Its population rose from 57,000 in the mid-1960s to 87,000 at the end of the 1980s. Accordingly, 21,000 new flats were built in this period, mostly in new housing development areas. With reunification, however, the economic decline of the city began. Most of the 8,100 jobs in the semiconductor works (VEB Halbleiterwerk) were lost. Considerable efforts were undertaken to build the city's future on its high-tech tradition through, for example, the establishment of research institutions like the Institute for Semiconductor Physics, subsidized by the European Union as well as by the national and federal government; however, the erosion of Frankfurt's economic base could not be prevented. Finally, in 2003, the establishment of a large chip producer failed. As a consequence of the economic decline, but also due to demographic developments and suburbanization, the city lost 22% of its population between 1990 and 2003. By 2025 the population of Frankfurt is expected to drop to no more than 59,000 people (more pessimistic prognoses estimate that between 45,000 and 50,000 inhabitants will be left). The decline of the population also indicates a severe brain drain of experts looking for work in other (German) regions. This is also a consequence of significant trends in suburbanization, largely due to the attractive rural landscapes surrounding Frankfurt (Matthiesen and Bürkner, 2004: 59ff.). An additional disadvantage of Frankfurt is its relative proximity to Berlin. The distance between the cities is only approximately 80 km. Thus, many high-income employees, for example, of the newly established university 'Viadrina', prefer to live in Berlin and commute to Frankfurt.

This shrinking process poses major problems for the networked infrastructure systems of the city. The case of the sewage treatment plant of Frankfurt, driven by the utility FWA (Frankfurt Water and Wastewater Company), may illustrate the dimensions of this problem. When its construction began in 1993, it was calculated for 20,000 cubic metres per day. In 1996, the real inflow was between 15,000 to 16,000 cubic metres. In 2004, it amounted to only 10,000 cubic metres (Interview FWA). As a consequence, the FWA incurred extra-investment to fight the malodour resulting from the facility operating under capacity. However, the amount of water necessary to flush the sewage pipes increased from 1,920 to nearly 12,000 cubic metres only between 1999 and 2001 (Koziol, 2004: 71).

The infrastructural problems caused by the shrinking process are the main drivers for the privatization of water companies in East Germany (see Figure 16.2).[5] In recent years, private companies like Veolia, Suez or Remondis have bought a range of shares in water companies in shrinking regions.[6] They may intend to enter the still predominantly public German water market while enlarging their supply areas step-by-step (even if the steps are small). In Frankfurt, the water and wastewater company FWA is still a public enterprise.[7] However, for several years there has been a fairly broad consensus among politicians, heads of administration and heads of utilities concerning the necessity for transformation in order to increase its efficiency, and to expand the supply and service provision area, for example, by exploiting its bridge position in respect to the Polish water market, thereby enabling the utility to contribute to improving the precarious financial situation of the city. In May 2005, the city council decided to transfer municipal shares in the FWA to Frankfurt Service Holding, a public enterprise which manages the city's shares in housing companies, hospitals and municipal services. This looked like a possible first step towards partial privatization. By selling a part of the municipal shares in FWA to a private investor, it was expected to earn €15 million, which should be used to consolidate the city's budget, reduce the debt and support the local economy. However, the particular form privatization would take was contested. One option was the integration of the FWA into Frankfurt's municipal services (Stadtwerke Frankfurt), a semi-privatized local company for energy supply and waste disposal, 39% of the shares of which are held by the E.on subsidiary E.dis and 10% by the former E.on subsidiary EWE. This option would have been combined

---

[5] The privatization of municipal companies in Eastern Germany is not only a concern of the water sector. Eastern German cities were among the forerunners in Germany for privatizing housing companies, hospitals and other social institutions.

[6] Veolia is involved in water companies in Görlitz, Weißwasser and Saxony-Anhalt; the Suez subsidiary Eurawasser serves the cities of Schwerin, Rostock and its surrounding area; Remondis recently signed a contract in the small town Senftenberg in the Land Brandenburg.

[7] The legal status of the FWA is that of a limited liability company: 90.5% of the shares are held by the city of Frankfurt, and the rest by two small neighbouring municipalities.

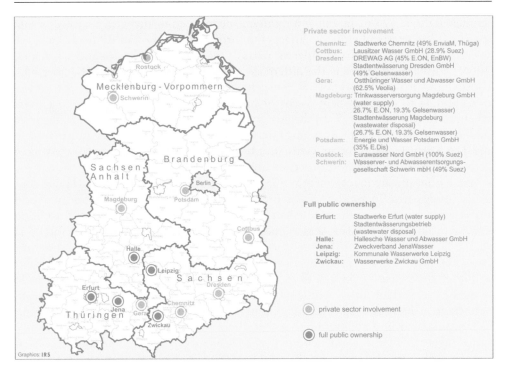

Figure 16.2 Water privatizations in Eastern German cities with more than 100,000 inhabitants (as of 2007)

Source: IRS.

with an unbundling of the FWA, expected to save costs and create synergies between the different branches of infrastructure supply. Because this would have meant dissolution of the FWA, it was heavily fought against by the management of the latter, which preferred another option: to locate a private water company to take over some of the municipality's shares in the FWA. This would have meant taking advantage of a situation in which private firms were prepared to pay high prices in order to get a foot into the regional and the Polish water market, both of which are being restructured.

In Western Germany, privatization frequently causes strong protests among different actors (trade unions, environmental groups, globalizations critics of ATTAC etc.), with privatization in some cases being blocked by local resistance. In contrast, privatization in Eastern Germany takes place in a rather 'silent way'. With a few exceptions, for example, Potsdam or Dessau, there are no campaigns against privatization of former municipal water supply and wastewater sewage companies. On the one hand, the lack of protest movements reflects the weakness of environmental groups and other NGOs in East Germany; on the other hand, it is the outcome of the strong belief of many municipalities that there is no solution but privatization. As far as Frankfurt is concerned, the 'inevitability' of further privatizing the infrastructure supply is not seriously questioned by any social group or political party in Frankfurt. Most actors have internalized the opinion that 'there is no alternative' (TINA). Only the form to be taken by a further privatization ('dissolution' of the water company into the municipal services or maintaining it as an independent company with a strong private 'partner') is contested because it affects the further distribution of influence over the infrastructure supply between different parts of the local state (Wissen and Naumann, 2006). Furthermore, there is the difference in motives among municipal actors who consider privatization inevitable. Whereas for some, privatization is an issue of conviction, leading to more efficient forms of infrastructure supply, others are just overwhelmed by the problems. Even the post-communists of the Left Party, who constitute the biggest fraction of the city council, have joined the TINA-coalition, in contrast to the Left Party in many other German cities and at the national level, where they actively support anti-privatization campaigns.

Aside from private involvement in the water sector, a more controversial issue concerns sewage fees. In rural areas of Brandenburg, households connected to central wastewater networks in the 1990s had to pay large sums for access, and in addition are now confronted by rising fees. Some attempted to reject 'connection by force', as they called it, in favour of building small decentralized wastewater treatment plants. However, these attempts failed in court. Nowadays, protest is articulated at the political level, for example, through the forming of civil initiatives and demonstrations in front of the Parliament of Brandenburg. In 2000, a hunger strike took place in Briesensee in Brandenburg. The inhabitants, including the mayor of the village himself, protested against connection to a central sewage system, which implied costs five times higher than the construction of decentralized systems. After four weeks of hunger strike, the Ministry for Rural Development, Environment and Consumer Protection of the Land Brandenburg agreed to consider a decentralized solution for the village.[8] But, because of resistance by the technical staff, house owners in 2004 organized a protest camp and hunger strike in front of the Parliament of the Land Brandenburg against forced connection to central sewage works. Private households as well as economic activities in rural areas are disadvantaged by the high wastewater fees, which are considered to be a severe negative location factor (Boecker, 2004) (see Figure 16.3).

To sum up, the shrinking of cities and regions has major consequences for networked infrastructure systems. The latter were previously designed to satisfy the demands of a growing population under conditions of economic growth. Now they have to cope with exactly the opposite conditions, paving the way for major transformations such as the privatization of utilities. In turn, infrastructural transformations can affect socio-spatial developments, if the deterioration or increasing prices of infrastructure provision reduce the profitability of economic activities and worsen the living conditions of significant parts of the population – thereby making a city or a region even less attractive. The degree as well as the manner of politicizing these issues varies. In places like Frankfurt, there is a heterogenous coalition that considers the further privatization of the water and wastewater utility a necessity without agreeing on the form that privatization should take. More intense struggles occur in rural areas where significant parts of the population fight against connection to central wastewater systems. These protests, however, are ideologically quite diffuse, ranging from populist campaigns (see Figure 16.3) to progressive forms of resistance which may have the potential to promote environmentally innovative infrastructural solutions at affordable prices.

---

[8] See http://www.eigenheimer.de/grogerhu.htm.

Figure 16.3 Poster of the private-house-owner association 'Haus und Grund'

Source: Haus und Grund Sachsen e.V.

16.5 LEARNING FROM SHRINKING REGIONS

The infrastructural transformations and conflicts analysed above have to be seen in the context of the specific constellations of problems that marked the situation in Eastern Germany after reunification, especially in the context of shrinking processes. However, some general conclusions can also be drawn. This is due to the fact that shrinking is a major tendency not only in Eastern Germany, but also in other parts of the world. While many cities in the developing and even in the developed world are still growing, old industrial centres and rural regions in the United States, Europe or Russia have suffered from economic decline and significant population losses (Oswalt, 2004). Thus, one can say that shrinking is not a general but an increasingly important mode of spatial development within the fragmented geographies of post-Fordism.

It takes on three forms (Hannemann, 2003). Firstly, it means deindustrialization. Old industries decline without being replaced by modern services, in other words, the economic base of a city or a region erodes. Second, there is a declining population, resulting from outward migration and a birth rate which is lower than the mortality rate. Thirdly, shrinking is characterized by suburbanization, leading to a socio-spatial differentiation between declining core cities and former residential areas, on the one hand, and growing or at least stable suburbs, on the other. Furthermore, shrinking processes are closely connected with vacancy and the deterioration of existing buildings (Bürkner et al, 2005: 12).

In spite of the various dimensions and increasing importance of shrinking processes, social science approaches to the transformation of networked infrastructures, until now, have not reflected sufficiently on the infrastructural impacts of shrinking. Therefore, in the following section we explore the conceptual consequences of shrinking for social science research on infrastructural change. In doing so, we refer in particular to the splintering urbanism approach (Graham, 2000; Graham and Marvin, 1996, 2001; Guy et al, 2001; Offner, 2000). Within this approach, urban development is investigated in relation to its interactions with the restructuring of networked infrastructures. The latter are seen to have a basic, often underestimated and taken for granted importance for urban life: they organize the 'perpetual process of transformation of nature into city' (Kaika and Swyngedouw, 2000: 120). Furthermore, utilities often belong to the largest local and regional enterprises, thus being important employers, real estate owners and investors (Marvin et al, 1999). A recent development is the unbundling of networked infrastructures, that is, the organizational and technical segmentation of formerly standardized systems, which often goes along with the privatization of profitable parts of infrastructure systems, whereas the less lucrative segments remain under control of the (local) state. On a symbolic level, this means that the promise of universal access, as represented in the 'modern infrastructural ideal', is no longer upheld.

This process is seen to have far-reaching spatial implications. Privileged places like the financial and business districts of global cities, known as 'premium network spaces' or 'hot spots' of infrastructure provision, are closely connected by highly efficient infrastructural systems, which 'bypass' marginalized places like poorer residential or old-industrial areas ('cold spots' of infrastructure provision), often situated in the same cities. Infrastructural transformations thus become a means of socio-spatial segregation or urban splintering (Gandy, 1997; Moss, 2003; Speak and Graham, 1999; Summerton, 2004).

The splintering urbanism approach sheds light on the 'invisible city' of networked infrastructures by focusing on the processes of their social production. It also shows how infrastructures themselves shape societal processes and power relations by their specific materialities and symbolic meanings. It is a path-breaking work that inspires empirical research on infrastructure and urban development. However, the experience of shrinking cities and regions in general and of Eastern Germany in particular suggests some conceptual revisions to the approach. These are summarized in the following:

First, there is a need to extend the scale of investigation. Whereas in many cases the urban scale may be appropriate to investigating processes of splintering and unbundling, in the case of infrastructural transformations under conditions of shrinking we have to focus on the regional scale in order to grasp the complexity of the picture. As the Eastern German experiences reveal, the degradation of core cities, caused by deindustrialization and population losses, and the resulting emergence of cold spots of infrastructure provision (marginalized network spaces) is accompanied by severe problems in peripheral rural areas within the same region and the development of rather stable or even growing suburban spaces (hot spots or premium network spaces). Against this background we see the necessity of a *shift from splintering urbanism to splintering regionalism.*

Second, the integration of rural areas into the analysis emphasizes the notion that the 'modern infrastructural ideal' was 'more a deeply symbolic construction than a tangible, achievable reality' (Graham, 2000: 185, see also Abegg and Thierstein, 2003; Thierstein et al, 2004). In rural areas of industrialized countries, let alone marginalized rural and urban areas in developing countries, this ideal infiltrated everyday life only as a vision of progress and emancipation. In contrast, the reality was often characterized by a lack of access to centralized infrastructure systems such as wastewater disposal. This doesn't mean that nothing has changed, because infrastructure provision has always been socially and spatially selective. However, it demands that future research on splintering urbanism and regionalism pay more attention to the *spatially and historically differing constellations of disparities and uneven development*, instead of focusing primarily on the transition from universal to uneven access to infrastructure provision. The distinctive feature of the current, post-Fordist constellation would then be a rise in the complexity of disparities, producing a fragmented (patchwork-like) socio-spatial and infrastructural landscape.

Third, the emergence of socio-spatial disparities of infrastructure provision in the water and wastewater sector does not necessarily take place via the bypassing of marginalized spaces in favour of premium network spaces, as one can observe in other networked infrastructure systems like telecommunications, roads and railroads. In Germany, the bypassing of 'cold spots' in the water sector is not possible, because water supply and wastewater disposal are part of the obligations of German municipalities. Thus, the question is not, *if* there is water supply and wastewater disposal, but *under which conditions* are these services provided. As far as these conditions are concerned, there are significant differences between German cities and regions. In contrast to the Eastern German shrinking regions with their high fees and prices and technical problems, booming cities such as Munich could be understood as 'premium network spaces' with high-quality services. Interestingly, in the shrinking regions of Eastern Germany it is the very attempt to universalize access to water and wastewater systems that becomes a medium for the development of disparities, as well as for a neoliberal shift in modes of governance. This is due to the specific materiality of centralized water and wastewater infrastructures: once established, they cannot be rapidly adapted to changing socio-spatial environments. Instead, their persistence imposes high costs on utilities, the local state and consumers in shrinking areas. They thus contribute to enhancing the problems of these areas, giving rise to claims for privatization and further depriving infrastructure provision of public control. In the long run, water infrastructures, originally an instrument of regional cohesion, might even turn out to be a medium of uneven development. This would be the case, if prices, quality and services of water supply and wastewater disposal not only reflect spatial developments such as shrinking, but also became a factor of deepening spatial differences in living quality and economic competitiveness. Our third proposal therefore is to pay more attention to *specific materialities of different networked infrastructures*, which in the case of water and wastewater systems can result in a dialectical rather than a dichotomic relation between universalizing and splintering.

The experience of Eastern Germany, analysed in this chapter, reveals a desideratum in infrastructure research. There is a need for more in-depth-analysis of infrastructural transformations in shrinking regions, taking into account urban as well as suburban and rural areas. Decreasing demand, need for modernization and increasing tendencies towards commercialization and privatization affect shrinking regions in Eastern Germany, but also take place in West European old-industrial or peripheral regions. The analysis of shrinking processes can therefore be viewed as highlighting future developments in other parts of the industrialized world. East European countries may be of special interest in this respect because their networked infrastructure systems can be expected to undergo major transformations after their accession to the EU. Of particular interest are comparative studies of shrinking and splintering. These promise to enrich the social science research on networked infrastructures with the experiences of an increasingly important type of socio-spatial development – the shrinking city and region – thus contributing to completing the picture of new geographies of infrastructure provision, drawn using the splintering urbanism approach. Furthermore, they could reveal general modes of infrastructural transformation, socio-spatial selectivities and governance forms in shrinking regions, whose identification is a precondition for shaping the transformation processes in socially and environmentally just ways.

REFERENCES

Abegg, C. and Thierstein, A. 2003. The liberalization of public services and their impact on the competitiveness of firms: a case study in the Alpine regions of Switzerland. Paper presented at 43rd Congress of the European Science Association, 'Peripheres, centers and spatial development in the new Europe', Jyväskylä, 27–30 August.

Bakker, K.J. 2004. *An Uncooperative Commodity. Privatizing Water in England and Wales.* Oxford: Oxford University Press.

Bakker, K.J. 2005. Neoliberalizing nature? Market environmentalism in water supply in England and Wales. *Annals of the Association of American Geographers*, 95(3): 542–65.

Berkner, A. and Spengler, R. 1991. Die hydrogeographischen und wasserwirtschaftlichen Bedingungen in den neuen Bundesländern. *Geographische Rundschau*, 43(10): 580–89.

Bernt, M. and Naumann, M. 2006. Wenn der Hahn zu bleibt. Wasserversorgung in schrumpfenden Städten. S. Frank and M. Gandy (eds) *Hydropolis. Wasser und die Stadt der Moderne.* Frankfurt am Main and New York: Campus, pp. 210–29.

Boecker, A. 2004. Die Mark im Schulden-Schlamm. *Süddeutsche Zeitung*, 18/19, September.

Bose, M. and Wirth, P. 2006. Gesundschrumpfen oder Ausbluten? *Aus Politik und Zeitgeschichte*, 21–22: 18–24.

Bürkner, H-J., Kuder, T. and Kühn, M. 2005. Regenerierung schrumpfender Städte. Theoretische Zugänge und Forschungsperspektiven. Working Paper. Erkner: Institute for Regional Development and Structural Planning.

BGW (Bundesverband der deutschen Gas- und Wasserwirtschaft). 2005. Marktdaten Info. Abwasserstatistik 2003. Available at: http://www.bgw.de/de/abwasser/marktdaten/article_2005_4_22.html.

Deutsche Bank Research. 2000. *Wasserwirtschaft im Zeichen von Liberalisierung und Privatisierung.* Deutsche Bank Research, Aktuelle Themen, 176.

Edeling, T., Stölting, E. and Wagner, D. 2004. *Öffentliche Unternehmen zwischen Privatwirtschaft und öffentlicher Verwaltung. Eine empirische Studie im Feld kommunaler Versorgungsunternehmen.* Wiesbaden: Verlag für Sozialwissenschaften.

Ewers, H-J., Botzenhart, K., Jekel, M., Salzwedel, J. and Kraemer, A.R. 2001. *Optionen, Chancen und Rahmenbedingungen einer Marktöffnung für eine nachhaltige Wasserversorgung. Endbericht eines Forschungsvorhabens im Auftrag des Bundesministeriums für Wirtschaft.* Berlin: Federal Ministry of Economics and Technology.

Gandy, M. 1997. The making of a regulatory crisis: restructuring New York City's water supply. *Transactions of the Institute of British Geographers*, 33(3): 338–58.

Graham, S. 2000. Premium network spaces: reflections on infrastructure networks and contemporary urban development. *International Journal of Urban and Regional Research*, 24(1): 183–200.

Graham, S. and Marvin, S. 1996. *Telecommunications and the City. Electronic Spaces, Urban Places.* London and New York: Routledge.

Graham, S. and Marvin, S. 2001. *Splintering Urbanism. Networked Infrastructures, Technological Mobilities and the Urban Condition.* London and New York: Routledge.

Guy, S., Graham, S. and Marvin, S. 1999. Splintering networks: the social, spatial and environmental implications of the privatisation and liberalization of utilities in Britain. O. Coutard (ed.) *The Governance of large technical systems.* London and New York: Routledge, pp. 149–69.

Halbach, U. 1992. Abwasserentsorgung in den neuen Bundesländern wird extrem teuer – Teil 1. *Wasserwirtschaft Wassertechnik*, 2: 62–63.

Hannemann, C. 2003. Schrumpfende Städte in Ostdeutschland – Ursachen und Folgen einer Stadtentwicklung ohne Wirtschaftswachstum. *Aus Politik und Zeitgeschichte*, B 28: 16–23.

Herbke, N., Pielen, B. and Kraemer, R.A. 2005. Europäische Liberalisierungstendenzen: Auswirkungen auf die kommunale Wasserwirtschaft in Deutschland. *GWF Wasser Abwasser*, 9: 660–64.

Jakubowski, P. 2006. Stadt ohne Infrastruktur heßt Stadt ohne Zukunft. Zur Agenda kommunaler Infrastrukturpolitik. *Informationen zur Raumentwicklung*, 5: 237–49.

Kaika, M. and Swyngedouw, E. 2000. Fetishizing the Modern City: the phantasmagoria of urban technological networks. *International Journal of Urban and Regional Research*, 24(1): 120–38.

Kluge, T., Koziol, M., Lux, A., Schramm, E. and Veit, A. 2003. Netzgebundene Infrastrukturen unter Veränderungsdruck – Sektoranalyse Wasser. netWORKS-Papers, 2. Berlin: German Institute of Urban Affairs.

Kluge, T. and Lux, A. 2001. Privatisierung in der Wasserwirtschaft. Sozial-ökologische Forschungsperspektiven. ISOE Diskussionspapiere, 17. Frankfurt am Main: Institute for Social-Ecological Research.

Koziol, M. 2004. Folgen des demographischen Wandels für die kommunale Infrastruktur. *Deutsche Zeitschrift für Kommunalwissenschaften*, 43(1): 69–83.

Krautzberger, M. 2001 Wohnungsleerstand und Rückbau: die ostdeutsche Stadtentwicklung. Raum. *Österreichische Zeitschrift für Raumplanung und Regionalpolitik*, 44: 40–43.

Kreistagsfraktion der PDS Barnim (ed.) 2003. Zur Entwicklung der Abwasserproblematik im Landkreis Barnim. Eine vorläufige Bilanz und Ausblick. Bernau.

Marvin, S. and Guy, S. 1997. Infrastructure provision, development processes and the co-production of environmental value. *Urban Studies*, 34(12): 2023–36.

Marvin, S., Graham, S. and Guy, S. 1999. Privatised networks, cities and regions in the UK. *Progress in Planning*, 51, Part 2: 89–165.

Matthiesen, U. and Bürkner, H-J. 2004. Wissensmilieus – Zur sozialen Konstruktion und analytischen Rekonstruktion eines neuen Sozialraum-Typus. U. Matthiesen (ed.) *Stadtregion und Wissen. Analysen und Plädoyers für eine wissensbasierte Stadtpolitik.* Wiesbaden: VS Verlag für Sozialwissenschaften, pp. 65–89.

Moss, T. 2003. Utilities, Land-use Change and Urban Development: Brownfield sites as 'Cold-spots' of Infrastructure Networks in Berlin. *Environment and Planning A*, 35(3): 511–29.

Naumann, M. and Wissen, M. 2006. Neue Räume der Wasserwirtschaft. Untersuchungen zur Trinkwasserver- und Abwasserentsorgung in den Regionen München, Hannover und Frankfurt (Oder). netWORKS-Papers, 21. Berlin: German Institute of Urban Affairs.

Offner, J-M. 2000. Territorial deregulation: local authorities at risk from technical networks. *International Journal of Urban and Regional Research*, 24(1): 165–82.

Oswalt, P. (ed.) 2004. *Schrumpfende Städte. Bd. 1: Internationale Untersuchung*. Ostfildern-Ruit: Hatje Cantz.

Reidenbach, M. 2002. Der kommunale Investitionsbedarf. Eine Schätzung für die Jahre 2000 bis 2009. DIFU-Beiträge zur Stadtforschung, Band 35. Berlin: German Institute of Urban Affairs.

Runge, M. 1994. Standortfaktor kommunale Infrastruktur. Organisation und Finanzierung der Wasserversorgung und der Abwasserentsorgung. *Mitteilungen der geographischen Gesellschaft in München*, 79: 339–58.

Schiller, G. and Siedentop, S. 2005. Infrastrukturfolgekosten der Siedlungsentwicklung unter Schrumpfungsbedingungen. DISP, 160: 83–93.

Sigismund, M. 2006. Regionale Disparitäten in der Wohnungsversorgung – noch ein Problem? *Informationen zur Raumentwicklung*, 6/7: 345–54.

Speak, S. and Graham, S. 1999. Service not included: private services restructuring, neighbourhoods, and social marginalisation. *Environment and Planning A*, 31(11): 1985–2001.

Statistisches Bundesamt. 1995. *Öffentliche Wasserversorgung und Abwasserbeseitigung*. Wiesbaden.

Statistisches Bundesamt. 2006. *Öffentliche Wasserversorgung und Abwasserbeseitigung*. Wiesbaden.

Summerton, J. 2004. The new 'energy divide': politics, social equity and sustainable consumption in reformed infrastructures. D. Southerton et al. (eds) *Sustainable Consumption. The Implications of Changing Infrastructures of Provision*. Cheltenham and Northampton (Massachusetts): Edward Elgar, pp. 49–64.

Tegner, H. and Rehberg, J. 2006. Öffentlich Private Partnerschaften – Lösungspotenziale für technische Infrastrukturen? *Informationen zur Raumentwicklung*, 5: 281–93.

Thierstein, A., Abegg, C., Pfister, B.G., Rey, M., Natrup, W. and Thoma, M. 2004. *Liberalisierung öffentlicher Dienstleistungen. Auswirkungen auf die Wettbewerbsfähigkeit der Unternehmen im Schweizer Berggebiet*. Bern, Stuttgart and Wien: Haupt Verlag.

Wissen, M. and Naumann, M. 2006. A new logic of infrastructure supply. The commercialisation of water and the transformation of urban governance in Germany. *Social Justice*, 33(3): 242–65.

INTERVIEWS

Administration for Strategy, Economy and development of the city of Frankfurt (Oder) (ASWE), 22 June, 2004.

Federal Association of the Gas and Water Sector/German Association of Gas and Water, Section Frankfurt (Oder) (BGW/DVGW), 16 September, 2004.

Frankfurt Water and Wastewater Company (FWA), 8 November, 2004.

Association of Infrastructure Development in East Germany (INFRANEU), 22 September, 2006.

Ministry for Agriculture, Environmental Protection and Spatial Planning of Brandenburg, (MLUR), 23 September, 2004.

Party of Democratic Socialism, Section Frankfurt (Oder) (PDS), 9 November, 2004.

Service Holding, Frankfurt (Oder) Service Holding, 26 October, 2004.

Chapter 17

Urban water conflicts and sustainability: An ecological-economic approach[1]

Antonio Massarutto

University of Udine, and Centre for Research on Energy and Environmental Economics and Policy (IEFE) Bocconi University, Milano, Italy

17.1 INTRODUCTION

The main idea behind this book, as stressed in Chapter 1, is the subtle but fundamental difference between 'water wars' and '(urban) water conflicts' (UWC). Within these pages, the theme has been addressed by scholars from different disciplinary backgrounds (economics, geography, social and political science) as well as different countries. The sole constraint was that the authors should examine cases in which the presence of organized water services based on physical infrastructure was decisive in framing the conflict, regardless of the kind of actors involved. Even the meaning of 'conflict' has been left deliberately vague, in order to encompass a wide range of situations: from violent opposition to political confrontation, from court litigation to social polarization, from competition and trade-offs (among models, ideas, cultures) to frontal opposition of parties and passions.

'Water scarcity' and related 'water wars' have dominated debate over the last twenty years. Implicit in this concept is the idea that humans are competing for access to water resources that are becoming increasingly scarce, either because demand is increasing or because the water capital is being dissipated. Privatization is interpreted as the attempt of greedy multinationals to monopolize supply of the 'blue gold', exploiting people's need for this fundamental good, and denying the right to universal access.

A few voices have been raised in recent times to counter this new conventional wisdom and present these phenomena in a different light. Among these, for example, are influential books such as Rogers et al (2006) or Saleth and Dinar (2004), which question the very 'myth' around water scarcity. They reveal the extent to which scarcity is a socially constructed issue, concerned more with bad management and poor allocation mechanisms than physical shortage. As Saleth and Dinar put it, 'water scarcity,

---

[1] Support for the FIRB research programme 'Evoluzione delle forme di mercato e delle modalità di regolazione dei servizi locali di pubblica utilità in Europa' and for the EU-5FP Euromarket project is gratefully acknowledged. The author wishes to thank Vania Paccagnan, Elisabetta Linares and Alessandro de Carli for their collaboration on both projects. Special thanks are also owed to Bernard Barraqué and Esteban Castro for their useful comments on a previous version of this chapter.

whether quantitative, qualitative or both, originates more from inefficient use and poor management than from any real physical limits on supply augmentation' (Saleth and Dinar, 2004: 1).

As such, is the solution just to improve and modernize water management practices? This answer is too simplistic. However, conflicts do exist, as the many examples collected in this book demonstrate. The engine that fuels them is far more complex than it is often assumed, as are the solutions. We might say that they are triggered by modernization. Modernization indeed constitutes a solution to traditional water wars, but it also triggers different kinds of conflict.

The hypothesis shared by the authors of this book is that the 'fault' generating these conflicts lies along the complex and never definitively resolved relation between water availability (natural capital) and water services (artificial capital). The former is abundant in general, but neither easily nor promptly usable (thus scarce); the latter is potentially unlimited, but costly and requires sophisticated management, thus implying complex institutional arrangements and collective enterprises. The mutual evolution of both systems, each constrained by its own 'limits to growth', is the ultimate source of conflicts, occurring either within systems or between them.

In this concluding chapter, we would like to go back to the initial hypotheses set out in Chapter 1 and provide a more general theoretical interpretation of UWC, from the perspective of ecological economics and, more broadly, the theory of sustainability. We will try to argue that UWC are, above all, useful indicators that allow us to understand the patterns of sustainability of water management, and the extent to which water represents a factor determining a 'limit to growth' for human development – a 'carrying capacity'. Far from being a universal issue, water sustainability will be presented as a local issue – where the size of 'local' largely depends on economic as well as technologic, geographic and institutional factors. Far from being a 'war' in the conventional meaning, water issues arise ultimately from a problem of collective action, namely that of finding an agreed solution to the question of how far water management systems should develop and who has to pay for them (Hanemann, 2006).

The chapter starts with a brief overview of the ecological economic approach to water sustainability, which enlightens the crucial importance of UWS. This is followed by an analysis of the present trend of UWS management and the crisis in the supply-side approach. Pressures arising from this change towards private sector involvement and full cost recovery are then discussed in general terms. Finally, this approach is used to frame the contributions in this volume and derive some general conclusions.

17.2 SUSTAINABILITY AND CARRYING CAPACITY IN THE WATER DOMAIN

The reason why water allocation might give rise to conflicts is not obvious. After all, in everyday life people have often to choose between alternatives that are mutually exclusive; this does not necessarily imply conflicts, provided that the rules followed are agreed and perceived as legitimate, fair and reasonable. Scarcity is not the answer either: there are very few paintings by van Gogh in the world, and many people would like to own one, but there is no 'conflict' around the possession of van Goghs. This is because everyone accepts the rule that the market price is an acceptable method, providing that the general public, who cannot afford to buy a painting, have collective access to a sufficiently high number of paintings shown in museums and exhibitions. Nor is it the fact that water is an essential good – at least as far as urban water use is concerned: 'in developed countries, the fact that water is essential for human life is almost certainly irrelevant when assessing the value of residential water supply, because the ways in which water is used are nowhere near the threshold level at which essentialness applies' (Hanemann, 2006).

Allocation of a scarce resource implies by definition that some demands will have to remain unsatisfied: this is intrinsic in the very meaning of the word 'scarcity'. Allocative choices imply trade-offs, whose solutions might or might not give rise to a conflict, depending on the capacity of the existing set of rules, property rights and management systems to resolve the related issues in a way that is both socially acceptable and sustainable over the long term. Conflicts arise, in turn, when the institutional setting is inadequate to the task of settling disputes and implementing the required technical solutions.

In pure mainstream neoclassical economics, conflicts do not exist, since the market is the universal engine providing solutions to trade-offs in a way that ensures the maximum of benefits. If a would-be user of a scarce good has to remain unsatisfied, this is a result of the fact that his or her demand is not sufficiently worthy compared to other competing demands. Welfare economic theory has provided a robust and coherent framework to deal with trade-offs when they entail an economic dimension – that is, when the values at stake are commensurable with each other. Cost-benefit analysis states that a given alternative is better than another (from a social welfare point of view) if its total costs are lower than its total benefits; where 'costs' mean 'opportunity costs', that is the economic value of the alternatives being sacrificed. Since society is an aggregate of individuals and groups that are affected by costs and benefits in different ways, this test can be refined by introducing compensation for the 'net losers': if total benefits are larger than costs, distribution of outcomes can be re-arranged in an order so that all concerned groups gain.

However, in the real world this does not necessarily take place: first, because redistribution of outcomes does not actually occur, or cannot occur within a given institutional framework (Johnston, 1996; Saleth and Dinar, 2004); second, because trade-offs are not always and not only economic – in other words, values cannot always be expressed in the same unity of measure (e.g. monetary values) allowing them to be traded-off in a coherent way and compensated (Martinez-Alier et al, 1998; O'Connor, 1998); third, because the net losers could be under-represented in political bargaining and decision-making, or even not represented at all, as in the case of future generations (Ekins, 2000).

Ecological economists have developed the concept of *strong sustainability*, according to which values that cannot be traded-off require the definition of specific constraints and threshold values in order to achieve sustainability. This occurs, in particular, whenever 'critical natural capital' is at stake – namely, environmental resources which cannot be substituted for the provision of valuable functions that society cannot afford to give up (Turner, 1993; Ekins, 2000; Ekins et al, 2003).

Water is an excellent case to apply these concepts, due to its multifunctionality. Water sustainability implies ecological dimensions (because it is a scarce and fragile natural resource, and essential for ecological life), an economic dimension (because it is a scarce valuable input with an economic value) and a social dimension (because it represents a basic need and should be shared in an equitable way). Water provides a large set of valuable environmental functions entailing economic as well as non-economic values (Hanemann, 2006; Young, 2005). Many of these functions are critical ('water needs' to be satisfied at any cost, 'social rights' to be guaranteed at the expense of the community, etc), while others can even be traded-off ('water demands' whose value can, in principle, be compensated by other economic values).

In a narrow sense, an economic approach to water policy issues deals with the concept of water as an economic good, in other words, it considers only those values that belong to the economic sphere (Dalhuisen et al, 1999). In a broader sense, the economics of water should be able to understand the multiple dimensions of scarcity that arise both from economic and non-economic spheres (Green, 2003). Following the latter perspective, an economic approach – 'applying reason to choices' in Colin Green's own definition – is not incompatible with the argument that certain demands should not be evaluated in terms of economic efficiency, but rather in terms of social justice, ecological soundness, political acceptability and so on.

If water unquestionably forms part of the planet's critical natural capital, the water domain is also characterized by degrees of substitution between man-made and natural capital: at a certain (high, though finite) cost, freshwater can be supplied in virtually any desired quantity. Desalination, which can be considered as backstop technology, has reached costs in the order of €0.5 – 1/m^3 (Lora et al, 2004). The same is true for other environmental functions, especially those for which economic values are predominant. The social choice problem here is whether this cost is worthwhile (i.e. if the value of water demand is higher), and if users are able and willing to pay for it. This consideration is fundamental, since it enlarges the spectrum of trade-offs (and potential conflicts). These concern not only appropriation and use of water as a natural resource, but also the inter- and intra-generational allocation of costs of artificial water systems. In the same way, sustainability concerns not only intergenerational aspects of natural resources management, but also includes the economic and financial viability of the water industry (de Carli et al, 2002). To put it another way, both natural and man-made capital belong to the category of 'critical capital' and therefore require strong sustainability criteria (Massarutto, 2005).

In terms of Figure 17.1, governance issues arise in a given system when there is a mismatch between social demands and supply. The system is initially defined in terms of natural patterns (geography, hydrology), as well as artificial assets and governance institutions. All these factors together determine how much water is available, how much can be used and by whom, and how costs and benefits are shared. On the opposing side, we have social demands for the environmental functions that water provides, taking into account the direct costs that each has to face and the price paid to resource owners and/or providers of water services. The dynamics of the system depend on natural factors (e.g. climate change), as well as social and economic factors (e.g. urban growth, industrial development, new emerging demands).

In the event of a mismatch, allocation rules may be able to settle the dispute: as in the case of the van Goghs, some demands will remain unsatisfied, but will accept the outcome of the allocation process. Or they may not, and this could mean that either the losers will not accept the outcome and will initiate a political struggle (eventually leading to a violent confrontation) to obtain what they want, or that some fundamental

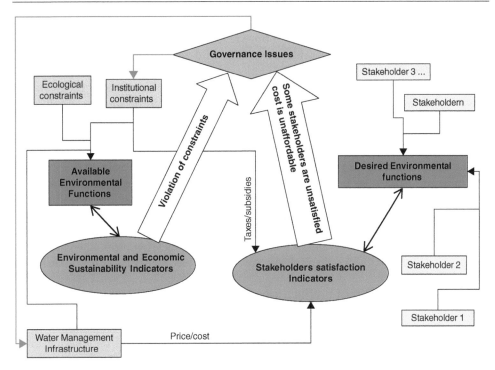

Figure 17.1 An ecological economic framework for understanding urban water conflicts

Source: Author.

demands will not be satisfied (e.g. many people will not have access to water); or finally that everyone is tempted to use more than the sustainable quantity (e.g. over-pump from aquifers), thus creating a 'tragedy of commons', leading in the long run to resource dissipation. In all cases, the institutional setting is unable to provide for sustainable use of the common-pool resource.

In this sense, sustainability is ultimately a governance issue: existing governance rules are not able to provide solutions to scarcity that are ecologically viable, technologically feasible, economically affordable and socially acceptable. For example, the existing set of water rights might be incapable of resolving a given problem (e.g. over-exploitation of an aquifer), but the creation of a new institutional framework is impossible because of inability to reach a political agreement.

The inability to develop appropriate governance can be traced back to the situation known in the public choice literature as 'tyranny of the status quo', which typically arises when the change implies unanimity of consensus (Mueller, 1990). North (1990) and Bromley (1991) emphasize that the process of institutional building does not occur mechanically, but rather through 'trial and error' in which communities that are able to develop appropriate institutional settings will be able to solve environmental problems. The process of institutional change is constrained by rent-seeking of economic actors and the transaction costs of change (Johnston, 1996).

Yet a 'tyranny of the status quo' might arise even if policymakers have the formal power to decide (with a majority rule), but are unable to implement and enforce their decision because of information asymmetries and lack of control over essential resources (Dente, 1995). For example, implementing state ownership over groundwater has proved to be very difficult in Italy or Spain, where hundreds of thousands of private wells are in place. This is even more probable when the solution implies not only 'negative' actions (preventing someone from doing something, which may eventually be forbidden by law and enforced), but also 'positive' actions, which cannot be precisely specified (e.g. developing cleaner technologies).

The substitution of natural capital with artificial systems may provide the solution. Yet once the process of modernization begins, artificial water management systems pose new and different constraints. First of all, water services are costly and require the diversion of adequate financial resources. This might be unfeasible for many different reasons. For example, a community might be simply too poor to afford water services; the cost of services could be higher than the value of additional environmental functions; or, finally, people might be unwilling to pay higher taxes or higher water prices.

Second, water services are technologically complex and require appropriate skills. The implementation of artificial water systems implies management by specialized organizations who should be able to mobilize adequate resources (human skills, technology, capital etc.) in order to recover the related costs and remunerate those resources so as to remain attractive. For example, water specialists should be paid enough to prevent them from going elsewhere, and capital should be adequately rewarded to prevent its reinvestment in other industries or countries. This is ultimately the reason behind the increasing role of professional water companies – possibly profit-oriented with an industrial standing – in replacing traditional management systems. This transformation is again not straightforward: it requires appropriate regulatory systems (the water industry is a natural monopoly with very limited opportunities for competition), as well as conducive social and political attitudes (acceptance of 'privatization'), institutional mechanisms and so on (Massarutto, 2006). The public good aspect of water infrastructure, due to the prevalence of fixed costs and the long duration of investment cycles, poses further problems in terms of business risk for investors and the need to secure long-term cash flows (Linares et al, 2008).

We can affirm that a carrying capacity threshold has been reached – and a conflict is likely to take place – when one of the above scarcity limits has been violated. This poses a 'limit to growth' to environmental functions that can be satisfied by the existing water management system. Reaching this limit means that a trade-off between social demands has to be managed and solved; and if it cannot be solved within the existing framework, this generates a conflict, whose existence can be interpreted as the sign that a carrying capacity threshold is being violated.

The limit is not set forever: an income improvement may enable the community to pay for UWS that would previously be unaffordable; technological improvements might render UWS provision possible (or simply cheaper); new governance institutions might cope with political incapability to decide; social learning processes might generate new cultural frameworks enabling the community to accept solutions that were previously rejected. But in the short term the individuation of a carrying capacity limit leads to the consideration of water resources – at that particular territorial scale – as a critical natural capital, and thus to pose an upper limit to their consumptive use.

This discussion also highlights the importance of UWS for sustainability. Water sustainability not only concerns natural water resources, but also the long-run viability of management systems, and the capacity to renovate artificial assets and finance their long-term maintenance and operation in a way that is politically legitimate and socially acceptable. For this reason, the assessment of cost recovery is crucial for sustainability. In neoclassical economic terms this is not necessarily the need to ensure optimal allocation of resources through marginal-cost pricing, but rather the need to assess the long-term capacity of the community to raise the financial means to ensure its viable operation, whether through prices or taxes, or to decide which equalization measures are put in place (Green, 2003; Massarutto, 2005).

This simple conceptual model helps us to understand the evolution of water policies in past and recent times. Economic development over the long term affects the relations between water and society in two different directions. On the one hand, it increases the capacity to finance WSS expansion and positively affects innovation; on the other hand it causes growth in demand (especially for those environmental functions that are income-elastic) and increasing external effects. The final effect on sustainability is thus ambiguous. Sustainability of actual patterns of water use requires that GDP grow more rapidly the cost of WSS, induced by demand expansion and/or depletion of natural capital.

In the literature, the optimist attitude is shared by those believing in an 'Environmental Kuznets curve' (namely, a positive relation between economic growth and sustainability). In the case of water, many indicators seem to show that the previous century's growth patterns were sustainable, as GDP grew more rapidly than deterioration of water capital, thus allowing for investment in WSS that ultimately improved both supply and water quality. As a result, in most developed countries, water quality and river ecosystems are now in much better condition than 100 years ago. On the other hand, the belief that this trend will not be reversed in the future is not based on empirical evidence. These features contribute to identifying water as a critical natural capital; therefore, a precautionary principle can be invoked (Arrojo et al, 2005; Ekins et al, 2003).

17.3 THE PARABOLA OF URBAN WATER SYSTEMS

17.3.1 The 'further from farther' crisis

Chapter 1 in this book synthesizes the modernization of water management in the shift from the *free access* world, where human demands mostly rely on raw water resources to which individuals and/or small communities have access in a dispersed manner, to *the urban water service* world, in which artificial systems allow organized access to natural capital.

While the former management regime entails the typical problems of a free-access good, that is rivalry (at least after demands have reached a critical threshold that impairs natural capacity to reproduce the resource) but not exclusion, the latter is typically a club-good (where rivalry is not a problem as far as sufficient economic resources are invested in the system, and in turn access can be more easily denied to those that do not fit the requirements imposed by other club members, e.g. do not pay their share of the cost).

The changing nature of economic problems with regard to the water resource and the water service domain is the subtle but fundamental reason behind the peculiar nature of UWC. While raw water – as a rival good – may trigger conflicts regarding property rights and access to the scarce resource, and eventually prime 'tragedies of commons', water services as public – though excludable – goods foster conflicts that concern fundamentally the size of the 'club', access to membership and the way costs and benefits are shared. Yet, the main problem is that development of water services also includes components of positive externality (e.g. public health, urban drainage etc.) and merit (willingness to guarantee a 'right to water' and connection to collective water and sanitation as a component of citizenship); while on the supply side, economies of scale, prevalence of fixed costs and the long economic life of assets define the industry as a natural monopoly.

Therefore, the dominant model has been the transformation of urban water services, initially driven by private companies on behalf of closed communities of individuals and entirely paid for by users, into publicly owned systems, intended as universal services managed collectively and financed by the public budget. In many cases (notably Northern Europe and most of the US), this never intended as a complete abandoning of user-geared finance. Users continued to pay a substantial share of the cost, eventually complemented by public subsidies, while the main issue became the way in which the cost had to be shared (on the basis of personal wealth or benefits received from the service), as well as who has the right or duty to join the collective system in place of provision on an individual basis. In other countries (e.g. Mediterranean Europe) central-state involvement has historically been stronger, and public subsidies have represented and still represent a substantial source of finance, while government policies have traditionally been dominated by infrastructure, not only for urban supply but also for irrigation and industry.

In all cases, decisions concerning water services have been advoked by governments and became, ultimately, a political issue. As a matter of fact, the political approach to urban water services has long been 'structuralist' (Arrojo et al, 2005), that is, dominated by a supply-side vision. The substantial growth of demand had to be compensated for by artificial systems, a useful indicator of which is the continuous increase of per-capita cumulated investment in water assets, made possible by the significant improvements in civil engineering works and economic growth. In the initial phase, social investment for expanding infrastructure has improved the supply of environmental functions, complimenting those furnished by natural capital; in the subsequent phase, the natural capital itself has deteriorated (because of increased pollution loads and over-abstraction), and expansion of the artificial system has been required to compensate for this loss.

Behind this approach we can easily recognize the dominance of a culture based on civil engineering and a coalition of interests composed of growing cities, large water users, construction industries and public water bureaucracies ('iron triangles', as analysed by Bressers et al, 1995).

Much of the supply increase has been made possible through the development of public services with substantial investment costs carried by the public budget. In the long run this has resulted in uncontrolled growth of the implicit public debt, represented by cash flows.

From a different angle, this approach parallels Boulding's 'cowboy economy', in the sense that solutions to the local mismatch of supply and demand has been found by gathering more freshwater from outside. 'Further from farther', as Barraqué (2003) characterizes it.

However, this solution is affected by many 'non-market failures', leading to a crisis that can be traced back to many factors:

The emergence of new social demands for water quality: while 'traditional' demand, focused on quantity, has reduced growth, demand for environmental quality cannot be easily satisfied through a simple expansion of WSS. While the coalition of interests behind traditional water policies was simpler and coherent, the actual structure of demands emphasizes trade-offs and complicates the task of decision-makers, who have to mediate between opposing stakes (Bressers et al, 1995).

The deterioration of natural capital, which has reached critical levels and hampered further critical ecological functions. Evidence of this previously neglected aspect forces water management to adopt more conservationist approaches and terminates the 'further from farther' model (Arrojo et al, 2005).

Evidence that at least some uses (such as irrigation and hydropower) have been greatly subsidized, and that this subsidization is unjustifiable at present since these uses no longer belong to the 'merit' category. The same applies to domestic supply, at least once the full cost-recovery price is affordable to nearly all.

Evidence that further expansion of infrastructure is often unjustified in economic terms, while alternative solutions are technologically feasible (Arrojo et al, 2005).

The crisis of public finance and the requirement for involving users in the financing of WSS, in the context of rapidly increasing costs of service provision.

While in principle, public decisions can inspire sound evaluation criteria aimed at maximizing social welfare, in fact these can be easily distorted by private interests wishing to derive the maximum of individual benefits at the expense of the community. As far as the water sector is concerned, we may see the dominance of the 'iron triangle' as a proof of the validity of Buchanan's theory of rent-seeking, especially if we consider the fact that many water works realized in the past would fail to pass cost/benefit tests (Anderson, 1983).

Once transferred into the analytical framework outlined earlier, the crisis of the structuralist model can be interpreted on the basis of the increasing costs that society has to face (and consequently share). This is particularly true for external costs.

As far as urban water is concerned, Barraqué 2004 has argued that the supply-side approach at some point becomes a vicious circle: increasing social demands call for further artificialization and further delegation of power and responsibility to the 'experts' and the technical structure. This creates further costs (both in terms of financial costs and externalities generated on other water users) and calls for further pressure to develop the artificial system. This vicious circle generates conflicts through the tension between social demand for water services and the need to allocate the underlying increasing costs (Figure 17.2).

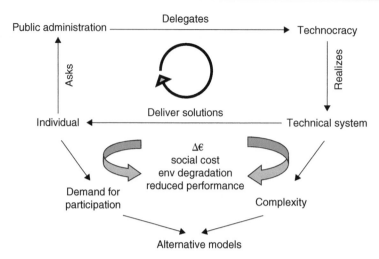

Figure 17.2 **The vicious circle of modernization**
Source: Author.

Further forces increase conflictual outcomes. This process also entails growth in territorial scale (and thus increasing *physical* distance between citizens and decision-makers), increasing professional specialization (and thus increasing *cultural* distance), and finally, increasing pressure towards economic efficiency, which usually culminates in water operators acquiring a business-like mentality, if not full privatization (and thus increasing the *motivational* distance between profit-oriented water companies and local political stakes).

17.3.2 Modernization and sustainability of urban water management

All factors outlined in the previous paragraph erode the basis of consensus and the legitimacy of the dominant model of managing urban water and generate pressure towards new solutions.

The Water Framework Directive (WFD) can be seen as a turning point that initiated a new European water policy focused on sustainability (Barraqué, 2004). Among its basic principles are: conservation of natural capital to a good ecological level; prevention of further deterioration; economic assessment aimed at implementing efficient allocation; full cost recovery aimed at avoiding the development of artificial water assets beyond social willingness and ability to pay; and public participation to ensure that all social stakes are duly taken into account. Similar principles frame water policies in other developed countries, while developing countries focus on ensuring complete accessibility.

Urban water services themselves have undergone a radical transformation in the last twenty years in most developed countries (de La Motte et al, 2005; Finger and Allouche, 2006; Massarutto, 2006). Among the most significant shifts are changes from local management to regional integration; from public subsidies and public finance to full cost-recovery and market finance; from direct management of local authorities to various forms of independent and professional water industry, often (although not solely nor necessarily) with the involvement of the private sector;[2] from the dominance of water supply and urban networks to water resources management at the river basin-level and the growing emphasis on cleanup, treatment and conservation of resources; from simple, discretional and benevolent regulation to more sophisticated, controversial and adversarial regulatory systems; and from sectoral water policy focused on infrastructure development to integrated management focused on sustainability and cooperation.

This transformation itself constitutes a fault line along which further conflicts arise: once the urban water management system develops into an autonomous entity, it becomes an actor on its own, creating further potential conflicts that go largely beyond the water management sphere. Rising costs of water management pose issues of affordability and challenge basic social rights. Integrated management forces local communities to give up control over their local territory in order to share resources (and problems) with other communities on a larger territorial scale. Water uses established over centuries may end under duress in order to allow space for new emerging societal demands. Increased corporatization (if not privatization) of service operation implies loss of control and substantial delegation of power to water professionals and profit-oriented organizations, whose faithfulness to the general interest of the community has to be proved and cannot be taken for granted.

To put it in a different way, sustainability of urban water systems is not only and simply a technical, managerial and economic problem; it requires instead a thorough adaptation of the institutional, political and social spheres. The outcome of transformation cannot be reduced to different technological solutions; it should be able to reproduce the basis of cohesion, trust, legitimacy and consensus. This on its own implies that patterns of allocation of costs and benefits (or, more generally, negative and positive outcomes) are perceived as fair. Furthermore, this perception also has its roots in common 'cultures' and 'rules of reason', which need to be developed and shared through open participation (Swyngedouw et al, 2002; Pahl-Wost, 2002). From this perspective, the transformation of urban water management can be described as a problem of governance (Hanf and Jansen, 1998; Picciotto, 1997), whose solution cannot be found through the interplay of individuals within a given set of cultures and institutional rules, but rather implies a reconstruction of both (Amin, 1997).

Ultimately, as suggested in some of the contributions to this volume and more effectively in Chapter 2 by Barraqué and Katko in their historical overview, the direction of the modernization process might reverse. Once the development of UWS along the traditional lines encounters limits (environmental, financial, social, technological) the challenge is to find a way back to the world of individual access to common property, via an institutional development that makes social control of such access actually feasible. The underlying paradigm is that of an innovative understanding of 'integrated water resources management'.

---

[2] Private sector involvement (PSI) is intended here in its widest meaning; it involves regulated private monopolies with full divesture (e.g. England and Wales), full or partial delegation via competitive tendering (e.g. France), outsourcing of activities from public entities, and various forms of private-public partnerships for operation.

17.3.3 Drivers of modernization

Figure 17.3 summarizes the main forces that drive the modernization of urban water systems.

Pressures for change arise when an external element alters the equilibrium. For example, new stakeholders appear on the scene with new demands that need to be accommodated; or the deterioration of natural capital reduces the supply of environmental functions. As policy analysts have pointed out, the existence of problems is not a sufficient condition for generating a crisis: this occurs only after the water policy network has filtrated 'problems' and constructs them as 'governance issues', namely problems that the society recognizes as such (Bressers et al, 1995).

On one hand, we can thus individuate those actors in the policy community with the capacity to put pressure on policymakers: water experts can identify problems and propose solutions, the scientific community provides the means to understand the issues, the water industry develops technologies and announces its capacity to resolve problems, water users exert pressure in order to increase water availability, environmental pressure groups demand improved water quality and so on. Of course, close interactions may exist between these actors: for example, scientific research can be influenced by water industry financing, and the appreciation of certain problems and

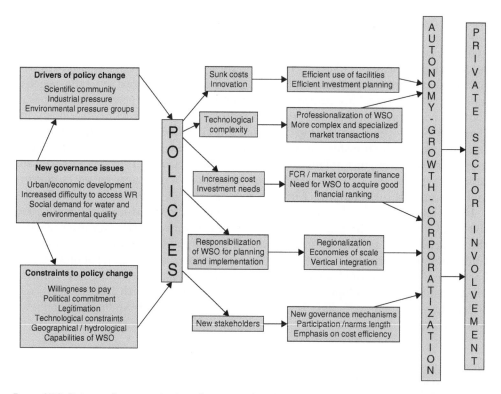

Figure 17.3 Drivers of corporatization of water undertakings and private sector involvement

Source: Author.

issues can be strongly conditioned by water industry investment and its perception of what can or should be done.

On the other hand, we can point to factors that constrain water policy action: lack of financial resources, lack of readiness to pay by consumers, as well as geographical and institutional factors (e.g. localism, inefficiency of public administrations, and the relative weight of water-related issues in the political arena).

Altogether, these factors shape and determine new water policies, placing pressure on urban water management, for example, by increasing quality standards and/or promoting new water investment. As we have argued before, a significant change has occurred in recent decades, causing a shift from traditional 'extensive' towards more 'intensive' water policies.

New water policies impact on the equilibria of the water industry basically through five channels, each putting in crisis traditional forms of public management and promoting innovative solutions that either imply the involvement of the private sector or at least the development of more independent and business-oriented public systems.

First, urban water management requires fixed costs that are for the most part sunk. These costs grow as technical complexity increases and demand (arising both from service users and environmental policy) becomes more exigent. This represents a formidable pressure on territorial growth of water undertakings (to share these costs across a wider base of customers) and outsourcing of activities, implying economies of scale or high labour costs (e.g. public works, customer services, billing, emergency management, project and design, RTD etc.).

Second, the increased technical complexity of water management forces local water management systems to delegate decisions to professionals with the necessary expertise. This leads to the creation of specialized organizations, which belong to either the private or public sector according to tradition and national policy. Specialization and complexity also mean that less critical input is produced 'on-site'; instead, input is incorporated into 'technological objects' that local operators select and purchase on the market. Water operators are thus required to position themselves to communicate with external suppliers of technology, and this can ultimately represent a strong incentive for corporatization of water service operators and increasing recourse to the market for provision of inputs. In this way, the value chain of water services includes more and more 'added value' produced by specialized firms in the marketplace.

This does not necessarily mean that water operators themselves will be privatized, at least not in the forms of delegation or full divesture: Italian and German municipally owned water companies, as well as the Dutch water companies or British Water Authorities prior to privatization are well-known examples of water industry developed within the public sector, albeit with some form of entrepreneurial autonomy. At a certain point, however, the pressure to privatize might become overwhelming, either because of a general policy trend in this direction, or because autonomy has become so strong that public control has weakened. The need to engage in heavy sunk costs (e.g. for R&D) also requires larger scales of operation, and thus weaker control of individual municipalities. The sale of shares on the market, often forced by budget constraints, completes the process, until companies finally behave as private companies, even if formally owned by the public sector.

In other countries, the need for professional skills has led quite early to delegation of operation to private companies, instead of developing the same skills in-house.

Peculiar institutional mechanisms may influence this development. For example, in the case of France, delegation to private companies often became necessary because of the legal difficulties encountered in the creation of local, publicly owned companies, and by the problems municipalities have in accessing credit (Linares et al, 2008).

The third channel is financial. The increasing cost of water services in the present phase is paralleled by the crisis in public finance. This has brought an end to the traditional way of financing through the public budget. Self-finance through full cost recovery has started to become necessary. However, tariffs and prices paid by consumers will serve as a cash flow for sustaining market-based finance, for which the water operator becomes the intermediary.

Once again, this does not necessarily lead to private operation of services, since public companies might also have the capacity to access financial markets. However, given the high capital intensity of the water industry and the very long depreciation schedule of water assets, the financial reputation of operators is a crucial factor in reducing the cost of capital and ultimately the cost of water to end users. In cases where water companies have remained under public control, many still engage in enlargement of capital via new investors, buyouts and other financial operations which open their property to external investors.

Moreover, the decision to privatize can be 'political', in the sense that governments and local authorities may wish to sell the property in order to give a signal that eventual losses in the future will no longer be transferred to the public budget. From this perspective, privately owned companies will need to achieve a good balance between revenues and costs, and therefore will not engage in new expenditure unless they are able to finance it through prices paid by consumers. The history of the British Water Authorities is enlightening. When they were created in 1973, they were required to operate on a cost-recovery base. However, charges used to be kept low for political reasons, while significant new investment continued. Consequently, the WA experienced a substantial rise in their debt, possible only because the financial markets treated this debt as if it was warranted by the central government. When the Water Authorities were privatized in 1989, the British Government had to write off this debt and absorb it into the public debt, otherwise the whole operation would have been a failure (Ballance and Taylor, 2005).

In some other cases, the decision to privatize operators may have been, in one way or another, imposed by financial institutions. This is particularly the case in developing countries, where the financial rating of local firms, the trust of international agencies such as the World Bank, and the lobbying capacity of Western transnational water firms were all factors conditioning the lending of capital to the delegation of operation to private companies (Hall, 2001; Finger and Allouche, 2002).

The fourth channel concerns the 'division of expertise' between service operation and regional water policy. Under the traditional model, water management could remain 'local' and 'simple' as long as external subjects (the state) carried out the necessary actions through public works, aimed at making water available at the local level. 'Water plans', carried out at the state, basin or regional level, were policy instruments in which the role of the public sector was realized.

As a result, the water industry is required to play a more important and responsible role, with substantial degrees of freedom. One important consequence is that local water management can no longer be considered as a 'simple' task within the framework of a publicly controlled water plan. The water operator itself has to become 'a planner', and therefore acquire both suitable geographical scale and vertical integration.

In other words, the task that water service operators have to deliver has more and more to do with decisions with higher degrees of autonomy. In the 'traditional' model, public planning provides solutions and decides facilities and investment; in the 'new' model, asset management, projected solutions and gathering of consensus is increasingly delegated to operators, while river basin plans mostly limit themselves to indicating the desired environmental functions (e.g. final quality of the river), leaving water users a high level of discretion regarding the means to reach these objectives.

Finally, the fifth channel concerns the need of water service management to interact with a larger number of conflicting stakeholders, with stronger emphasis placed on the demand side rather than the displacement of new assets and technology. In the 'new' water policy network, the ranking of public water services, though still in some way predominant against other water uses, has lost the centrality it could have in the past. The maxim 'what is good for water services is also good for water policy' is no longer self-evident. Urban water users thus become one among several stakeholders of water policy.

Governance in the water sector becomes an increasingly complex exercise, and the meaning of 'public interest' is less clear and univocal. In many countries, the trend towards privatization also means the search for a new model of regulation, based on an 'arm's length' distance between water service regulation, environmental planning and water service operators, seen as in opposition rather than as partners.[3]

17.4 TOWARDS A GENERAL UNDERSTANDING OF UWC

17.4.1 Conflicts as indicators of urban water sustainability

The case studies presented in this volume have followed different approaches. In spite of this, all the conflicts analysed can be easily related to the modernization process. While 'water wars' typically oppose users against one other, eventually supported by political institutions, the character of UWCs can be much more variegated, including: 'club members' (those connected to the system who expect it to fulfil their needs), other water users (either willing to join the club or not), political institutions, professional water administrations and system operators.

We have argued that water management systems and water institutions represent the most significant limiting factor for urban and economic development, in terms of carrying capacity. Urban growth requires increasing quantities of water and generates impacts on the water environment. This in turn creates increasing costs in terms of sacrificed environmental functions and impacts on other communities and uses. In one way or another, this pressure implies a vicious circle that puts into question long-term

---

[3] This does not necessarily imply that independent regulatory authorities are created, as in the British case. Rather, it underlines the fact that as UWS undertakings become more autonomous and concerned with water service provision, definition and enforcement of water policy requires a more adversarial and formalized regulatory system.

sustainability: through overuse of natural capital and the loss of ecological functions, dilapidation of the value of capital, and erosion of the bases of social consensus.

The alternative is represented by increasingly sophisticated forms of WMS and the related need to face financial costs and delegate decisions. Yet this transformation is not straightforward. It requires the mobilization of adequate economic resources, the achievement of social consensus, and political agreement. The process also requires appropriate governance institutions. Until these prerequisite goals are achieved, the transformation remains unfeasible, and the system will continue to exist at the expense of its critical capital.

In this sense, the existence of urban water conflicts can be interpreted as a signal that something is going wrong or is missing, such that the transformation cannot take place and sustainability is not guaranteed. An understanding of UWCs, their roots and dynamic causes, and the possible ways out, is crucial to making a judgement about the long-run sustainability of a given water system. Existing water assets, both natural and man-made, physical and immaterial, pose a carrying capacity threshold that cannot be surpassed until further assets are put in place.

The current experience in developed countries shows that the way in which modernization has been attempted to date – namely, the 'further from farther' approach – is giving rise to new and previously unknown conflicts, since its costs (both financial and external) are growing higher and the capacity to recover them increasingly difficult. The different forms of sustainability (ecological, economic, social, financial, institutional) involve many trade-offs, and it is increasingly difficult to pursue any one without hampering the others. In the end, the persistence of UWCs might imply that the trajectory of modernization is itself unsustainable. A carrying capacity threshold might reside not only in the limits of natural capital, but also in the fact that societies are not ready to accept and metabolize change, and that institutions and cultures are incapable of evolving to dominate and govern it.

17.4.2 Interpretative frames: actors in conflict

Here, we attempt to generalize by using the framework summarized in Table 17.1, based on the typology of actors involved.

A first typology of conflicts arise among 'club members' (i.e. among urban water service users) and concern the definition of membership rights and duties, with special concern for charging criteria. The source of this conflict resides in the increasing difficulty of public budgets to follow the requirements of the system, both because past policies have disregarded the need to set aside cash flows for maintenance and replacement and because of the need to invest further to achieve quality standards.

In developing countries, the latter issue is prevalent, and prevention of self supply for public health purposes, extending sanitation and wastewater collection, potentiate the infrastructure and enable it to meet substantially increased demand, expressed not only in quantitative but also qualitative terms. This is exemplified by the case studies on India, Indonesia, Brazil and Argentina.

In Western countries, the issues are quite different. In East Germany, for example, the main issue concerns the over-sizing of infrastructure built during the Soviet era and the unwillingness of many people to continue paying, while new self-management options are feasible (e.g. wastewater recycling), making the possibility of disconnection

Table 17.1 Typologies of UWC according to the actors involved

Actors of conflict	Example
Club members	Cost-sharing and tariff-setting (polluter-pays, user-pays, rich-pays)
	Admissibility of cross-subsidies (among user categories, territorial areas, rich and poor, among services etc.)
	Degrees of freedom (connect or not, opt out or not, integrate w/self-supply or not)
	UWS as providers of individual services vs. collective utilities
Club members vs. would-be members	Extension of networks to suburbs
	Extension of sanitation networks to suburbs and rural areas
	Universal service definition (e.g. quality and continuity of supply)
Club vs. other clubs/users	Deterioration of resources induced by urban growth and industrial pollution
	UWS subtracting water to other users
	UWS segregating land use in catchment areas
	UWS affecting water quality for downstream uses
Club members and managers/operators	Privatization
	Governance of UWS management, regulation and public participation
	Commercialization of UWS
	Financing of investment
Club members vs. state	Affordability of impositions concerning environmental quality
	Inter-governmental conflicts
	Degree of integration and meaning of solidarity

more concrete. In Italy, as well, the shift from state subsidies to full cost recovery seems problematic.

It seems likely that full cost recovery through volumetric charges will have dramatic redistributive outcomes (either among regions or among rich and poor customers). What is conflictual at this stage are the alternative theories of justice and equity sub-tended by the polluter-pays principle, on the one hand, and the 'progressive taxation principle', on the other. The polluter-pays principle, first introduced by environmental economics as an efficient way to allocate external costs, becomes increasingly difficult to accept on an equity basis once the cost to be considered concerns infrastructure and basic services. Water as a social right is clearly in opposition to water as an economic good and a scarce resource.

On the other hand, full cost recovery makes clearer and more explicit what the costs are, why they are incurred, and who pays what to whom. As a result, citizens and politicians engage more easily in discussions about the desirability of service improve-ments, while the costs induced by past decisions become visible. It is easier to build a dam or a treatment plant using taxpayers' money than having to raise water prices.

Sometimes this might also create affordability problems, either at the individual level (water service demand not correlated with income, resulting in the poor paying substantially more) or at the collective level (low-density areas paying more). This trend is likely to be exacerbated by the high cost of capital implied by the recourse to the private market, as simulations in the Italian case study show.

The reaction of customers might differ from country to country: bill strikes in Barcelona or Buenos Aires; self-supply and opt-out in Germany; court cases in France; violent campaign and appeal to media in Italy; intense inter-institutional confrontation, as in England and Wales, where consumers and Ofwat often conflict with environmental and quality regulators about the desirability of service improvements. Yet all of these cases bear witness to the need to quickly develop a more robust institutional framework to deal with this issue.

Again solutions might differ according to national specific features; yet they all involve, in one way or another, the following:

1. a more progressive tariff structure (e.g. distinguishing connection charges from volumetric charges and applying the latter only to marginal cost components, while the allocation of infrastructure fixed costs might follow other criteria, including local taxation)
2. more effective cost-sharing mechanisms that allow some cross-subsidies among areas
3. an approach that distinguishes 'normal' service costs from resource management costs, the latter perhaps including extraordinary actions aimed at achieving basin-relevant targets
4. smart financial institutions that enable the provision of capital at a lower cost, through a more efficient allocation of business risks between the private and the public sector.

A second typology of conflict occurs between members and would-be members, concerning the criteria for being admitted into the club. This conflict is particularly acute in developing countries, such as India, Indonesia, sub-Saharan Africa and South Africa. The basic challenge is to expand access to UWS networks in the context of rapidly growing cities. Some case studies, and notably sub-Saharan Africa, also demonstrate the necessity to deal with collective affordability as well as individual affordability: in order to remain sustainable, investment should not surpass a threshold in relation to national income; most probably, this means that over-ambitious plans that aim to connect people in terms of Western standards should be put aside, and more realistic solutions found (e.g. investing in low-cost community schemes instead of connecting each individual dwelling).

Yet, large metropolitan areas in developing countries also demonstrate the necessity of coping with the many externalities raised by urban water management, and the unsustainability of maintaining suburbs outside the 'club'. As European and American history show, this situation occurred in Western cities 100 years ago, and could be solved only through the development of a local public finance scheme, be it financed by general taxation or perhaps by cross-subsidies among different services and economic categories. This trend can be easily seen in the historical overviews of Barraqué and Katko on Europe and Melosi on the US.

In developed countries, this issue concerns, in particular, sanitation and treatment in rural areas. As the Italian case shows, expanding centralized sewerage systems to rural areas looks prohibitively costly and incompatible with the narrower version of territorial solidarity (declined at the district level only) and a strictly commercial vision of service provision. Perhaps this difficulty will put into question choices made by regional and basin plans, and instead favour softer technologies. A similar debate is taking place in France and Western Germany with special focus on rural sewerage and rainwater management: here as well, the preferred option may be the development of soft services (e.g. compulsory maintenance of septic tanks).

A third typology occurs between members and other water users, whose competing management systems may be impaired by the expansion of urban systems, raising an 'appropriation' conflict over rights of access to water resources and/or compensation in cases where such rights are reduced; or, more generally, other areas where options are 'segregated' – namely, limited and conditioned – by expanding urban water systems.

Appropriation and segregation conflicts have been managed so far through state involvement in large waterworks and, at the basin level, shifting the focus from appropriation of the natural resource to that rights and priority access to artificial water systems. This occurs, for example, in Southern Europe and Brazil, where typically governance of appropriation conflicts is the task of water management institutions, and confrontation takes place predominately within administrative and political procedures rather than in other arenas.

An alternative mechanism, practised in the US, is based on direct negotiation among uses. While initially successful for the purpose of reallocating water rights, this 'market' mechanism has shortcomings, as illustrated by the Imperial Valley case: conflictual outcomes following the purchase of water rights emerged due to the lack of an integrated vision and a failure to internalize the many externalities occurring at the river basin level.

The Indian case visualizes, in turn, the opposition between exigencies of urban growth and the existence of customary water rights in the (not yet) modernized surrounding rural communities. This case resembles a modern version of the 'enclosure' conflicts that took place in Europe during the eighteenth century.

In the European cases we often find a conflictual relation between urban areas and other land uses, especially agriculture, where externalities occur on both sides (e.g. over drinking water protection from pesticides and nutrients, but over discharge of polluted wastewater into irrigation networks).

Further examples encourage a more optimistic vision. In Europe and the United States, in particular, the emergence of conflicts among urban systems and other water uses may lead to innovative solutions based on cooperation and bargaining, for example, concerning drinking water catchment protection, wastewater reuse and rainwater management.

A fourth typology concerns opposition between members and those who administer the system on their behalf (with a focus on governance rules and rewards for managers). Multiple trade-offs are visible here, both in developed and developing countries, but the majority of these can be traced to Karen Bakker's scheme: on one hand, the extent to which the 'club' of users is absorbed into the wider social and political community, and on the other hand, the degree of technical and economic complexity of the system, leading to a greater degree of delegation to experts at arm's length from users. The source of the conflict is the opposition between the requirements of specialists – in terms of remuneration and securitization of cash flows for repaying sunk costs – and that of avoiding monopoly rents. The latter factor exacerbates information asymmetries between suppliers and customers, while the former implies delegation to a third party (the municipality or the regional government) of the power to define service content and represent users, thus adding a further information asymmetry linked to the formation of political choices.

Here, again, it is interesting to note that some countries have developed more robust institutional responses – for example, in France, where most of the conflicts can be settled within the judicial system – while in other countries the lack of adequate regulatory institutions trusted by the public is one of the main reasons behind the violent reaction against private sector involvement (e.g. Mexico, India, Latin America). On the other hand, the French case itself gives rise to the argument that one of the factors that kept the volume of the conflicts low was the substantially uncompetitive system that prevailed before the *loi Sapin* (where contracts were based on the *intuitu personae* and renegotiation was free and opaque). A more adversarial confrontation might ultimately persuade municipalities to revert to direct management in order to avoid complicated judicial cases and overly complex and rigid contractual arrangements.

Clearly, an important source of conflicts here is privatization of utilities and commercialization of water services. The source of the conflict is mixed: it concerns the actual capacity to ensure adequate service governance and induce open competition, but there are also other less straightforward reasons, ranging from ideological premises to local history. Paradoxically, it seems that privatization is more accepted and raises less conflicts where the quality of the public sector as a regulator is higher and the tradition of good administration is better rooted. Meanwhile, where the public sector has failed to perform adequately, opposition to privatization is stronger and harsher.

Also, it might be noted that in some cases privatization is contrasted per se, regardless of the outcome in terms of efficiency and service costs. However, in other cases the private sector is opposed on the evidence of dramatic price increases – the latter quite often being simply the result of implementation of the full cost-recovery principle and the quest for a more sustainable financial basis for water management undertakings. In other words, in many cases the private company becomes a sort of a scapegoat that catalyses conflicts, whose actual origins lie in changed patterns of financing services or willingness to invest to maintain the value of assets or improve services.

Finally, we can consider conflicts that create opposition among different levels of government. This type of conflict has a vertical (state, regions, municipalities) as well as a horizontal dimension (boundary regions, municipalities that have to merge in order to collectively provide the service). In some countries, we can notice the conflictual relation between politically elected institutions and special-purpose institutions (independent authorities, collective boards, basin organizations).

In some countries (notably, Brazil, Portugal and Italy) the vertical dimension is important because of the traditional central government role in the mobilization of resources at the inter-regional basin level. In other countries the opposition is more concerned with regulatory aspects, such as central/regional governments imposing requirements that local water management have to achieve at their own cost. In both cases, the source of conflict is first of all concerned with power (power to decide, but also to allocate money and own information).

In other cases, the focus rests on symbolic issues (e.g. countryside 'resistance' against being absorbed by urban management systems with consequent loss of autonomy and the obligation to share in their problems). In Italy, for example, the decision to compulsorily merge the over 10,000 municipal water undertakings into less than 100 inter-municipal agencies has given rise to a struggle concerning both the 'demand' side – decision-making within the inter-municipal boards concerning aspects such as solidarity levels, cost-sharing, resource optimization, location of facilities etc. – and the 'supply' side, this time concerning the ownership structure of the former municipal companies, which are increasingly becoming private-law companies with public shareholders.

17.4.3 Interpretative frames: categories of conflict

An alternative framework for interpreting urban water conflicts considers conflict categories and the main dimensions that they raise (Table 17.2).

A hypothesis that emerges from the contributions presented in this volume is that the first two typologies are more frequent in developing countries and wherever economic growth is related to massive urbanization processes. In a more advanced phase, urban growth becomes slower or even negative; suburbanization and connected sprawl-effects generalize the requirements for urban water services, while the possibility and political feasibility of further expanding the 'water ecological footprint' of the city on the rest of the basin reduces. Thus, priorities change (the aim becomes making the best and most efficient use of available resources instead of searching for new ones), complexity of management and technical sophistication increases, and financial costs become higher.

An important policy lesson can also be derived. In the first phase, costs are represented by externalities (on other communities), and by the construction of network systems; while in the second phase, these are mostly related to improving the quality of the management system. In the first phase, typically, local resources are left to deteriorate and water ecology is constrained, as enlarging the footprint looks easier and cheaper. In the second phase, in turn, costs of supply-side approaches grow suddenly higher or even infinite. The UWS has to adapt its strategy, once the deterioration of its assets becomes significant and thus very difficult to invert. This lesson learned by developed countries – now having to adapt their development trends at high cost in order to keep urban water use sustainable – is that indulging in supply-side approaches is essentially myopic; it would have been far more efficient, even in economic terms, to make provision early on for the maintainence of local water environmental functions.

Finally, the contributions to this book highlight the importance of social, political and institutional dimensions of urban water management as a limiting factor, although one which has been neglected to date. In a climate where water policy is dominated by engineering and economics, solutions may appear straightforward, once technologically feasible, and economically sound, providing that the market is in a condition to implement them. This very outcome, however, is facing unexpected reactions and difficulties that cannot be simply dismissed as displays of Nimby syndromes, anti-market ideological stakes and Luddism. How water institutions deal with these kinds of conflict is rapidly emerging as the foremost issue facing water management. Since institutional change and social learning require time and are based on trust, the focus switches to legitimacy and ultimately 'social capital'. This gives rise to a further important lesson: while investment in these areas might seem of little importance in the short term – through public participation and direct involvement – returns on this investment will appear much more valuable in the future.

Table 17.2 UWC categories, main dimensions and critical aspects

Category	Description	Dimension	Critical aspects
Access	Who has the right to use the resources? What criteria are used for allocating water rights? Are those that cannot have access to water being compensated in some way? And by whom?	Sectoral Territorial Economic	Property rights allocation/tradability Historical rights Planning criteria
Segregation of uses	Some users' actions make water unusable or inaccessible to other users Some users generate externalities that are suffered by other (potential) water users or by society as a whole Some water uses entail restrictions on other activities (land use, regional development)	Sectoral Territorial Economic	Externalities Regulatory institutions Planning criteria
Status and identity	Local communities fear becoming 'the garbage can' of other communities or being forced to share resources and problems with other communities. Local communities lose control over what they perceive as 'their' territory and resources.	Territorial Social Political Cultural	Balance of power (central/local) Planning dimension Finance of new infrastructure
Governance	The increased technical complexity of decisions forces communities to delegate decision-taking to professionals and 'experts', and to subtract power from the individuals and the community. Users with large political or market power have more weight in political decisions.	Social Cultural Institutional Ideologic	Participation Transparence Representation of interests Tradition Social conflict Regulatory institutions

		Type	Keywords
Ecology	Conservation of NC for the next generation is not guaranteed because conflicting demands tend towards overuse and next generations are not represented. Resilience of water management systems and capacity to adapt to future challenges. Basic ecological functions of water systems threatened by economic and urban development.	Intergenerational	Research Investment Collective understanding of water as a constraining factor
Artificialization	Is development of water infrastructure worthwhile? Allocation of costs should be fair and not jeopardize affordability. Water management systems should remain economically and financially viable. Deterioration of assets should be compensated by depreciation.	Intergenerational Fiscal Economic	Hidden public debt Efficiency of financial markets Public/private relationship and patterns of PSI Is economic growth rapid enough to guarantee affordability of WSS in the future?
Allocation of costs and economic risks	Is the allocation of costs (between individual users/sectors of use/territorial regions) fair? Are some users excluded because they cannot afford the water bill? Is the allocation of costs between users and taxpayers compatible with the constraints on public budget and with the need to avoid perverse incentives? Is the allocation of economic risks coherent with the remuneration received?	Fiscal Economic Regulation Equity	Instruments for cost equalization Economic balance between 'users' and 'taxpayers' Risk-sharing mechanisms

REFERENCES

Amin, A. 1997. *Beyond Market and Hierarchy: Interactive Governance and Social Complexity*. Cheltenham, UK: Elgar.

Anderson, T.L. 1983. *Water Crisis: Ending the Policy Drought*. Baltimore: Johns Hopkins UP.

Arrojo, P., Assimacopoulos, D., Barraqué, B., Bressers, J.T.A. and Castro, E. et al. 2005. *European Declaration for a New Water Culture*. Zaragoza: Fundaciòn Nueva Cultura del Agua. http://www.unizar.es/fnca/euwater/

Ballance, T. and Taylor, A. 2005. *Competition and economic regulation in water: the future of the European water industry*. London, IWA Publishing.

Barraqué, B. 2003. Past and future sustainability of water policies in Europe. *Natural Resources Forum*, 27: 200–11.

Barraqué, B. 2004. *Water policies in Europe after the Water Framework Directive*. E. Cabrera and R. Cobacho (eds) *Challenges for the new water policies in the XXI century*. Amsterdam: Balkema.

Bressers, H., O'Toole Jr., L.J. and Richardson, J. (eds) 1995. *Networks for Water Policy*. London: Frank Cass.

Bromley, D. 1991. *Environment and Economy: property rights and public policy*. Oxford, UK: Blackwell.

Dalhuisen J., De Groot R. and Nijkamp, P. 1999. The economics of water: a survey of issues, Research Memorandum, 36. Amsterdam: Faculty of Economic Sciences and Econometrics, Free University.

de Carli, A., Massarutto, A. and Paccagnan, V. 2002. La valutazione economica delle politiche idriche: dall'efficienza alla sostenibilità. *Economia delle fonti di energia e dell'ambiente*, 1.

de Groot, R., van der Perk, J., Chiesura, A. and van Vliet, A. 2003. Importance and threat as determining factors of critical natural capital. *Ecological Economics*, 44: 187–204.

de la Motte, R. 2005. WaterTime National Context Report – UK, WATERTIME Project. Greenwich: University of Greenwich.

Dente, B. (ed.) 1995. *Environmental Policy in Search for New Instruments*. Amsterdam: Kluwer.

Ekins, P. 2000. *Economic Growth and Environmental Sustainability*. London: Routledge.

Ekins, P., Simon, S, Deutsch, L., Folke, C. and De Groot, R. 2003. A framework for the practical application of the concepts of critical natural capital and strong sustainability, *Ecological Economics*, 44: 165–85.

Finger, M. and Allouche, J. 2002. *Water privatization: trans-national corporations and the re-regulation of the water industry*. London: Taylor & Francis.

Finger, M. and Allouche, J. (eds) 2006. *The Evolution of the Water Sector in Europe*. IWA Publishing, forthcoming.

Green, C. 2003. *A Handbook of the Economics of Water*. Chichester: John Wiley.

Hanemann, W.M. 2006. The economic conception of water. D. Rogers, R.M. Llamas and L.M. Cortina (eds) *Water Scarcity: myth or reality*. Amsterdam: Balkema.

Hall, D. 2001. *Water in public hands*. PSIRU: London. Available at: www.psiru.org.

Hanf, K. and Jansen, A. (eds) 1998. *Governance and Environment in Western Europe*. Harlow: Addison Wesley Longman.

Johnston, J. 1996. *Nature, State and the Economy: a political economy of the environment*. Chichester: John Wiley.

Linares, E., Massarutto, A. and Paccagnan, V. 2008. Private management and public finance of Italian water services: a marriage of convenience? *Water Resources Research*, forthcoming.

Lora, J., Rancho, M. and Soriano E. 2004. The future of desalination as a water source. E. Cabrera and R. Cobacho (eds) *Challenges for the new water policies in the XXI century*. Amsterdam: Balkema.

Martinez-Alier, J., Munda, G. and O'Neil, J. 1998. Weak comparability of values as a foundation of ecological economics, *Ecological Economics*, 26: 277–86.

Massarutto, A. 2005. *Competition and the water sector: a review of the economic literature*, Working paper series in Economics, 05-05. University of Udine, Dept. of Economics. Available at: http://web.uniud.it/dse/working_papers/working_papers_eco.htm

Massarutto, A. 2006. *Liberalization and private sector involvement in the European water industry*, Working paper series in Economics, University of Udine, Dept. of Economics 06-01. Available at: http://web.uniud.it/dse/working_papers/working_papers_eco.htm

Mueller, D. 1990. *Public Choice*. Cambridge, Mass: Cambridge University Press.

North, D. 1990. *Institutions, Institutional Change and Economic Performance*. Cambridge, Mass: Cambridge University Press.

O'Connor, M. (ed.) 1998. Social processes for environmental valuation: procedures and institutions for social valuation of natural capitals in environmental conservation and sustainability policy, VALSE project, final report, European Commission, DG 12, Environment and Climate Programme.

Pahl-Wost, C. 2002. Towards sustainability in the water sector: the importance of human actors and processes of social learning. *Aquatic Sciences*, 34: 394–411.

Picciotto, R. 1997. Putting institutional economics to work: from participation to governance. C.K. Clague (ed.) *Institutions and Economic Development: growth and governance in less-developed and post-socialist countries*. Baltimore: John Hopkins University Press.

Rogers D., Llamas, R.M. and Cortina, L.M. (eds) 2006. *Water Scarcity: myth or reality*. Amsterdam: Balkema.

Saleth, R.M. and Dinar, A. 2004. *The Institutional Economics of Water. A cross-country analysis of institutions and performance*. Cheltenham: Edward Elgar.

Swyngedouw, E., Page, B. and Kaika, M. 2002. Sustainability and policy innovation in a multi-level context: crosscutting issues in the water sector. H. Heinalt, P. Getimis, G. Kafkalis, R. Smith and E. Swyngedouw (eds) *Participatory Governance in a Multi-level Context*. Opladen: Leske & Budrich, pp. 107–31.

Turner, R.K. 1993. Sustainability: principles and practice. R.K. Turner (ed.) *Sustainable Environmental Economics and Management*. London: Belhaven Press.

Young, R.A. 2005. *Determining the Economic Value of Water: concepts and methods*. Resources for the Future series. Washington DC: RFF books.

Index

T - #0493 - 071024 - C346 - 254/178/16 - PB - 9780415498630 - Gloss Lamination